Ho Math Chess	Pre-K and Kindergarten Math

何数棋谜　棋谜式幼儿健脑思维趣味数学

© 2012 – 2021 Frank Ho, Amanda Ho, Canada copyright 1095661, Trademark 771400

Ho Math Chess

何数棋谜

Ho Math Chess Learning Centre

fho1928@gmail.com

www.homathchess.com

Copyright © 2012 – 2021 by Frank Ho, Amanda Ho

All rights reserved. No part of this book may be reproduced in any form or by any means whatsoever without written permission from the Publisher.

If you have any comments or problems on this workbook, please email them to fho1928@gmail.com.

一个做数学练头脑的高档培训中心

兒童趣味游戏数学專科培训中心

（專教资优兒童3歲或以上）

全球唯一独特发明,兒童智能数学

国际象棋谜题奥数融合教材

资优数学　SSAT　儿童趣味数学

数学IQ智能健脑思维　奥数加强

Ho Math Chess   Pre-K and Kindergarten Math
何数棋谜 棋谜式幼儿健脑思维趣味数学

© 2012 – 2021 Frank Ho, Amanda Ho, Canada copyright 1095661, Trademark 771400

Table of Contents

Preface 8
Ho Math Chess workbooks are good for children. (何数棋谜教材对儿童的好处) 9
Ho Math Chess worksheet simulating an internet screen (何数棋谜教材与互联网紧密结合) 10
Shapes and classification (图形的种类及配对) 11
Classification 12
Filling in next object 14
Classification 15
Learning numbers 0 to 9 (数字的写法) 17
Circling digits 0 to 9 23
Images of numbers 33
Tracing numbers 35
Tracing and copying 45
Tracing 1 to 9 48
Oral practice 52
Finding answers from numbers 1, 2, 3, 4, 5, 6, 7, 8, and 9. 54
Matching 55
Circling the correct writing to match the top number 56
Finding missing number. 57
找遗失的数字 57
Counting forward and backward (向前数, 向後数) 58
Saying numbers forward 63
Saying numbers backwards 64
Index counting 65
Circling the odd one 69
Right-left circle, opening up-down 73
Circling objects 74
Circling larger number of objects 79

Ho Math Chess   Pre-K and Kindergarten Math
何数棋谜   棋谜式幼儿健脑思维趣味数学
© 2012 – 2021 Frank Ho, Amanda Ho, Canada copyright 1095661, Trademark 771400

Circling the different parts 80

Circling the odd one 81

Writing numbers 85

Different meanings of numbers 89

数的不同意思 89

Tracing images of numbers 90

Tracing image 91

One to one correspondence 92

Match and connection 95

Matching 96

Counting and writing numbers 102

Dotted number showing its value 103

Linking numbers to dots 105

Unmatched numbers 115

Comparing 116

Length, weight, height 117

長度，重量，高度 117

Direction 118

方向 118

Clockwise ↻ and counter clockwise ↺ 122

顺时針，反时針 122

Traditional chess setup and Ho Math Chess setup 124

传统棋及何数棋谜教学棋 124

Ho Math Chess pieces' moves 125

Ho Math Chess chessboard setup 126

Chess moves 127

Rook's moves 128

車的走法 128

Directions and chess moves 129

Ho Math Chess	Pre-K and Kindergarten Math

何数棋谜　棋谜式幼儿健脑思维趣味数学

© 2012 - 2021 Frank Ho, Amanda Ho, Canada copyright 1095661, Trademark 771400

棋谜式数学 129

Directions and Frankho ChessMaze 133

Matrix reasoning 137

矩阵规律思考 137

Curve or straight line 144

曲线或直线 144

Handwriting of number 146

Figures made by connecting numbers 147

Number and its writing 148

More than (>), larger than (>) or less than (<) 150

多於，大於，少於 150

More than (>) or less than (<) 151

Depth of shapes 154

Checking sets matching the number of objects 155

Circling identical parts 156

Circle the following figures which cannot be traced by one stroke without lifting the pencil. 157

Counting 1, 2 or 3 158

Comparing objects 159

Matching objects 160

Connecting numbers 161

Counting objects 162

Logic connection 164

Review of numbers writing 165

Counting and writing numbers 175

Counting and writing numbers using table 179

Counting and writing numbers 184

Subitizing 188

知数感，一见即知数 188

Counting and sequencing 209

www.homathchess.com

Ho Math Chess Pre-K and Kindergarten Math	
何数棋谜 棋谜式幼儿健脑思维趣味数学	
© 2012 – 2021 Frank Ho, Amanda Ho, Canada copyright 1095661, Trademark 771400	

计数及数列 209

Counting beans 211

Copying numbers 212

Tracing 1 2 3 213

Filling in the missing number using 1, 2, and 3... 214

Counting the number of smiling faces 216

Pattern of 123 219

Figures and shapes 220

图及形狀 220

Basic Shapes.. 221

基本形狀 221

Weight.. 225

重量 225

Length.. 226

長度 226

Time 227

时刻或时间 227

123 writing.. 228

Adding more circles to match each number above.. 231

Writing a number to match its height 232

Filling in 1, 2, or 3 from inner circle to outer circle 233

Replacing each "?" by a number.. 234

Correcting any of the following incorrect counting numbers................................ 235

Filling in 1, 2, or 3 by pattern.. 236

Counting number and writing number on each number line 238

Colouring 1, 2 and 3 241

Finding the missing number 244

Weight.. 245

What is row and column? Matching by drawing... 247

| Ho Math Chess | Pre-K and Kindergarten Math |

何数棋谜　棋谜式幼儿健脑思维趣味数学

© 2012 – 2021 Frank Ho, Amanda Ho, Canada copyright 1095661, Trademark 771400

列及行 247

Circling the missing part 248

Sizes 249

Connecting numbers 251

Connecting shapes 253

Connecting numbers 254

Column and row 255

Left, right, top, down 256

　左右上下 256

Pattern 257

Maze 259

　数迷 259

Similarities 264

　類比求答案 264

Tracing each number 265

Counting the number of circles in each square 266

Counting numbers 271

Maze 273

　数迷 273

Matrix reasoning 275

Pattern 276

Sudoku 数独 277

Multi-task, multi-step, multi-concept 278

Sudoku 数独 280

Memory and computation training 281

　记憶及計算訓練 281

Knight moves 311

马的走法 311

Ho Math Chess   Pre-K and Kindergarten Math
何数棋谜　棋谜式幼儿健脑思维趣味数学
© 2012 – 2021 Frank Ho, Amanda Ho, Canada copyright 1095661, Trademark 771400

Mark an "X" on a square to show where each (knight) can move to. 311

Adding 1, 2, 3 314

Subtracting and adding up 9 337

Memory and computation training 356

Number ranking puzzles 366

数的大小比较 366

Ho Math Chess   Pre-K and Kindergarten Math
何数棋谜　棋谜式幼儿健脑思维趣味数学
© 2012 – 2021 Frank Ho, Amanda Ho, Canada copyright 1095661, Trademark 771400

何数棋谜　学习中心

只见棋谜不见题　　劝君迷路不哭涕

数学象棋加谜题　　健脑思维真神奇

Preface

We have found that many pre-k or kindergarten math workbooks do not seem to be suitable for those children who have encountered learning math difficulties. This workbook is written to help children as young as four years old learn to count or write numbers.

It is difficult for students with math dyscalculia to transfer math counting skills by using concrete objects or their fingers to calculate by using abstract symbols such as our number system 1, 2, 3 ... and even the magnitude of each digit does not come to be easy for them to understand. This workbook trains students in understanding abstract number sense without relying on counting fingers.

We developed this workbook by observing what difficulties our students have encountered and then revised our workbooks to tailor their needs. Thus, this workbook will be particularly helpful to those who have encountered challenges in mastering number writing, counting, or developing number sense since it is field-tested on children who have learning difficulties. This workbook can be considered as a remediation workbook for those children who have math disabilities or dyscalculia. Of course, it could be used as a regular math workbook by young beginners who have not experienced any math disabilities.

Frank Ho

February 2015

| Ho Math Chess   Pre-K and Kindergarten Math |
| 何数棋谜　棋谜式幼儿健脑思维趣味数学 |
| © 2012 – 2021 Frank Ho, Amanda Ho, Canada copyright 1095661, Trademark 771400 |

Ho Math Chess workbooks are good for children. (何数棋谜教材对兒童的好处)

使用何数棋谜教材的好处

加拿大 何数棋谜 培训中心
創辦人何数棋 (Frank Ho)
www.mathandchess.com

今天儿童面对的世界是学习如何处理数字,图形,资料搜寻,音影上下载,资讯比较,分类等资讯.这些活动实际已成為儿童生活的一部份.所以如果说学数学就是计算数字就错了.学数学的另一个目的就是学习如何利用数字资讯去解决问题及培养创造力.但是**传统式数学的计算练习题却完全没跟上科研已经改变了儿童面对的世界.**

儿童想要的计算题已经不是单纯的从上到下,从左到右的纯计算.**儿童需要的是他们情愿的而又快乐地做不枯燥的计算题.**所以如何将传统式数学计算题变得有趣而且又好玩,并且还可以增强儿童的计算及解决问题能力及培养创造力,同时还可以增进儿童记忆的能力达到全脑开发的目的?

何数棋谜首创已申情商标的几何棋艺符号并利用此符号发明了世界第一無二的何数棋谜教材及教学棋具.何数棋谜教材让儿童能利用几何棋艺符号进行数学的运算.

何数棋谜与传统式,数学教材不同的是小朋友不但要发掘题目,而且还要依国际象棋棋子的走法去发掘谜题与计算题（見下图）及答案. **只見棋谜不見题　劝君迷路不哭涕　数学象棋加谜题　健脑思维真神奇**

何数棋谜是将国际象棋融入数学以达到寓教於乐的教学理念.学生不但可以增强计算能力并且还可以增强解题能力及培养全脑开发创造力.

详细资料请上网查询 www.mathandchess.com.

Ho Math Chess   Pre-K and Kindergarten Math

何数棋谜　棋谜式幼儿健脑思维趣味数学

© 2012 – 2021 Frank Ho, Amanda Ho, Canada copyright 1095661, Trademark 771400

Ho Math Chess worksheet simulating an internet screen (何数棋谜教材与互联网紧密结合)

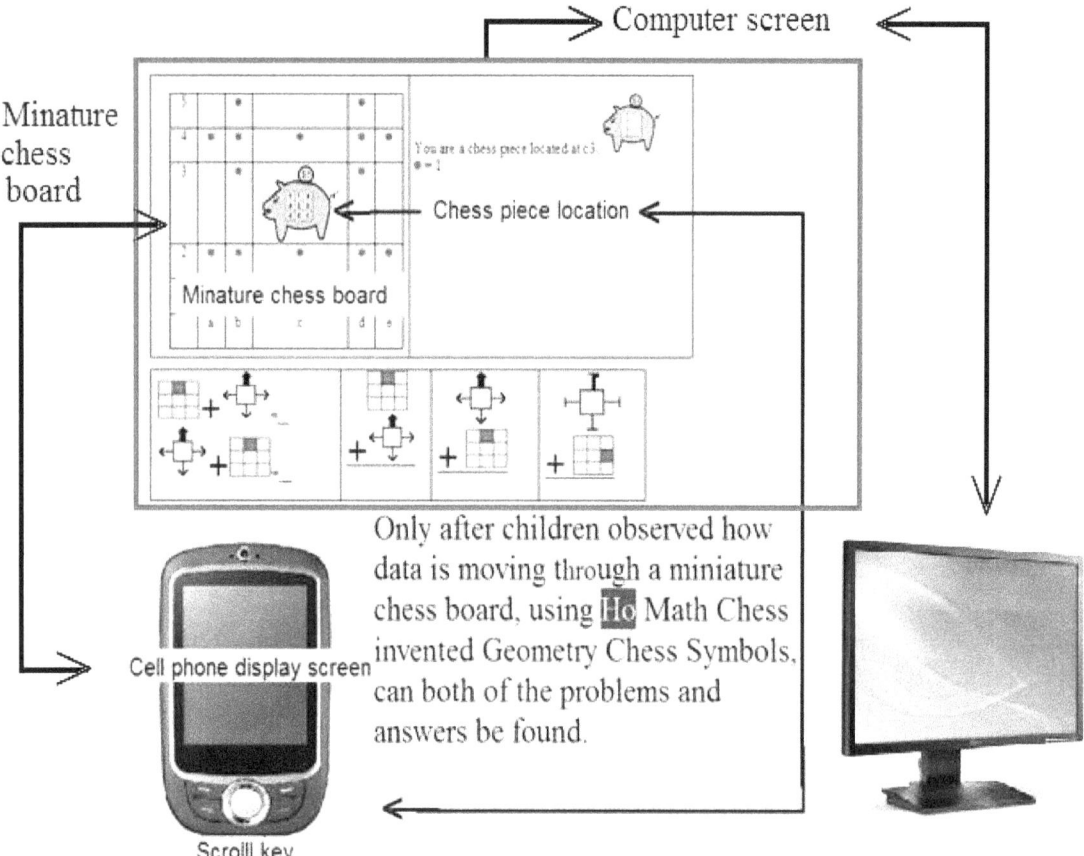

Ho Math Chess — Pre-K and Kindergarten Math

何数棋谜　棋谜式幼儿健脑思维趣味数学

© 2012 – 2021 Frank Ho, Amanda Ho, Canada copyright 1095661, Trademark 771400

Shapes and classification (图形的种类及配对)

Draw a line to match the left shape to the right object.

Circle	○	(face)
Oval	○	(beach ball)
Triangle	△	(hand mirror)
Square	□	(penguin)
Rectangle	▭	(envelope)

www.homathchess.com

Classification

Circle the following objects that you might use when you write or draw.

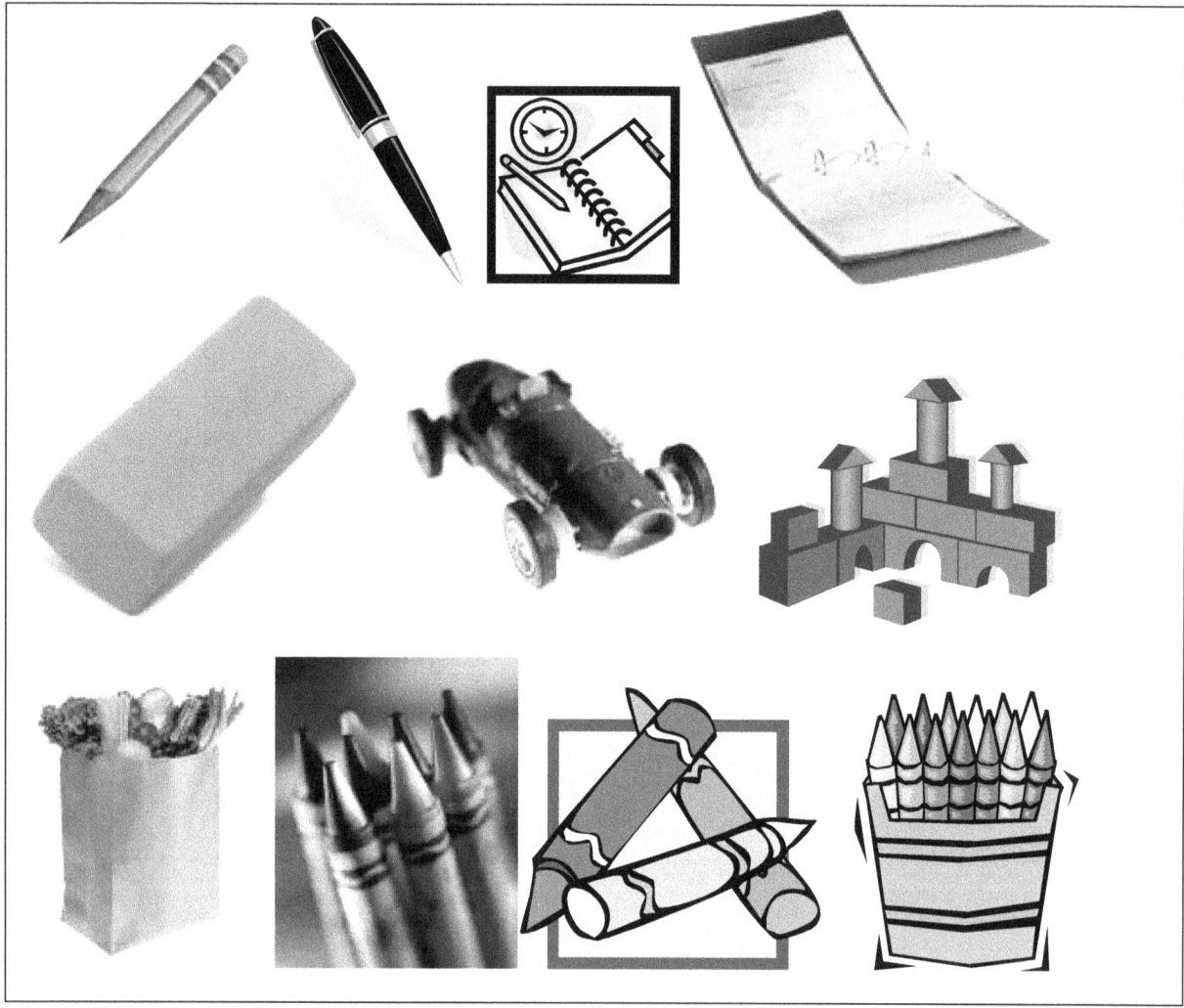

Do not circle 2 toys in the middle row.

Classification
Circle the following objects which you can ride on.

Do not circle the chair and the bed.

Filling in the next object

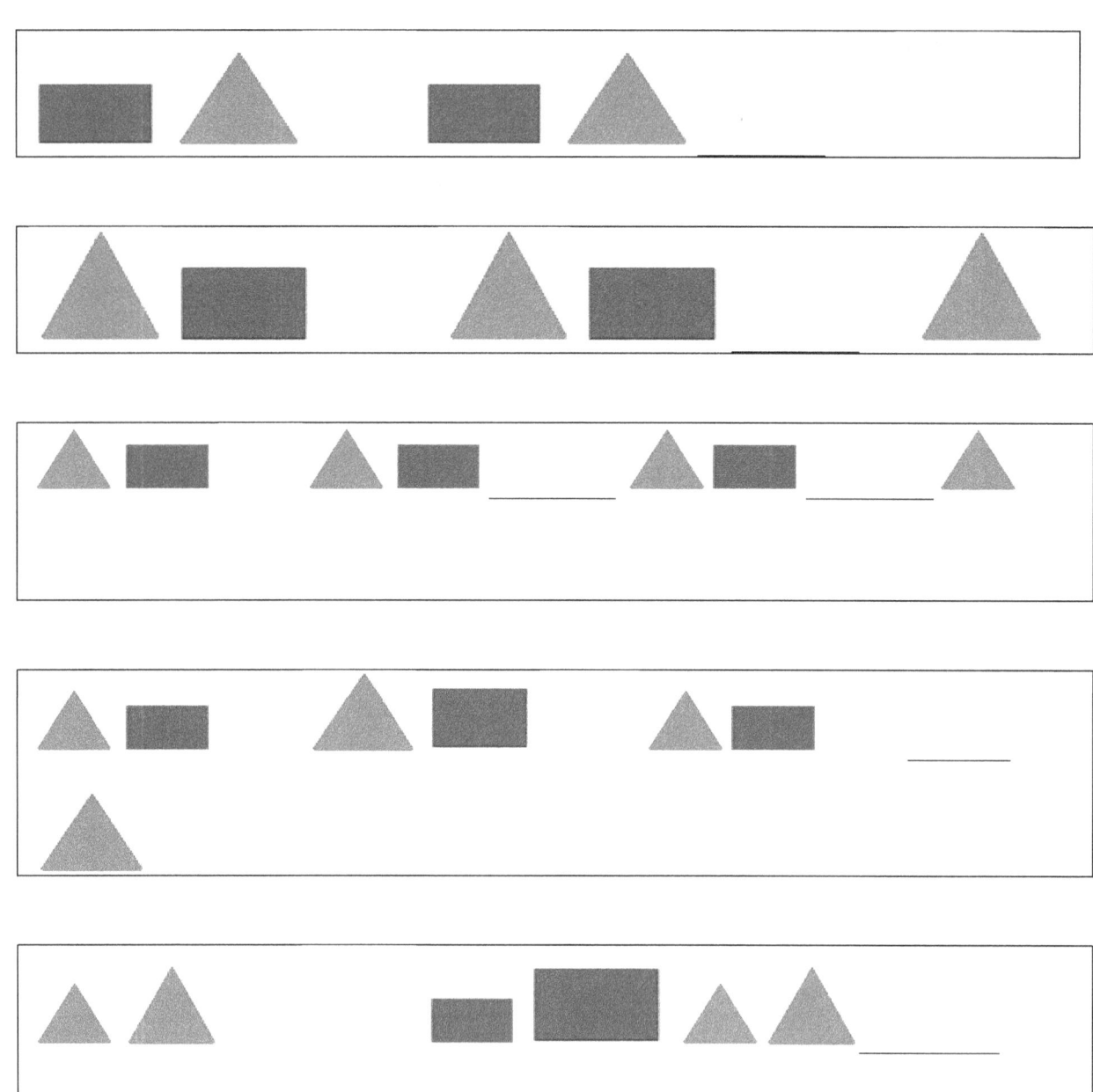

Classification

Circle the following shapes, which are made of curves.

Ho Math Chess — Pre-K and Kindergarten Math

何数棋谜　棋谜式幼儿健脑思维趣味数学

© 2012 – 2021 Frank Ho, Amanda Ho, Canada copyright 1095661, Trademark 771400

Classification

Circle the following shapes, which are made of straight lines.

Ho Math Chess Pre-K and Kindergarten Math

何数棋谜 棋谜式幼儿健脑思维趣味数学

© 2012 – 2021 Frank Ho, Amanda Ho, Canada copyright 1095661, Trademark 771400

Learning numbers 0 to 9 (数字的写法)

(The number writing symbols used here are designed so that there is a very minimal difference between the style of numbers and its popular print fonts.
The most noted number is the writing symbol of four, which is the same as the way it is printed in math textbooks.)

Counting objects	Original number representation (number of acute or right angles)	English words	Trace each of the following numbers.	**Copy each number.**
No bears	0	Zero	0	
(1 bear)	1	One	1	
(2 bears)	Z	Two	2	
(3 bears)	3	Three	3	

www.homathchess.com 17

Ho Math Chess   Pre-K and Kindergarten Math

何数棋谜　棋谜式幼儿健脑思维趣味数学

© 2012 – 2021 Frank Ho, Amanda Ho, Canada copyright 1095661, Trademark 771400

Learning numbers 0 – 9

Counting objects	Original number representation (number of acute or right angles)	English words	Trace each of the following numbers.	Copy each number.
(four teddy bears)	4	Four	4	
(five teddy bears)	5	Five	5	
(six teddy bears)	6	Six	6	

www.homathchess.com

Ho Math Chess   Pre-K and Kindergarten Math

何数棋谜   棋谜式幼儿健脑思维趣味数学

© 2012 – 2021 Frank Ho, Amanda Ho, Canada copyright 1095661, Trademark 771400

Learning numbers 0 – 9

Counting objects	Original number representation (number of acute or right angles)	English words	Trace each of the following numbers.	**Copy each number.**
(7 bears)	7 shape	Seven	7	
(8 bears)	8 shape	Eight	8	
(9 bears)	9 shape	Nine	9	

www.homathchess.com

Ho Math Chess   Pre-K and Kindergarten Math

何数棋谜   棋谜式幼儿健脑思维趣味数学

© 2012 – 2021 Frank Ho, Amanda Ho, Canada copyright 1095661, Trademark 771400

Learning numbers 0 – 9

Number writing	An object resembling each number	Visual presentation
(0)	A cookie It starts to write the letter "C".	(cookie image)
(1)	A pencil or a pen	(pencil image)
(2)	A swan's head and neck	(swan image)
(3)	A butterfly's wing	(butterfly image)

www.homathchess.com   20

Ho Math Chess   Pre-K and Kindergarten Math

Learning numbers 0 – 9

Number writing	An object resembling each number	Visual presentation
4	A triangular flag	
5	A teapot	
6	A whistle or a toothbrush or a golf club	

Ho Math Chess — Pre-K and Kindergarten Math

何数棋谜　棋谜式幼儿健脑思维趣味数学

© 2012 – 2021 Frank Ho, Amanda Ho, Canada copyright 1095661, Trademark 771400

Learning numbers 0 – 9

Number writing	An object resembling each number	Visual presentation
7	A cane or a rectangle cut in half.	(cane, 7, rectangle)
8	A pretzel or a chain fence. It starts to write a letter "S".	(chain fence, 8, pretzel)
9	A whistle or a toothbrush or a golf club	(golf club, 9, whistle, toothbrush)

www.homathchess.com

Circling digits 0 to 9

Circle all digits 0.

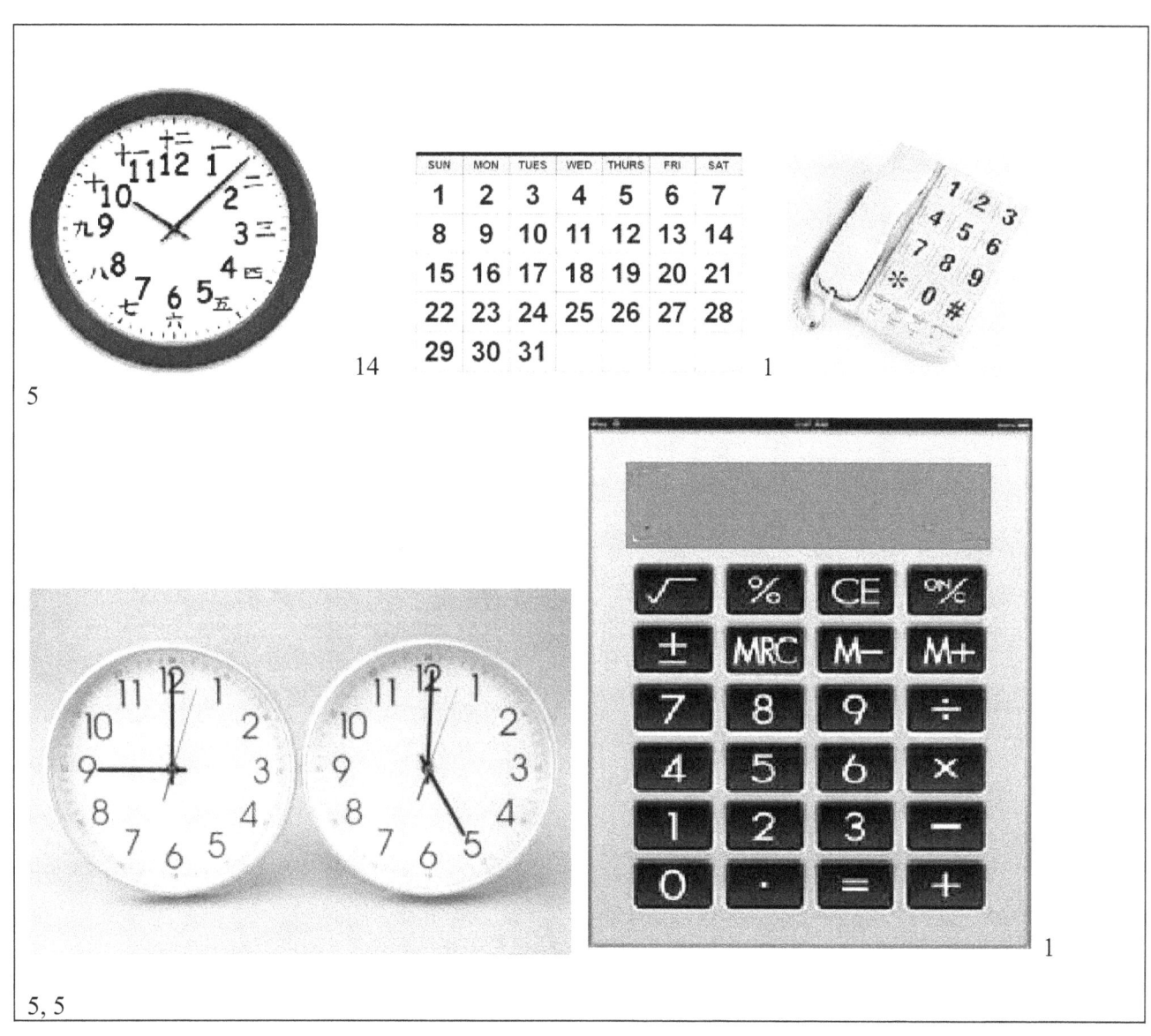

Circle all digits 1.

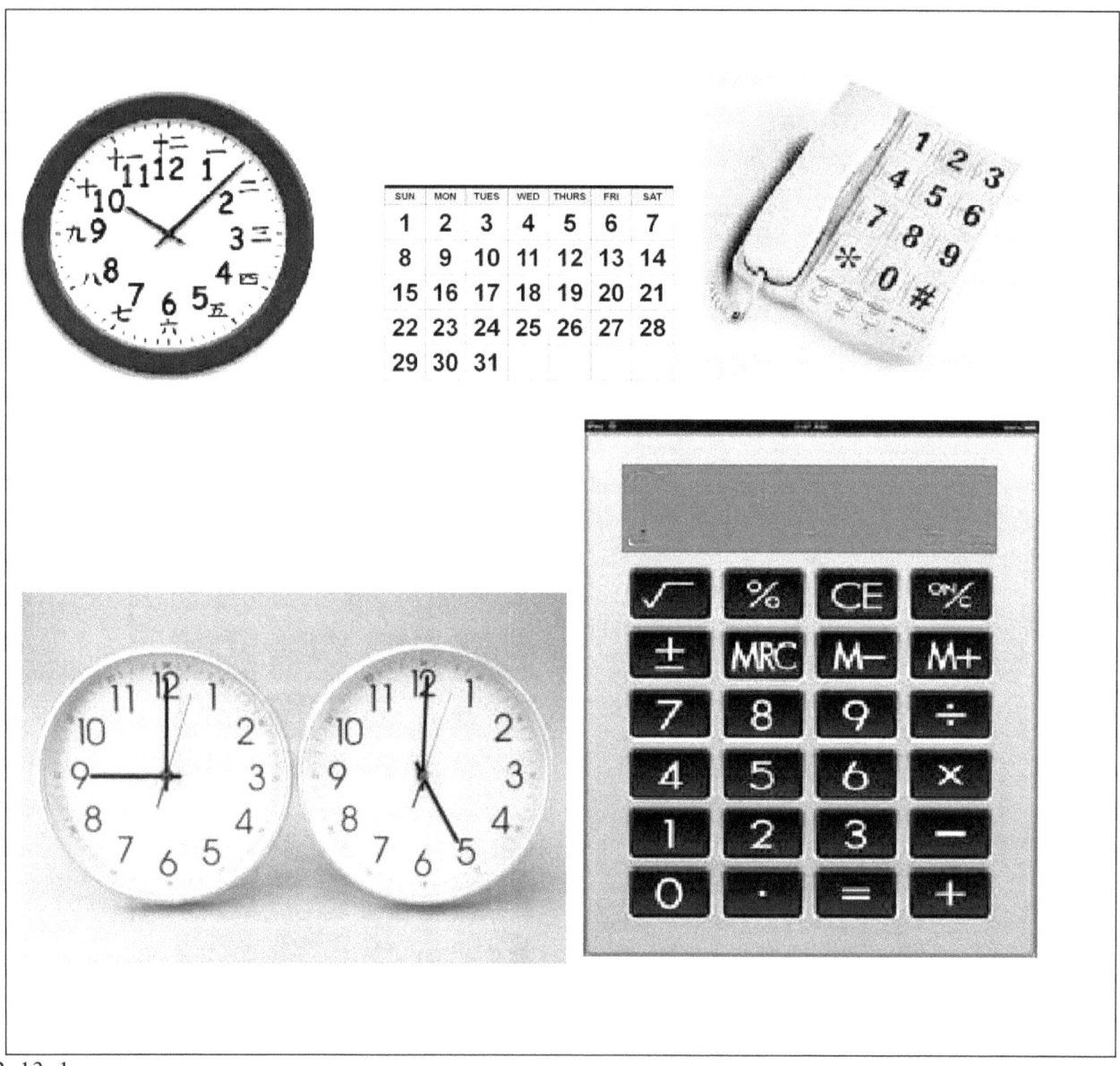

2, 13, 1,
2, 2, 1

Circle all digits 2.
Total 21

1, 5, 1
1, 1, 1

Circle all digits 3.
Total 10

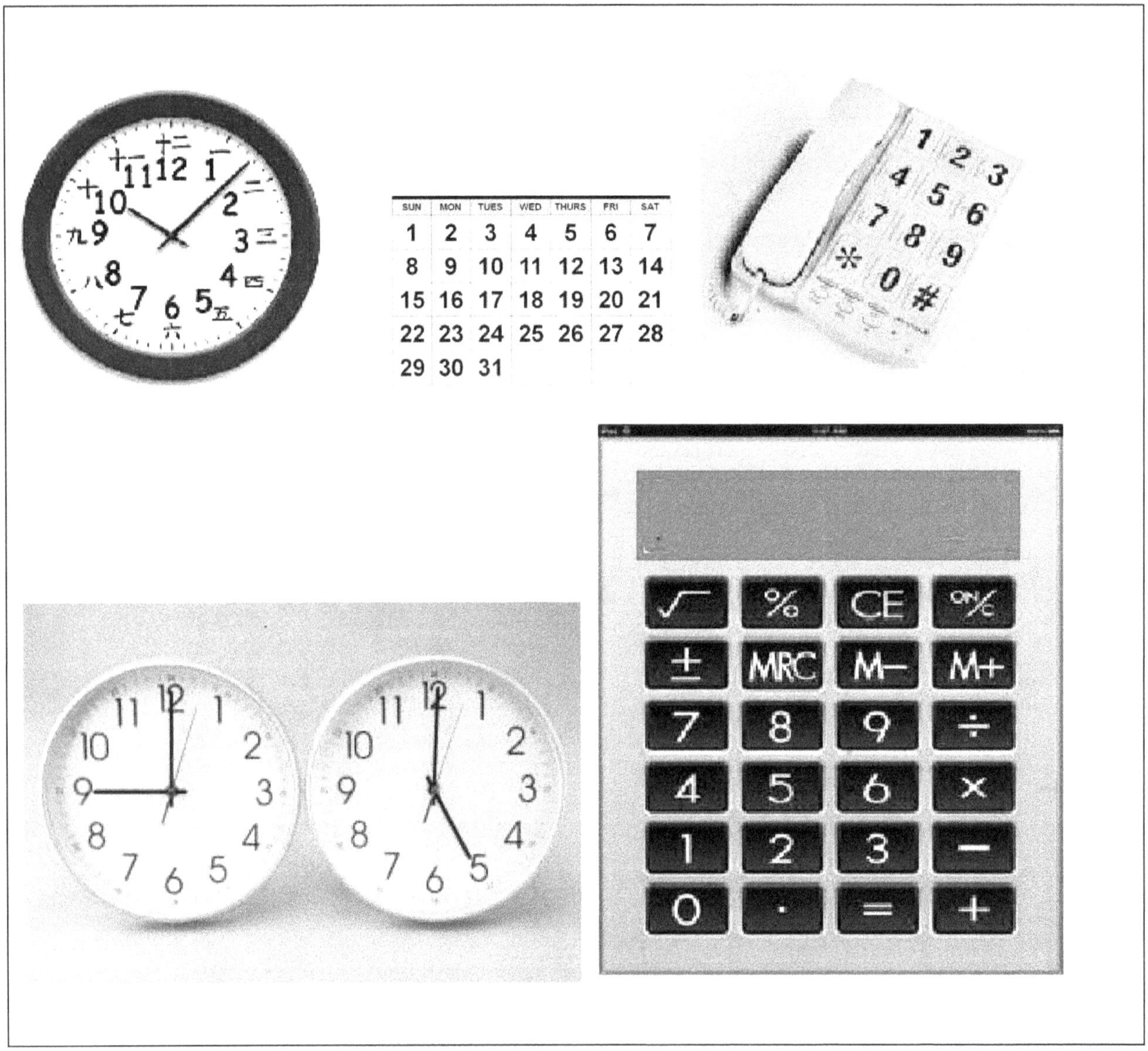

Circle all digits 4.
Total 8

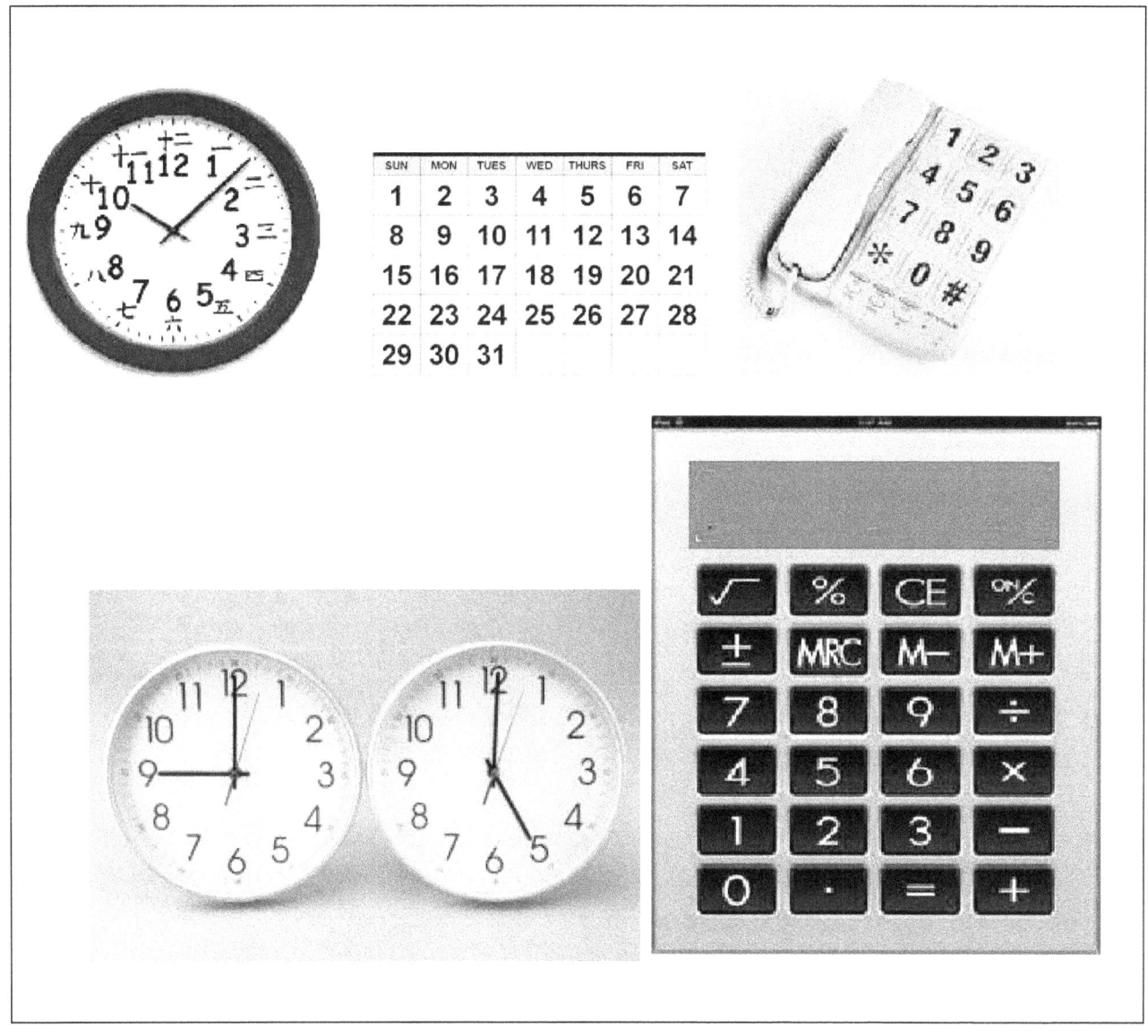

Circle all digits 5.
Total 8

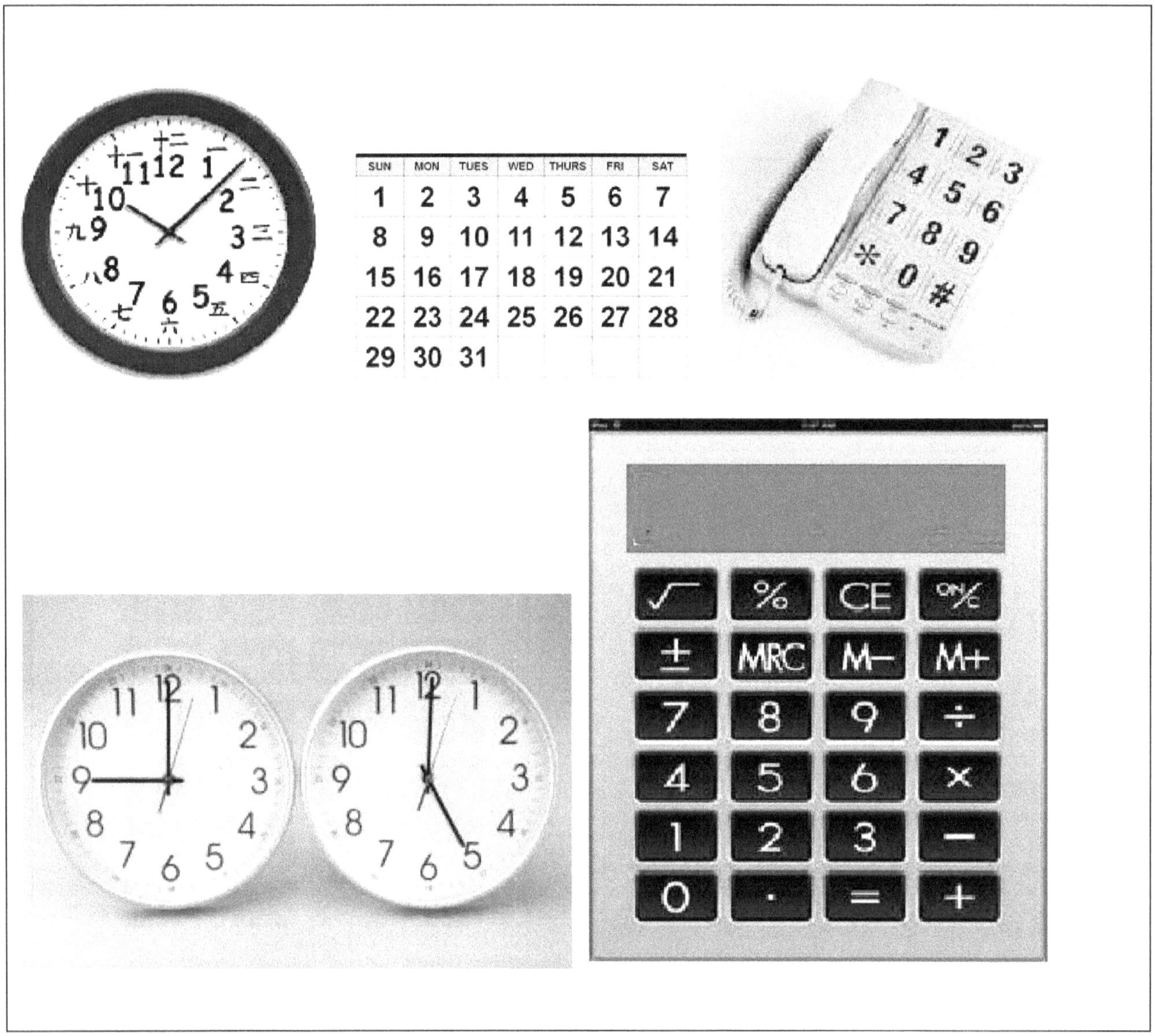

Circle all digits 6.

Total 8

Circle all digits 7.
Total 8

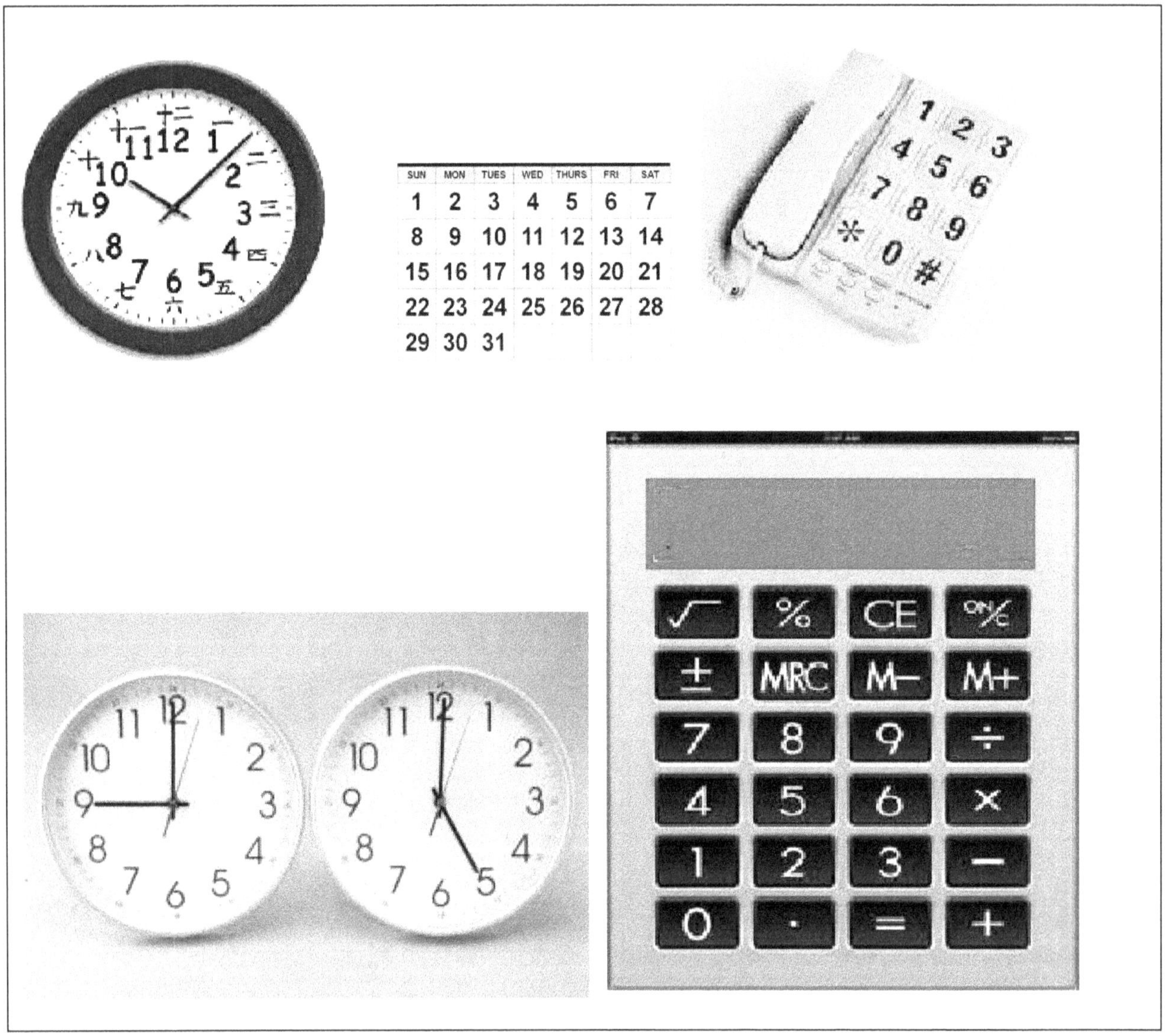

Circle all digits 8.
Total 8

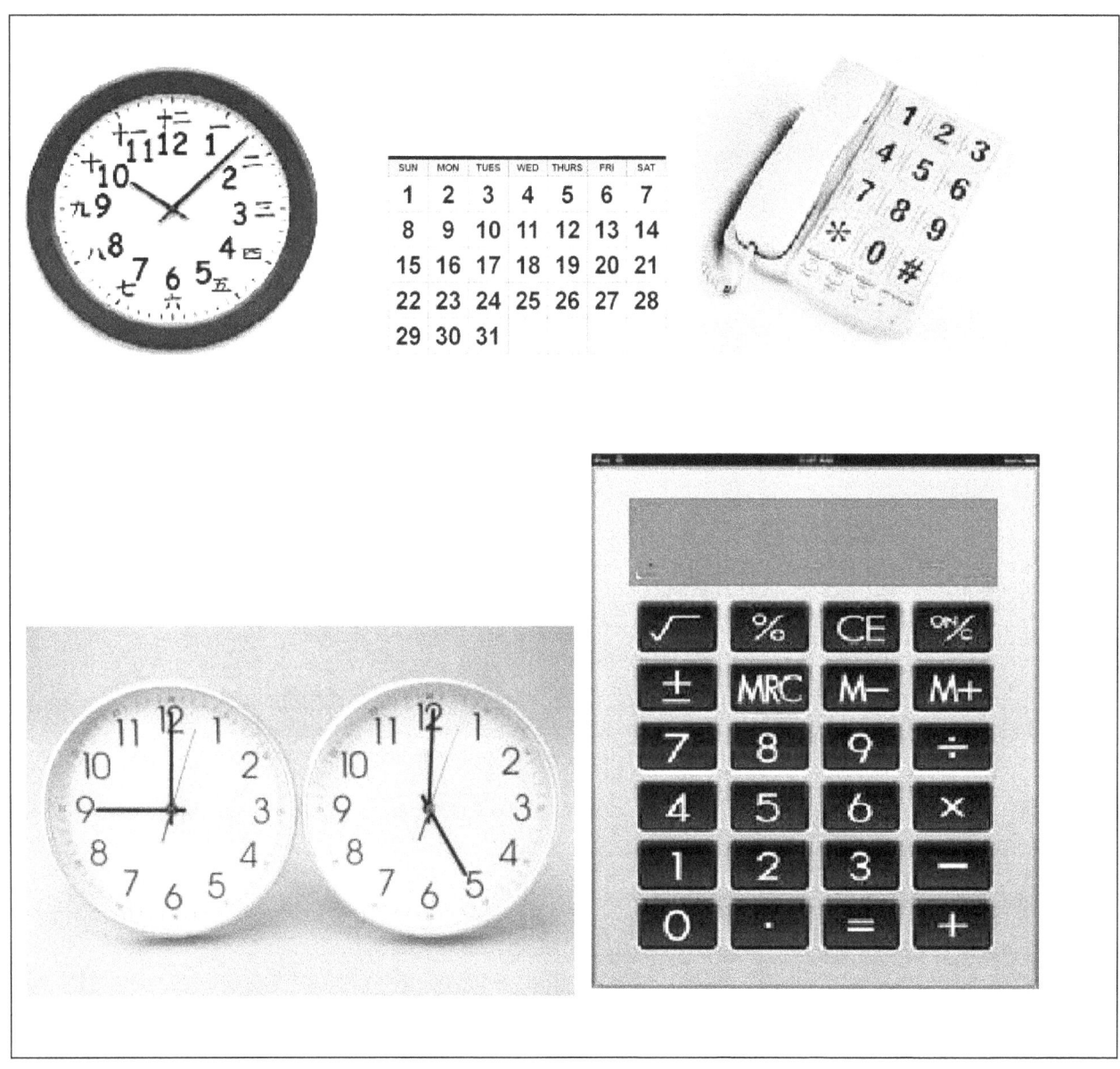

Circle all digits 9.
Total 8

Ho Math Chess — Pre-K and Kindergarten Math

Images of numbers

Numbers	Description	Visualized image
1	One is a pencil.	
2	Two is a swan's head and neck.	
3	There is a heart.	
4	Four is a triangular flag.	
5	Five is a teapot.	
6	Six is a whistle,	

Images of numbers

Numbers	Description	Visualized image
7	Seven is a cane. A half of a rectangle forms seven.	
8	Eight is a pretzel.	
9	Nine is an upside-down 6. 9 is 6 turned clockwise 180^0.	

Tracing numbers

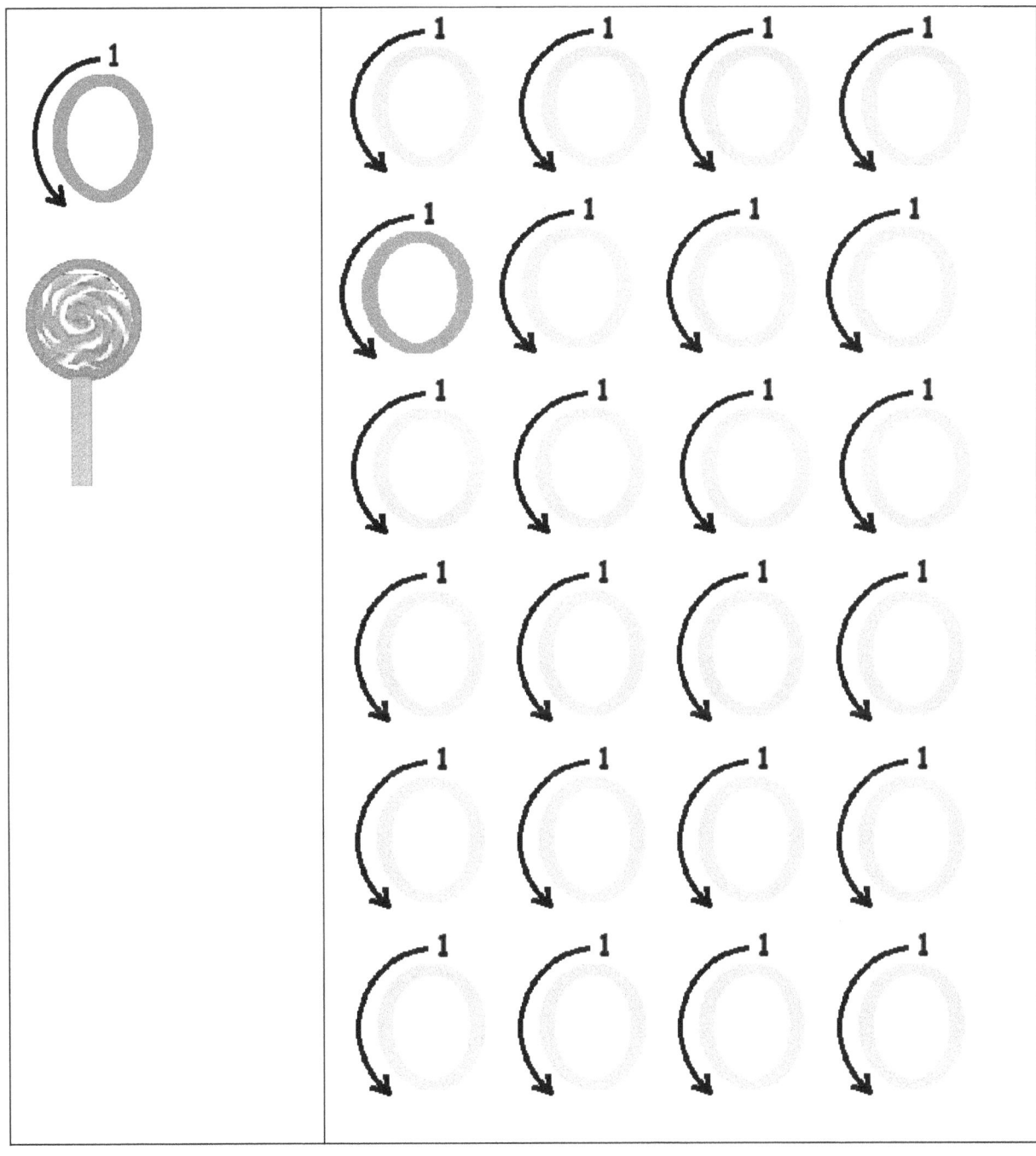

Tracing numbers

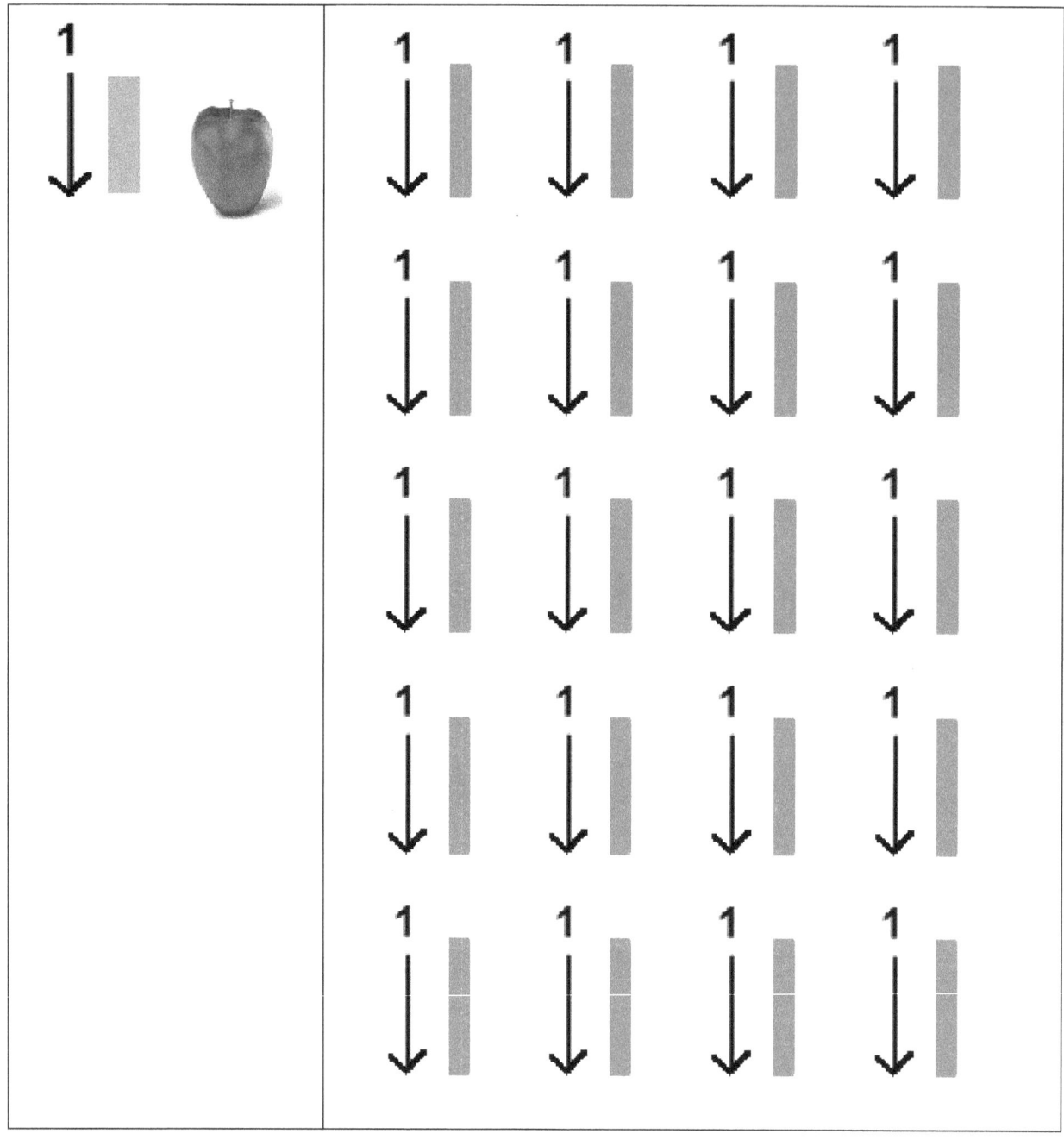

Tracing numbers

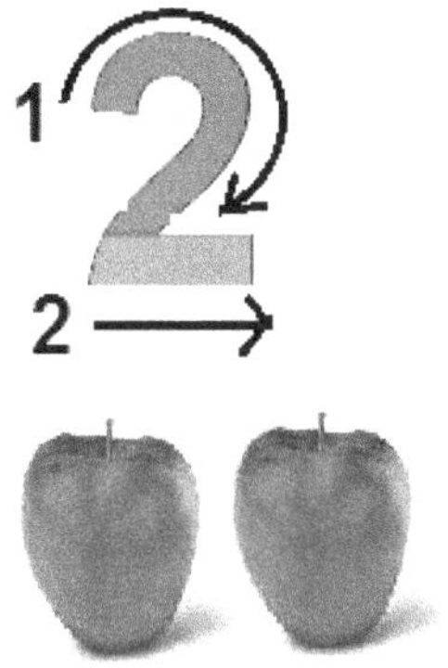

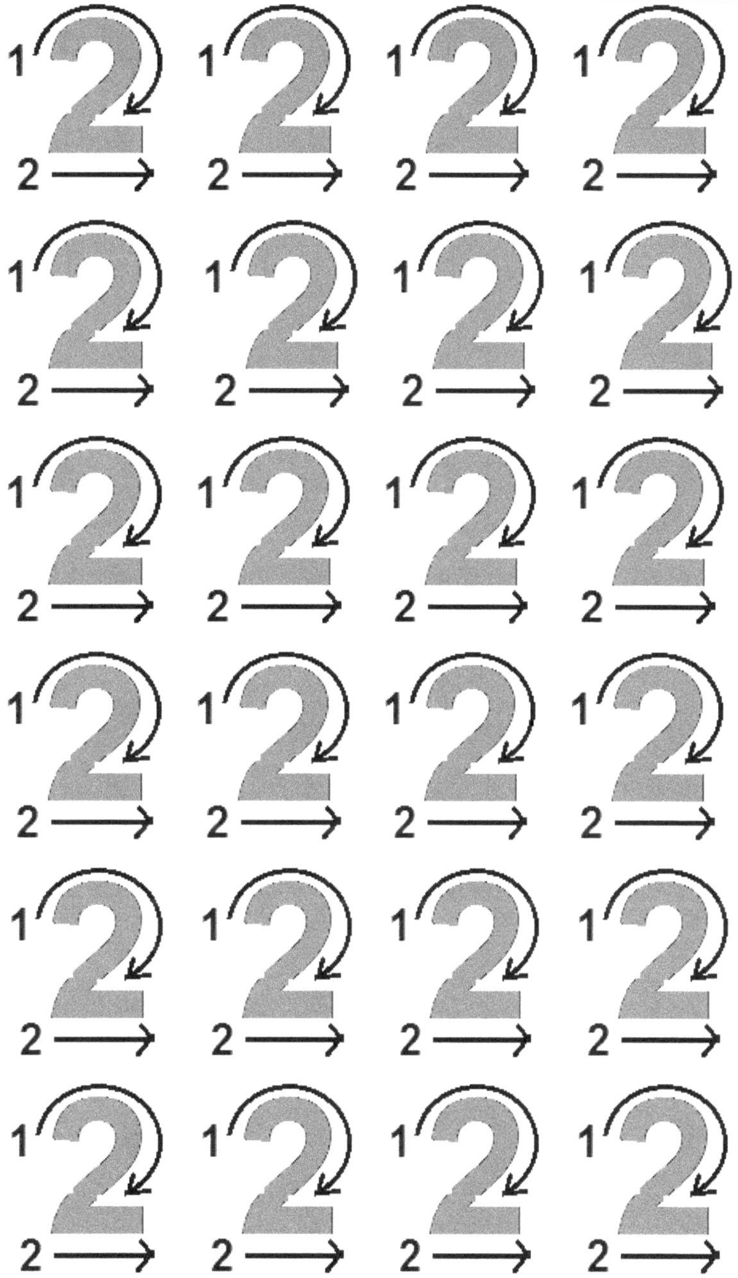

Tracing numbers

Tracing numbers

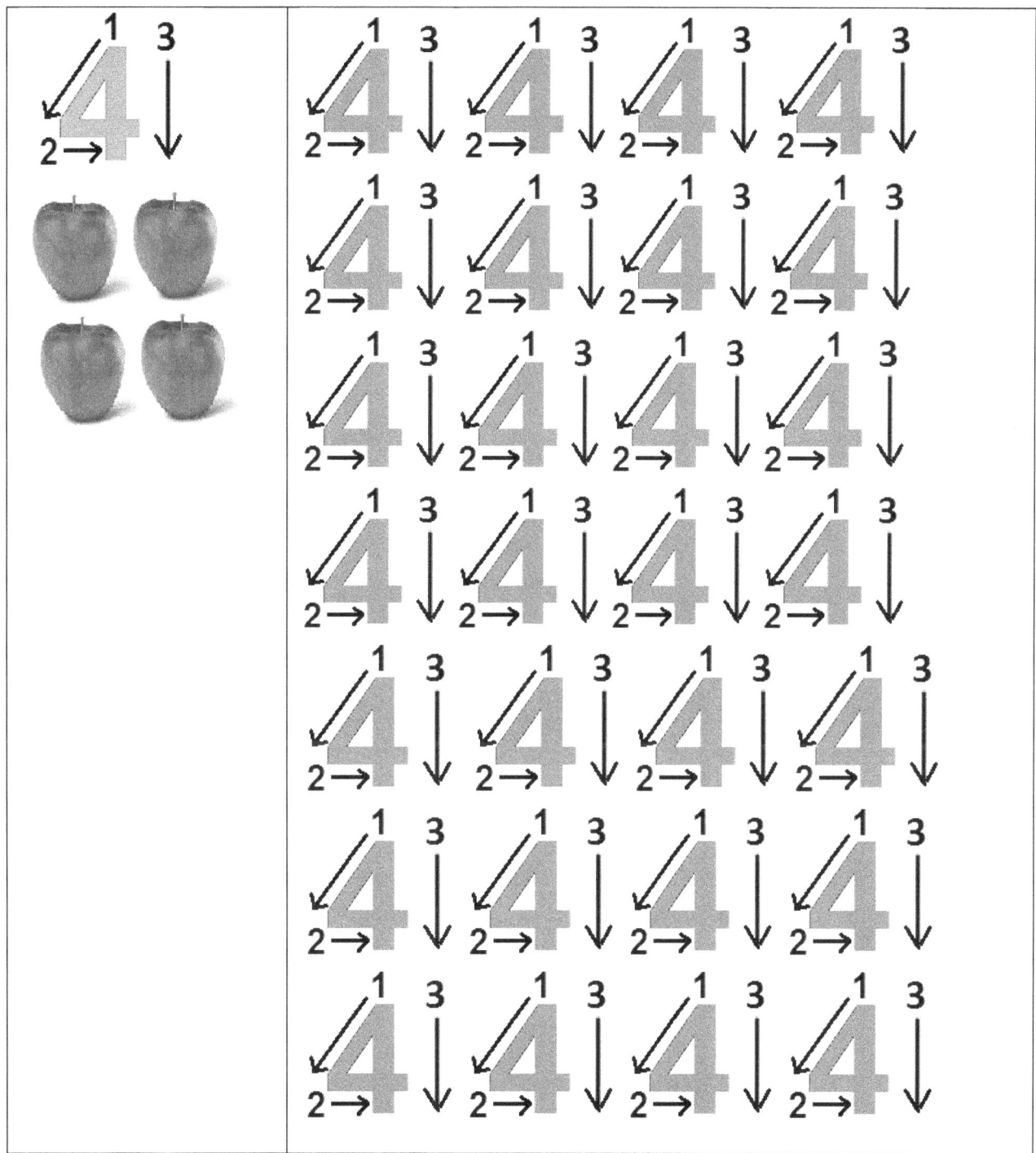

Tracing numbers

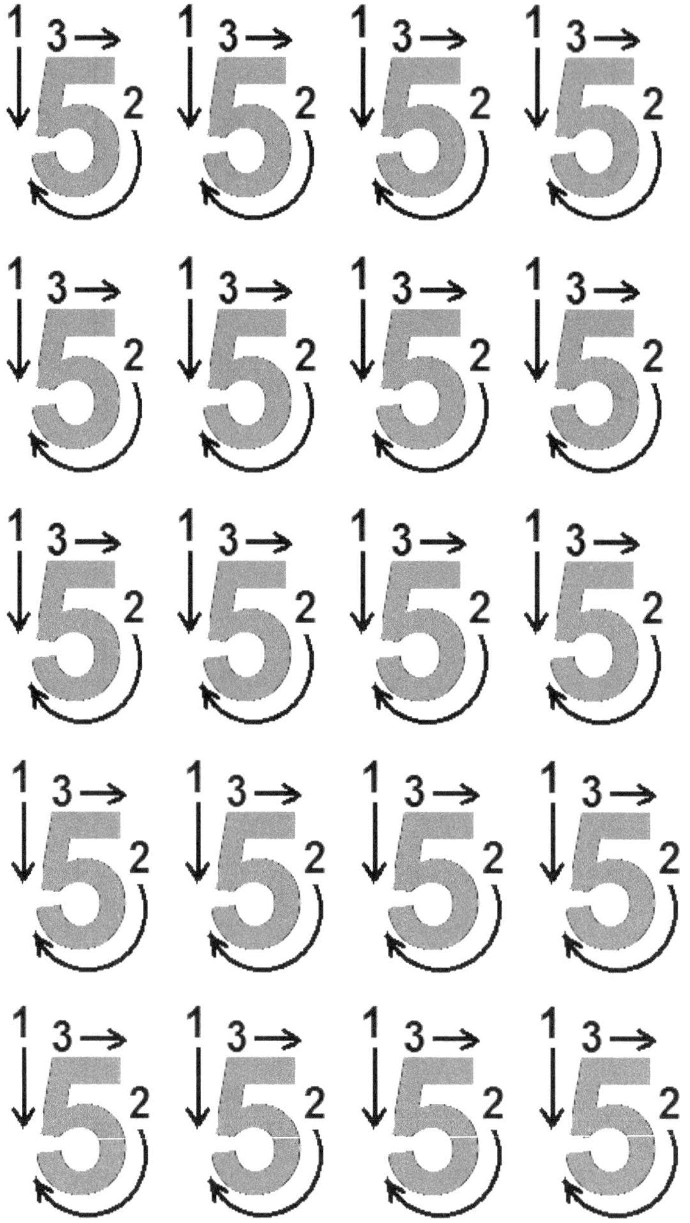

Ho Math Chess — Pre-K and Kindergarten Math

Tracing numbers

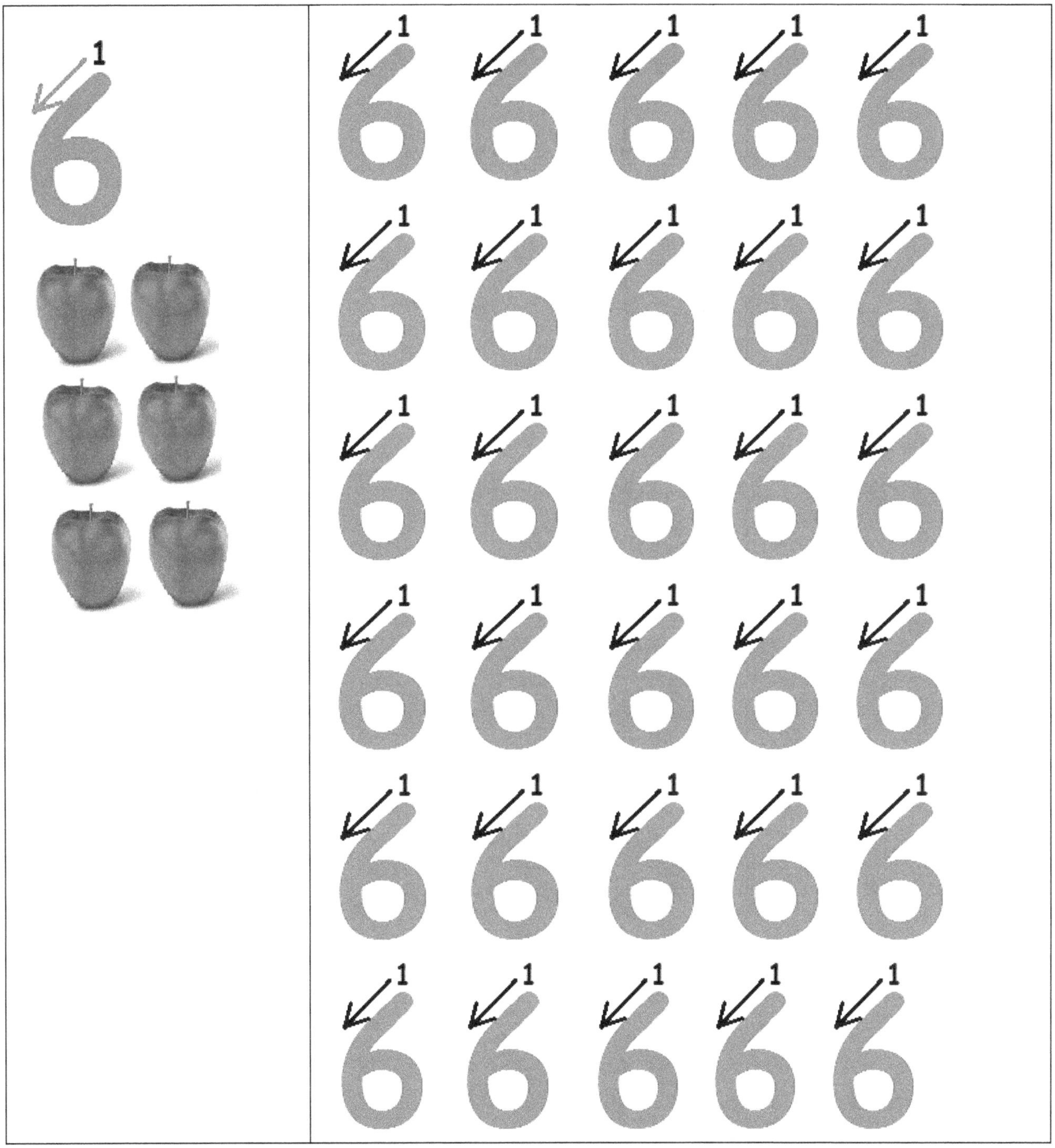

Tracing numbers

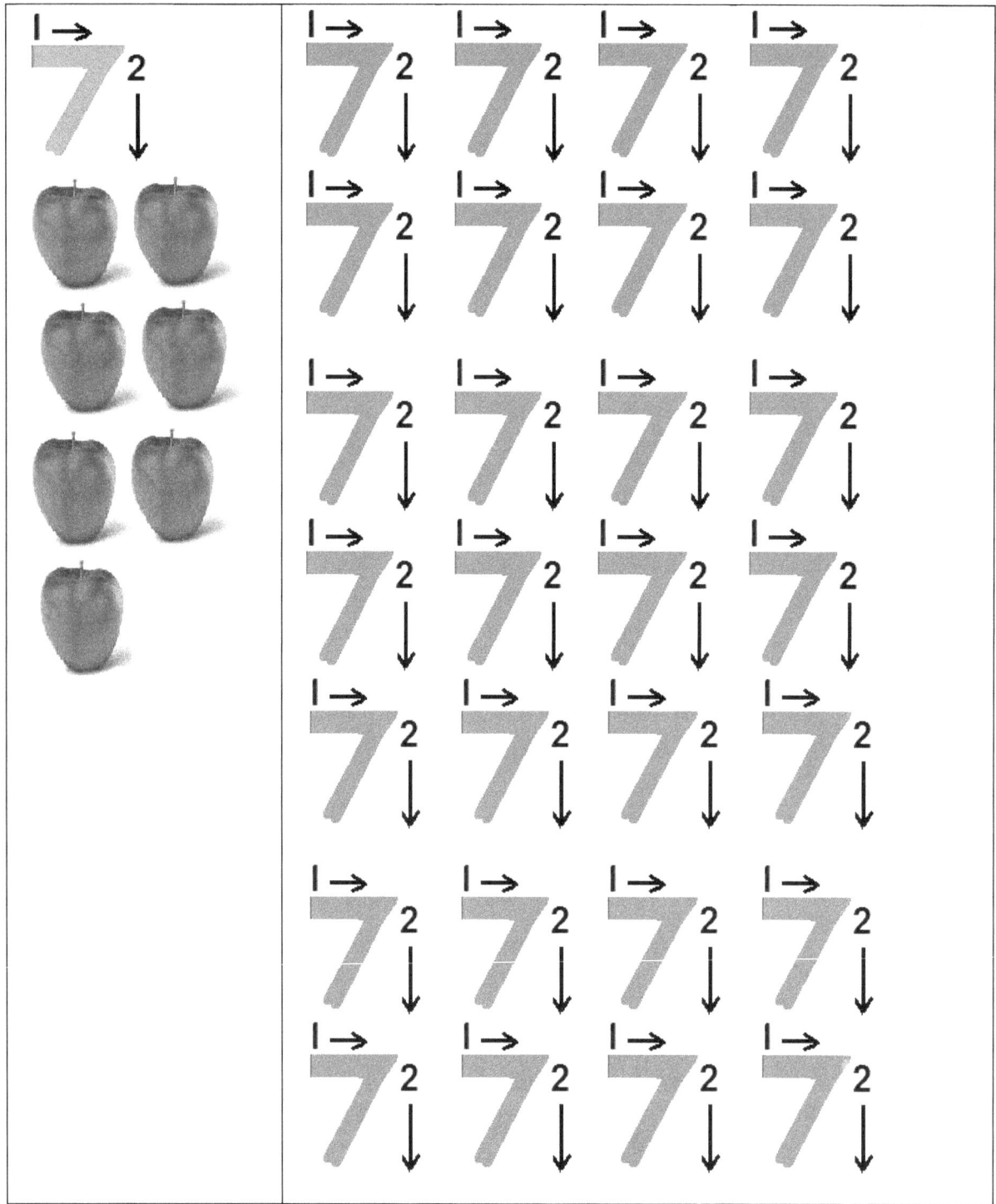

Tracing numbers

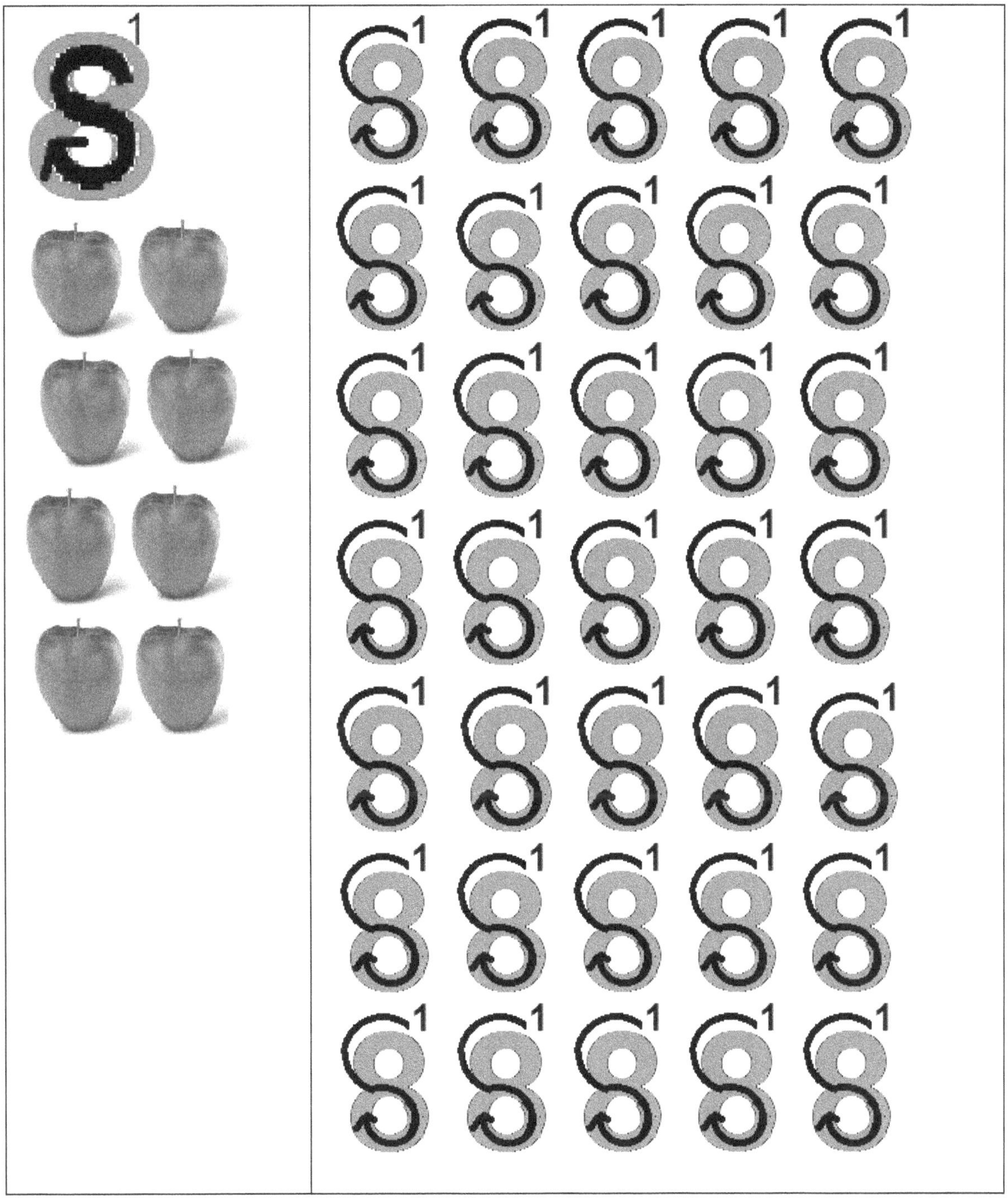

Tracing numbers

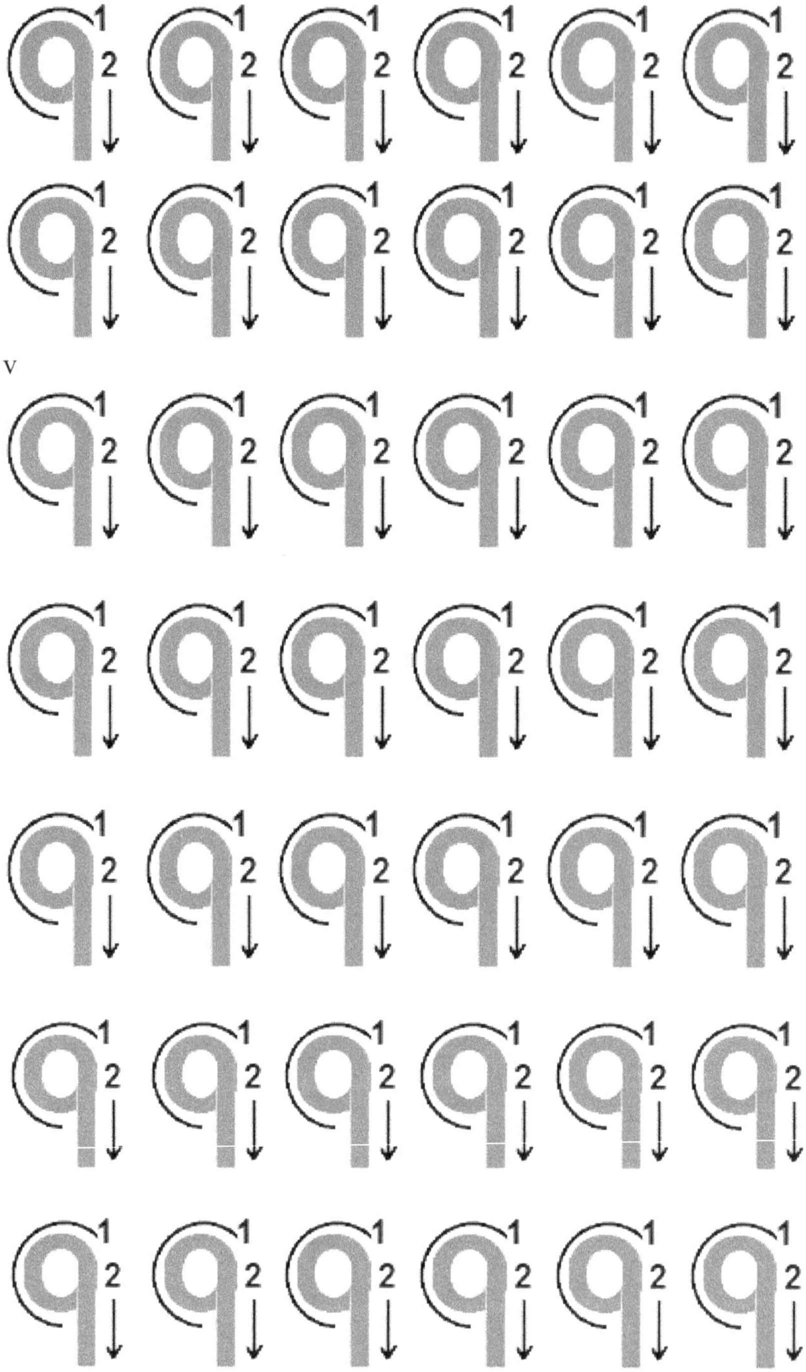

Tracing and copying

Print font (Times New Roman)	Trace handwriting	Copy handwriting number	Copy handwriting number
0	0		
1	1		
2	2		
3	3		
4	4		
>	>		
<	<		
=	=		

Ho Math Chess — Pre-K and Kindergarten Math

何数棋谜　棋谜式幼儿健脑思维趣味数学

© 2012 – 2021 Frank Ho, Amanda Ho, Canada copyright 1095661, Trademark 771400

Tracing and copying

Print font (Times New Roman\)	Trace handwriting	Copy handwriting number	Copy handwriting number
5	5		
6	6		
7	7		
8	8		
9	9		
&	&		
$	$		

Tracing and copying

Print font (Times New Roman\)	Trace handwriting	Copy handwriting number	Copy handwriting number
{	{		
}	}		
[ ]	[ ]		
%	%		
?	?		
@	@		
#	#		
!	!		

Tracing 1 to 9

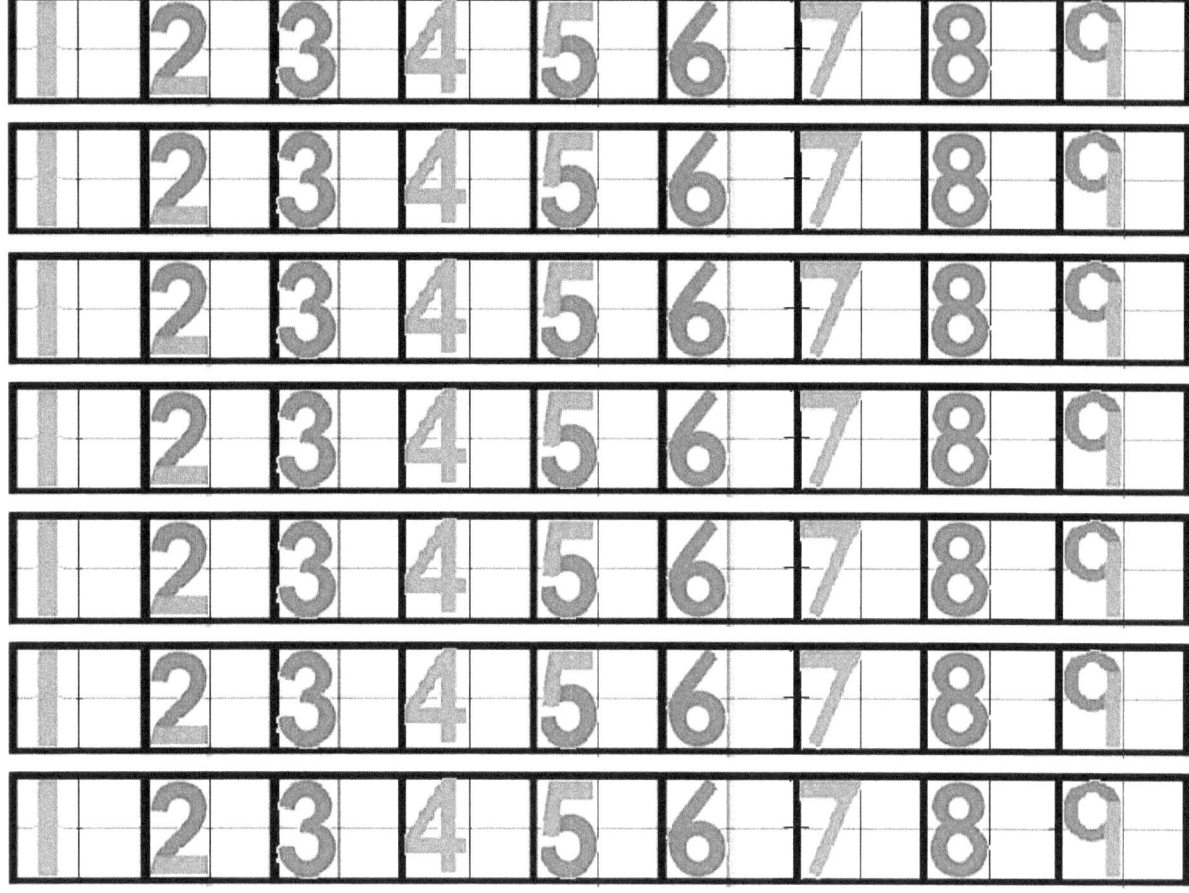

Ho Math Chess   Pre-K and Kindergarten Math
何数棋谜　棋谜式幼儿健脑思维趣味数学
© 2012 – 2021 Frank Ho, Amanda Ho, Canada copyright 1095661, Trademark 771400

Tracing 1 to 9

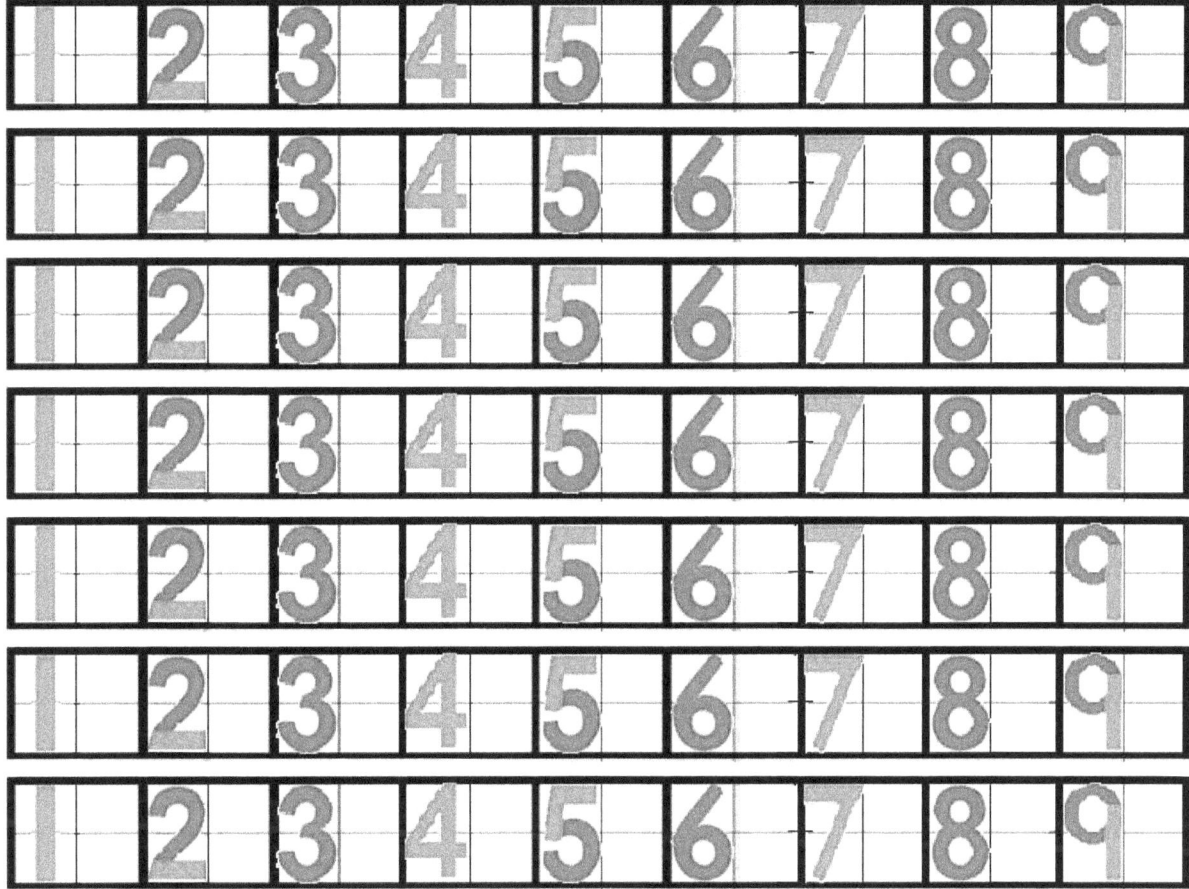

Tracing 1 to 9

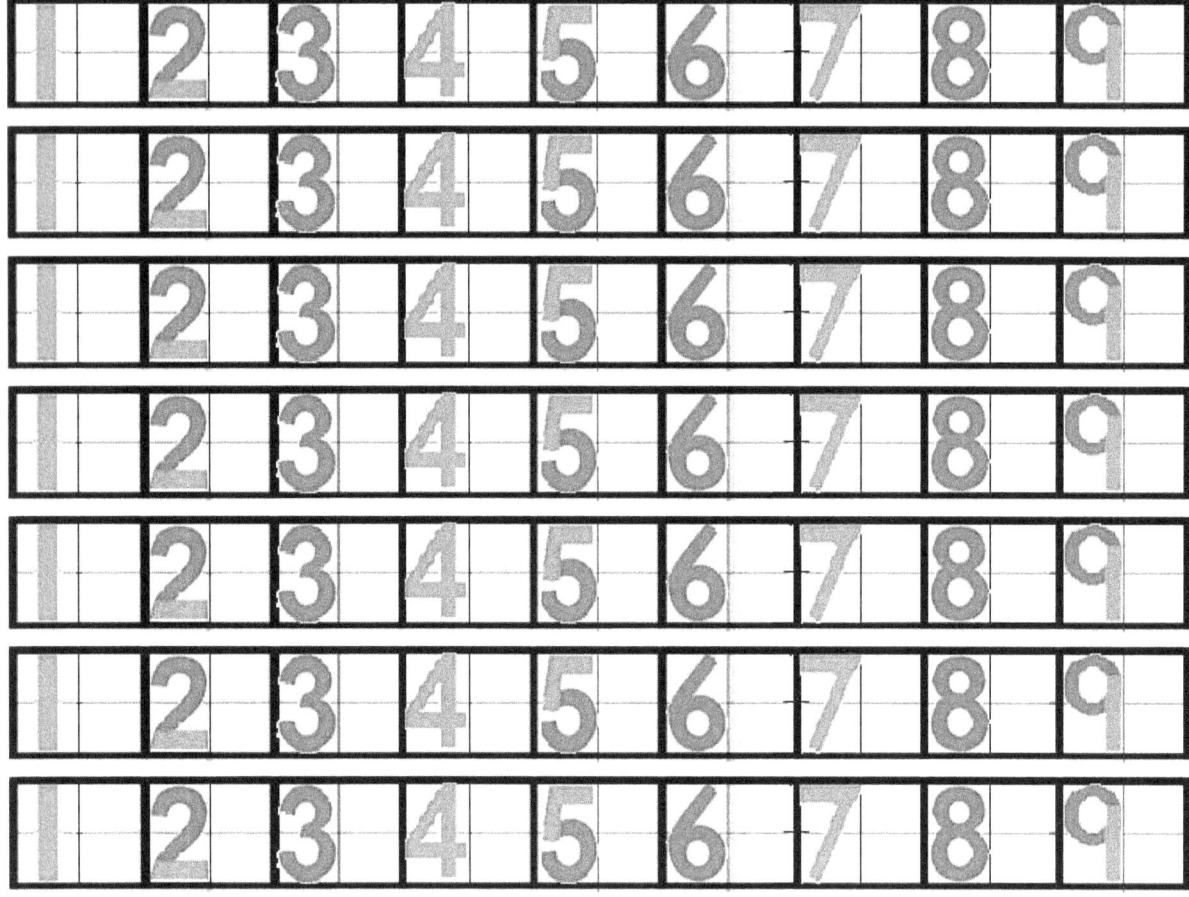

Tracing 1 to 9

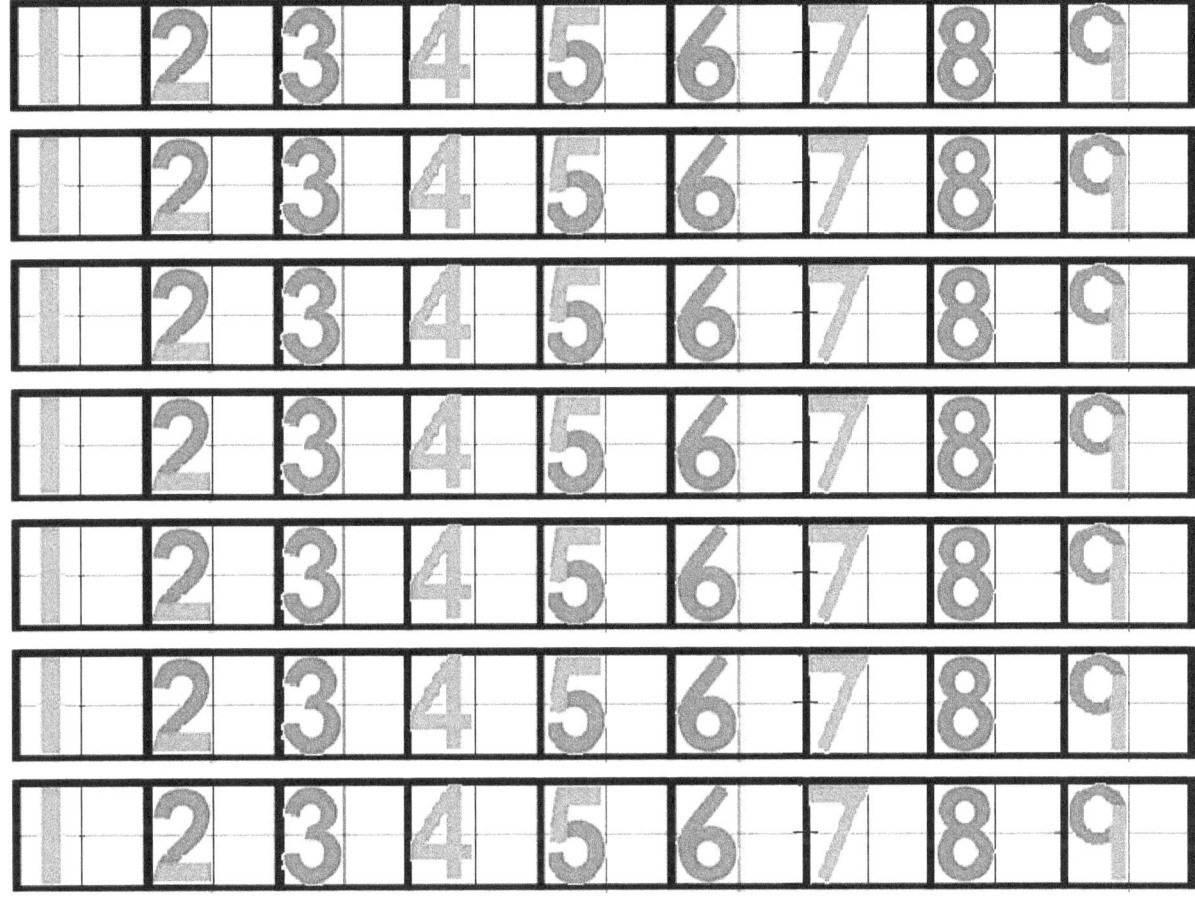

Ho Math Chess — Pre-K and Kindergarten Math

Oral practice

1. How many people are there in your family? Write down the number you answered. _____
2. How many eyes do you have? Write down the number you answered. _____ 2
3. How many fingers do you have on one hand? Write down the number you answered. _____ 5
4. How many legs does a dog have? Write down the number you answered. _____ 4
5. How many pairs of shoes do you have? Write down the number you answered. _____
6. How many books do you have? Write down the number you answered. _____
7. How old are you? Write down the number you answered. _____
8. How many cats do you have in your house? Write down the number you answered. _____
9. What is your home phone number? Write down the number you answered. _____
10. What is your house number? Write down the number you answered. _____

Oral practice

Write numbers from 1 to 9.
Write numbers from 9 to 1.
Write numbers from 2 to 9.
Write numbers from 9 to 2.
Write numbers from 3 to 8.
Write numbers from 8 to 3.
Write numbers from 4 to 7.
Write numbers from 7 to 4.
Write numbers from 3 to 6.

Ho Math Chess — Pre-K and Kindergarten Math

何数棋谜　棋谜式幼儿健脑思维趣味数学

© 2012 – 2021 Frank Ho, Amanda Ho, Canada copyright 1095661, Trademark 771400

Finding answers from numbers 1, 2, 3, 4, 5, 6, 7, 8, and 9.

1 2 3 4 5 6 7 8 9

1. Which three numbers each is at least made of one circle? _____   689

2. Which 6 numbers each is made of at least one straight line? _____   124579

3. Which three numbers each is at least made of one half circle? _____   235

4. Which two numbers each is made of only one-half circle? _____   25

5. Which numbers are written without lifting your pencil? _____   1236789

6. Which umbers each has at least two straight lines? _____   457

7. Which number is made of two circles? _____   8

8. Which numbers are only made of curves? _____   368

9. Which numbers only have straight lines? _____   147

Matching

Matching parts of lines or curves to the whole number

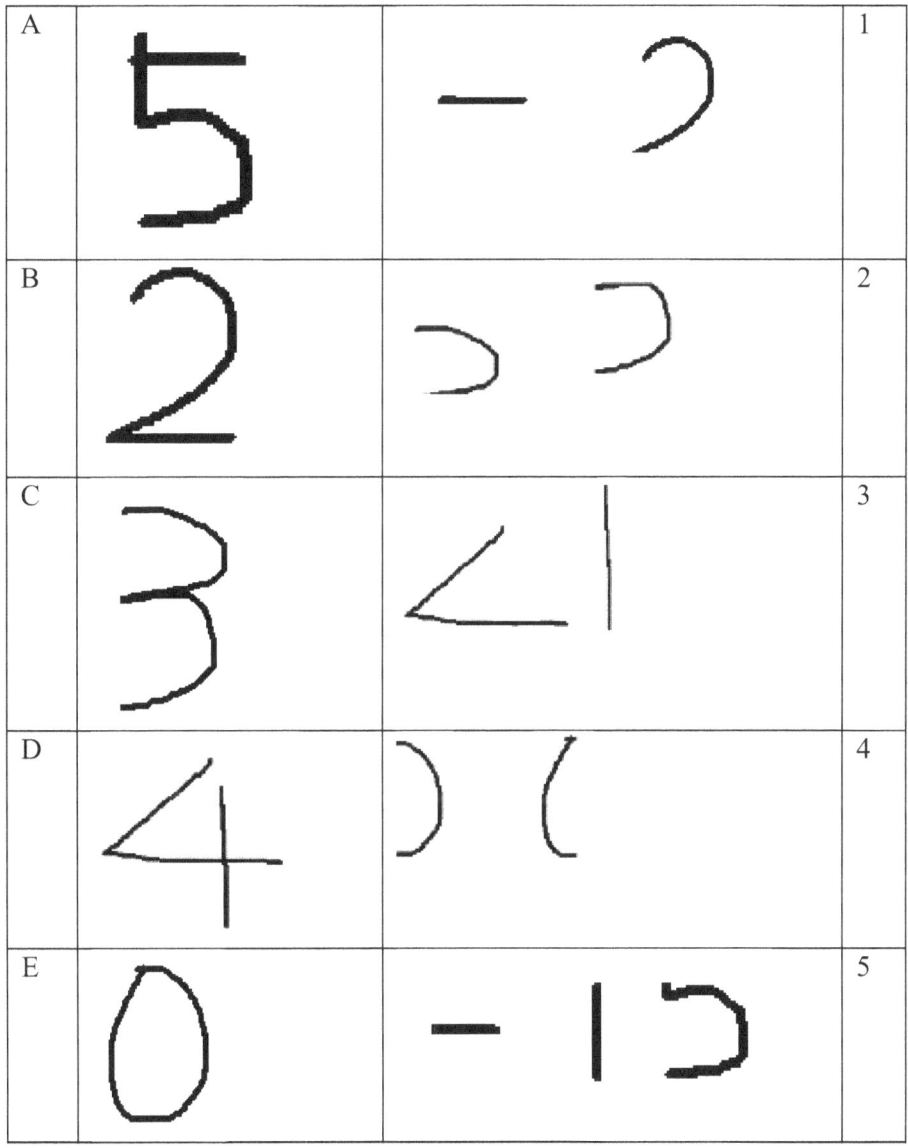

A5, B1, C2, D3, E4

Circling the correct writing to match the top number

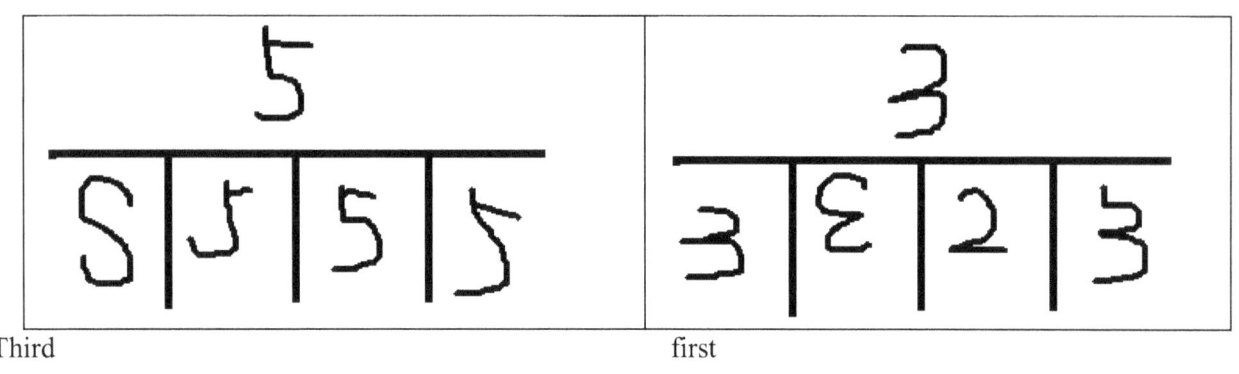

Third   first

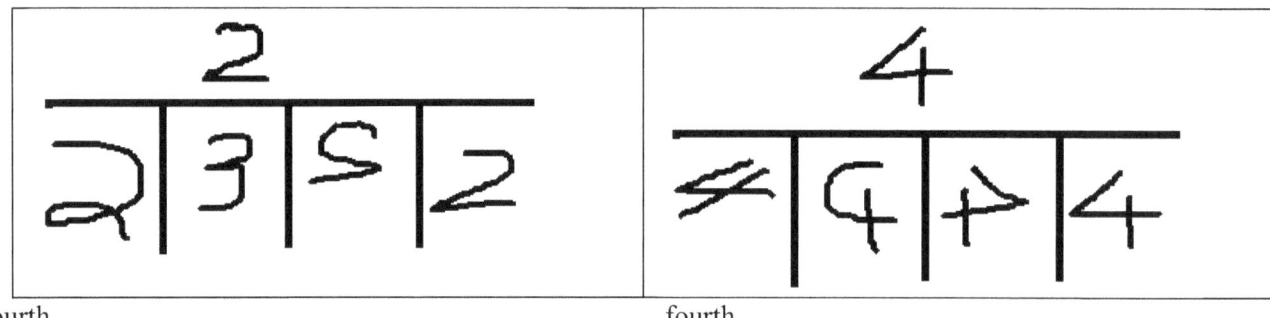

fourth   fourth

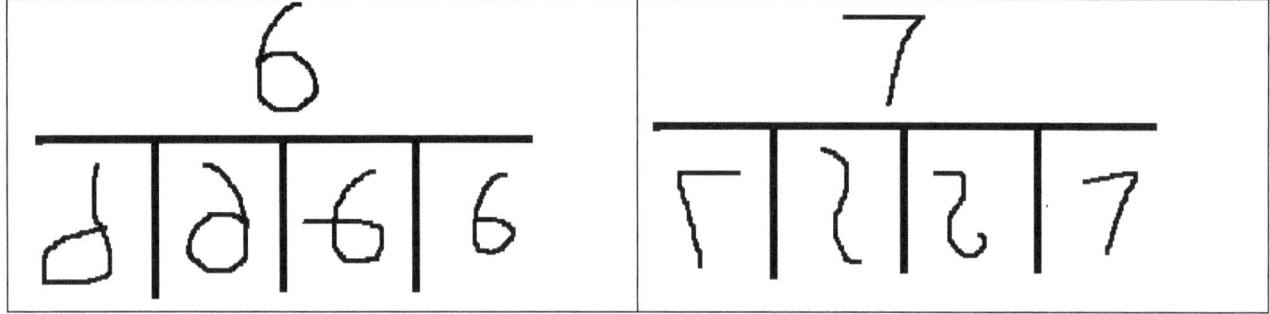

fourth   fourth

Ho Math Chess — Pre-K and Kindergarten Math

何数棋谜　棋谜式幼儿健脑思维趣味数学

© 2012 – 2021 Frank Ho, Amanda Ho, Canada copyright 1095661, Trademark 771400

Finding missing number.

找遗失的数字

Each number sequence has numbers from 0 to 9, but one of them is missing, find the missing number and write it.

1 3 5 7 9 2 4 6 8	Answer ____ 0
2 5 4 9 1 3 7 8 0	Answer ____ 6
5 3 9 7 4 1 0 6 2	Answer ____ 8
5 8 0 6 9 2 4 7 1	Answer ____ 1
4 3 5 7 0 2 8 6 1	Answer ____ 9
0 8 3 5 1 9 2 4 6	Answer ____ 1
3 1 8 0 2 5 6 9 5	Answer ____ 4
4 3 5 8 9 0 1 6 7	Answer ____ 2
4 6 5 7 9 2 0 8 3	Answer ____ 1
1 8 0 7 9 5 4 3 2	Answer ____ 6
9 3 5 7 1 2 0 6 4	Answer ____ 5
5 3 0 7 9 2 4 6 1	Answer ____ 6
1 3 0 7 9 2 8 6 5	Answer ____ 4
6 1 3 5 9 2 4 0 7	Answer ____ 8
1 0 3 7 5 2 4 9 8	Answer ____ 6

www.homathchess.com

Counting forward and backward (向前数, 向後数)

You are going upstairs. Write numbers from 1 (the bottom stair) to the top stair.

How many numbers are in the picture?
_____ 6
The largest number is _____ 6.
The smallest number is _____ 1.
The two middle numbers are _____ 3, 4.

You are going downstairs. Write numbers from 9 to 1 on steps from top to down.

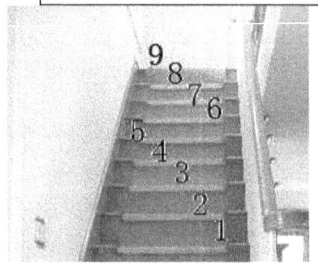

Ho Math Chess   Pre-K and Kindergarten Math

何数棋谜   棋谜式幼儿健脑思维趣味数学

© 2012 – 2021 Frank Ho, Amanda Ho, Canada copyright 1095661, Trademark 771400

Counting forward and backwards

You are going downstairs. Write numbers from 1 (the top stair) to the bottom stair.

How many numbers are in the picture?
_____   6
The largest number is _____ 6.
The smallest number is _____ 1.
The middle number is _____   3,4

You are going upstairs. Write numbers from 1 to 9 on stairs from bottom to top.

Counting forward and backwards

You are going upstairs. Write numbers from 1 (the bottom stair) to 9.

How many numbers are in the picture?
_____   9
The largest number is _____   9.
The smallest number is _____   1.
The middle number is _____   5.

You are going upstairs. Write numbers from 1 to 9 on stairs from bottom to top.

Counting forward and backwards

You are going upstairs. Write numbers from 2 (the bottom stair) to 9.

How many numbers are in the picture? _____
The largest number is _____
The smallest number is _____
The two middle numbers are _____

You are going upstairs. Write numbers from 1 to 9 on stairs from top to bottom.

8, 8, 2, 5 6

Counting forward and backwards

You are going upstairs. Write numbers from 1 (the bottom stair) to 9.

How many numbers are in the picture? _____
The largest number is _____
The smallest number is _____
The middle number is _____.

You are going upstairs. Write numbers from 1 to 9 on stairs from bottom to top.

9, 9, 1, 5

Ho Math Chess — Pre-K and Kindergarten Math

Saying numbers forward

Say numbers from 1 to 3. (Do not say "1 to 3".)	Write numbers separated by "." requested on the left column.
Say numbers from 1 to 3.	
Say numbers from 2 to 3.	
Say numbers from 2 to 4.	
Say numbers from 3 to 4.	
Say numbers from 3 to 5.	
Say numbers from 4 to 5.	
Say numbers from 4 to 6.	
Say numbers from 5 to 6.	
Say numbers from 1 to 3.	
Say numbers from 2 to 5.	
Say numbers from 1 to 3.	
Say numbers from 3 to 5.	
Say numbers from 4 to 4.	
Say numbers from 3 to 6.	

Saying numbers backwards

Say numbers from 3 to 1.
Say numbers from 4 to 1.
Say numbers from 4 to 2.
Say numbers from 5 to 3.
Say numbers from 6 to 4.
Say numbers from 3 to 5.
Say numbers from 6 to 3.
Say numbers from 5 to 2.
Say numbers from 4 to 2.
Say numbers from 3 to 1.
Say numbers from 4 to 1.
Say numbers from 4 to 2.
Say numbers from 5 to 3.
Say numbers from 6 to 4.
Say numbers from 3 to 5.

Index counting

Mr. Ho has left his footprints along the path he travelled.

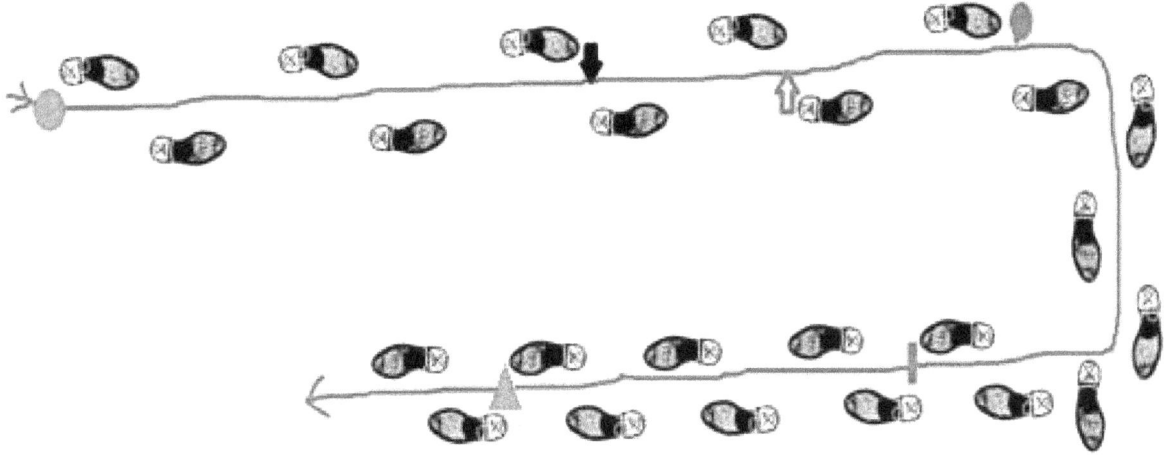

Follow Mr. Ho's path and find the number of footprints between two of the following marks.

marks	The number of footprints between each of the two marks
● ⬇	5
● ▮	16
⬆ ▲	15
▮ ▲	6
⬆ ●	2
● ▮	7
▮ ▮	10

One to one connection (数的对比及配对)

One to one connection from left to right by the same colour.

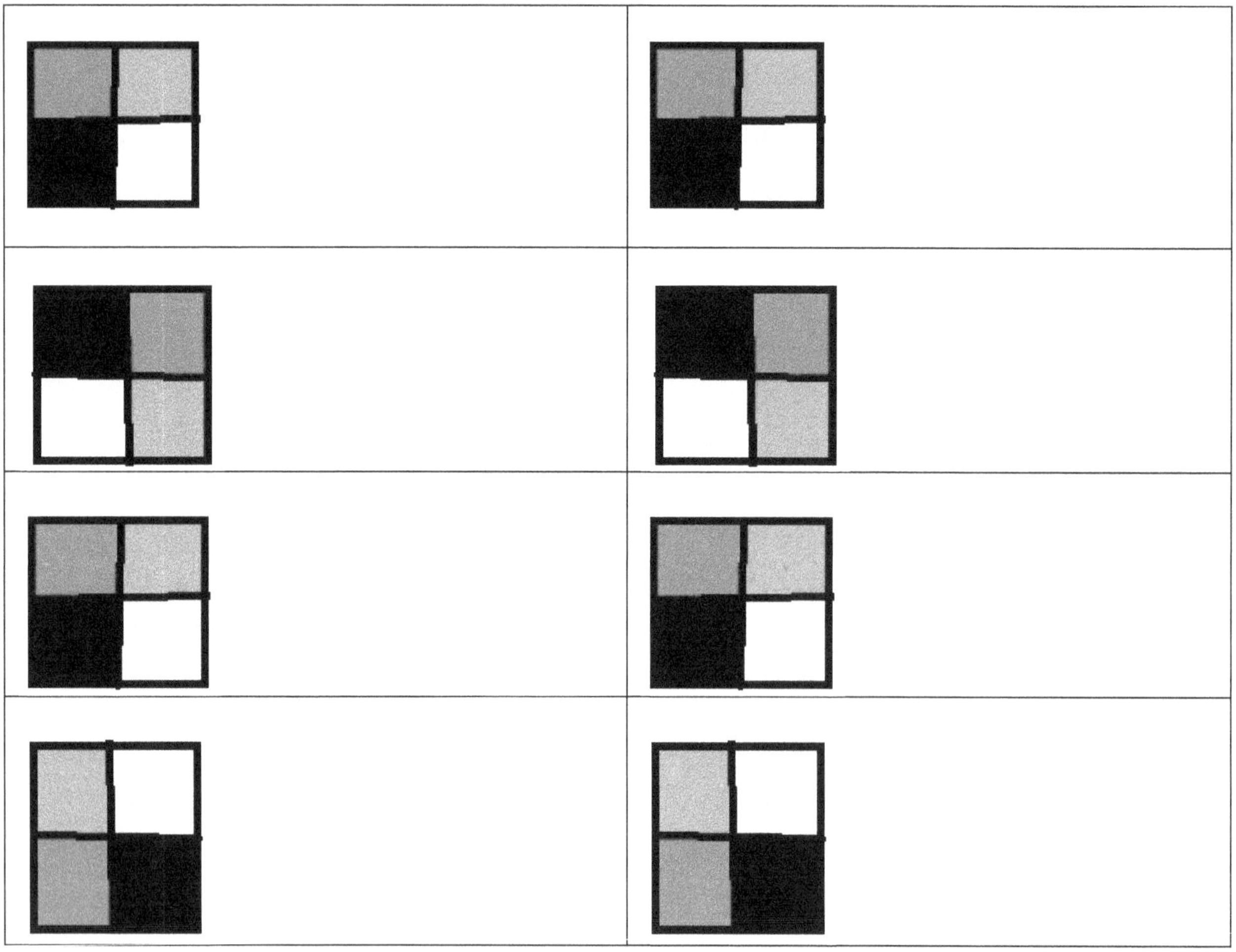

One to one connection from left to right by the same number.

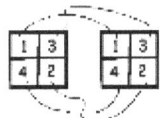

Ho Math Chess   Pre-K and Kindergarten Math

何数棋谜   棋谜式幼儿健脑思维趣味数学

© 2012 – 2021 Frank Ho, Amanda Ho, Canada copyright 1095661, Trademark 771400

From left to the matching number of dots on the right

Circling the odd one

The fourth of the bottom row.

Circling the odd one

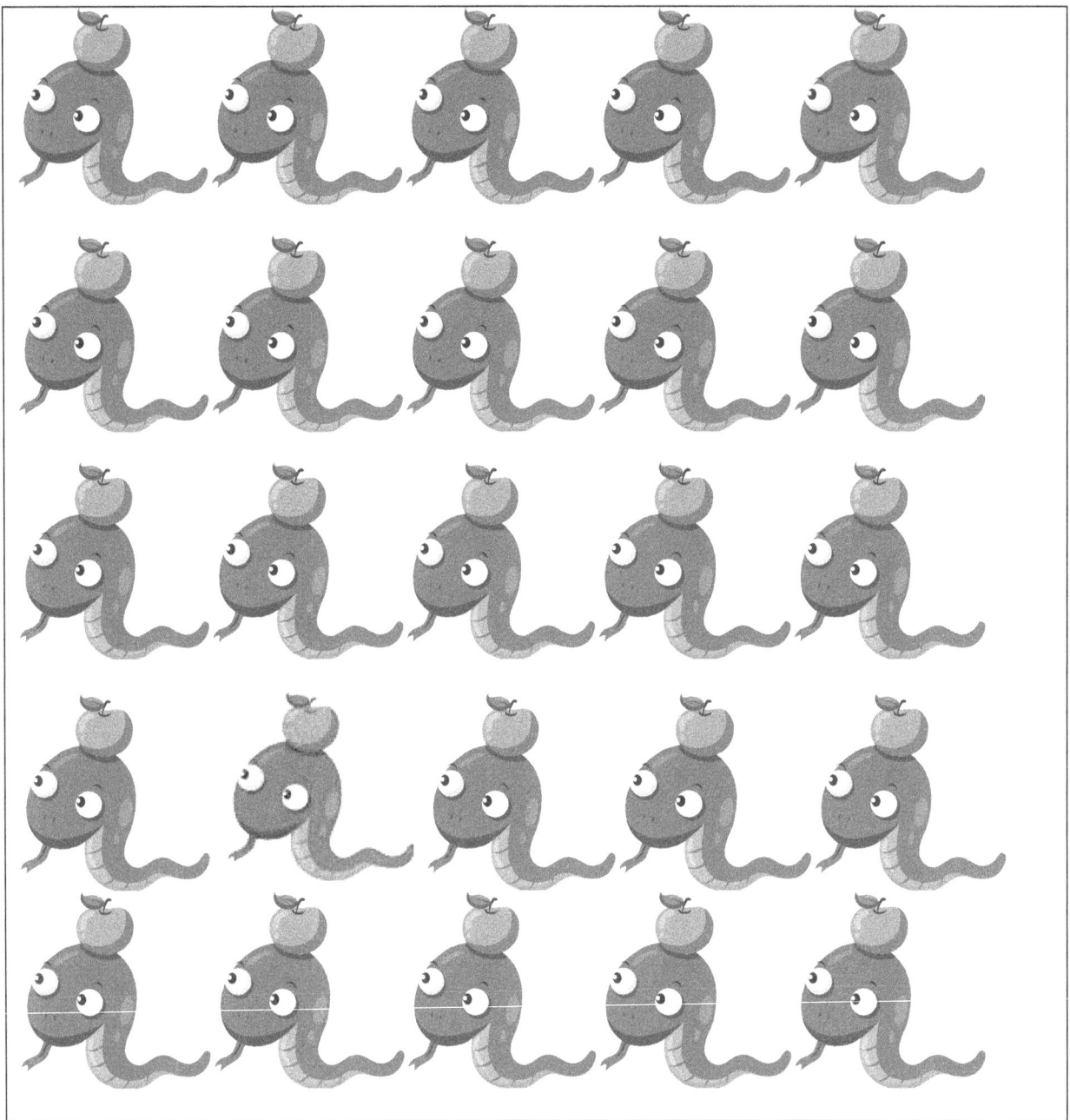

The second of the fourth row.

Circling the two odd ones

The second of the fourth row.

Circling the odd one

Circle the table.

Ho Math Chess — Pre-K and Kindergarten Math

何数棋谜　棋谜式幼儿健脑思维趣味数学

© 2012 – 2021 Frank Ho, Amanda Ho, Canada copyright 1095661, Trademark 771400

Right-left circle, opening up-down.

Circle those circles which are on the right sides of the straight lines.

d b p q

Circle those circles which are on the left sides of the straight lines.

d b p q

Circle those letters which have openings up.

u n m w

Circle those letters which have openings down.

u n m w

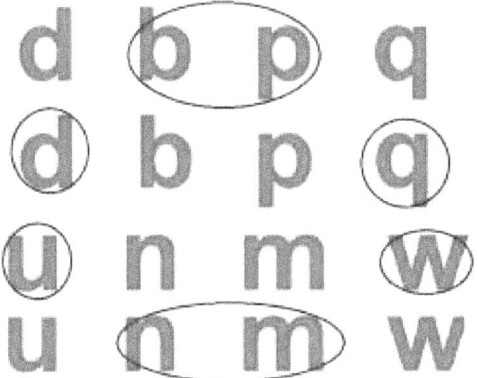

73

Circling objects

Circle anyone object only.

Circle all.
Circle the number "1".

Circle one set of any two objects.

Just circle any 2 plates.
Circle the number "2".

Circle two of 2's.

Circle one set of any three objects.

Circle any 3 plates.

Circle the number "3".

Circle three of 3's.

Circle one set of any four objects.

Circle all of 4 plates.

Circle the number "4".

Circle one 4.

Circle one set of any five objects.

No answer.

Circle the number "5".

Three of 5's.

Ho Math Chess — Pre-K and Kindergarten Math

Circling a larger number of objects

Circle the basket which has more apples.

Problem 1

Problem 2

Problem 3

Circle right, right, left.

Circling the different parts

The right side picture has 6 changes, which are different from the left side picture. Circle the 6 different changes on the right side picture.

The right side picture has 7 changes, which are different from the left side picture. Circle the 7 different changes on the right side picture.

  answer

Circling the odd one

Circle the odd one.

Apple

Circle the odd one.

Cement truck

Circle the odd one.

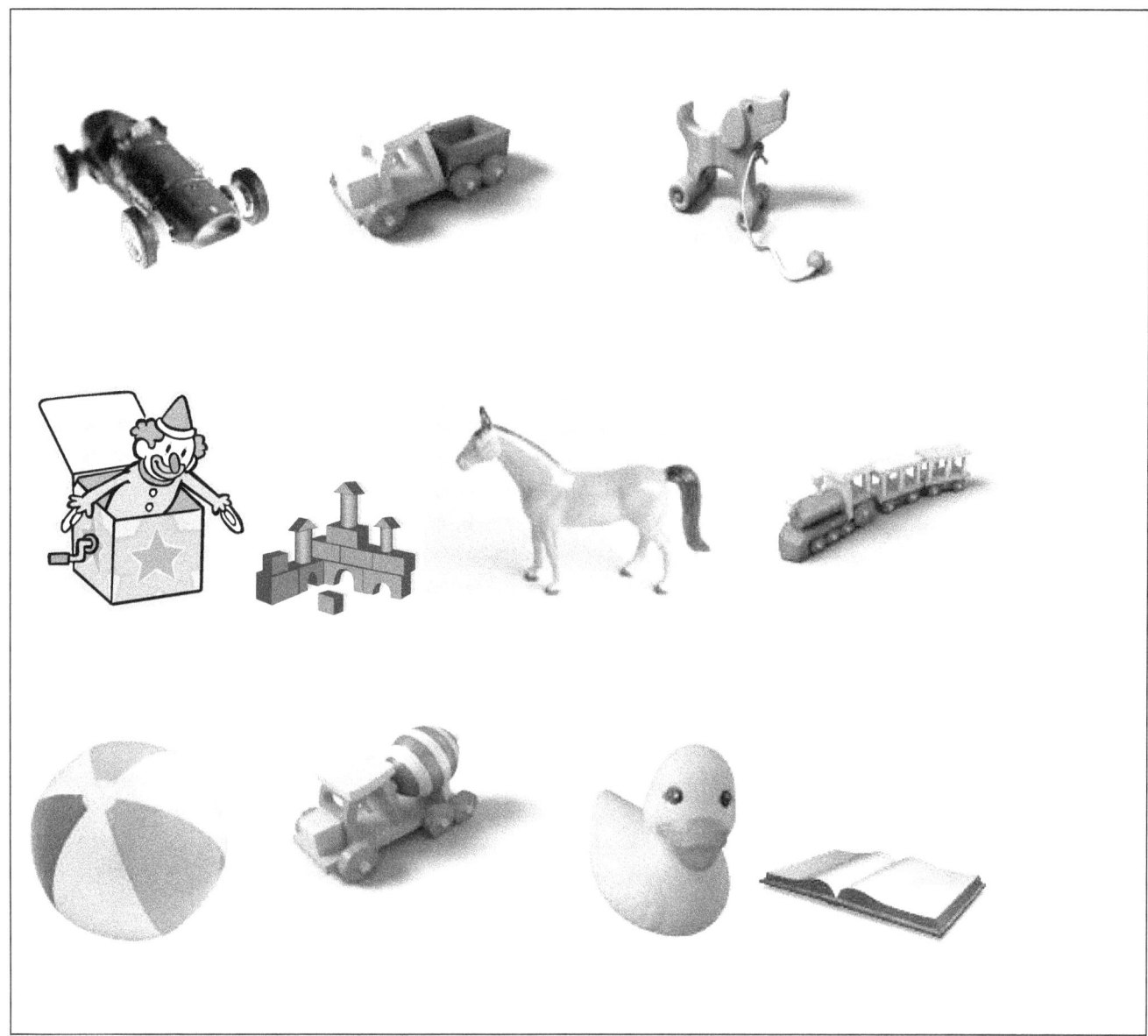

Book

Circle the odd one.

Wrench

Ho Math Chess — Pre-K and Kindergarten Math

何数棋谜　棋谜式幼儿健脑思维趣味数学

© 2012 - 2021 Frank Ho, Amanda Ho, Canada copyright 1095661, Trademark 771400

Writing numbers

The blue colour shows the straight line, and the red colour shows the curve.

	Lollipop	
	Lollipop stick	

	Duck neck	
	Dragonfly two eyes (half)	

www.homathchess.com

(kite image)	Kite	4
(teapots image)	Teapot	5
(teapots image)	Teapot	6

87

	Walking cane	
	Chained fence	
	hook	

Ho Math Chess   Pre-K and Kindergarten Math

何数棋谜   棋谜式幼儿健脑思维趣味数学

© 2012 – 2021 Frank Ho, Amanda Ho, Canada copyright 1095661, Trademark 771400

Different meanings of numbers
数的不同意思

Counter (Circle all beans which add up to the total counts.)	Value	Order or rank
1 🫘🫘🫘🫘🫘🫘🫘🫘🫘	1 bean = 🫘	1st
2 🫘🫘🫘🫘🫘🫘🫘🫘🫘	2 beans = 🫘🫘	2nd
3 🫘🫘🫘🫘🫘🫘🫘🫘	3 beans = 🫘🫘🫘	3rd
4 🫘🫘🫘🫘🫘🫘🫘🫘	4 beans = 🫘🫘🫘🫘	Circle the 4th bean on the left.
5 🫘🫘🫘🫘🫘🫘🫘	5 beans = 🫘🫘🫘🫘🫘	Circle the 5th bean on the left.
6 🫘🫘🫘🫘🫘🫘🫘🫘	6 beans = 🫘🫘🫘🫘🫘🫘	Circle the 6th bean on the left.
7 🫘🫘🫘🫘🫘🫘🫘🫘	7 beans = 🫘🫘🫘🫘🫘🫘🫘	Circle the 7th bean.
8 🫘🫘🫘🫘🫘🫘🫘🫘	8 beans = 🫘🫘🫘🫘🫘🫘🫘🫘	Circle the 8th bean on the left.
9 🫘🫘🫘🫘🫘🫘🫘🫘	9 beans = 🫘🫘🫘🫘🫘🫘🫘🫘🫘	Circle the 9th bean on the left.

Tracing images of numbers

Trace each part of the following image, which looks like the image of a number.

Tracing image

Trace each part of the following image, which looks like the image of a number.

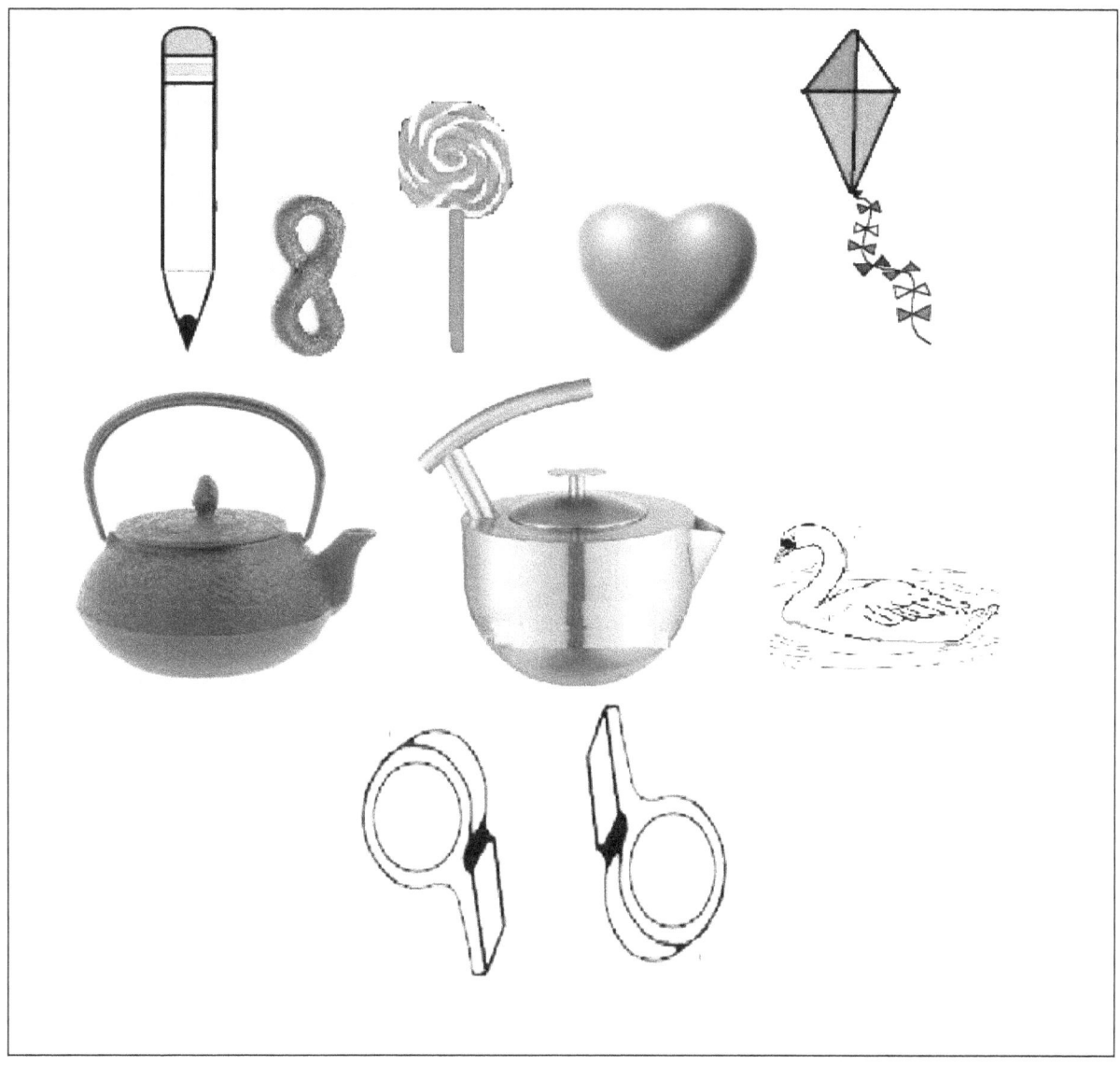

One to one correspondence

Use a line to make a one to one connection by the same colour.

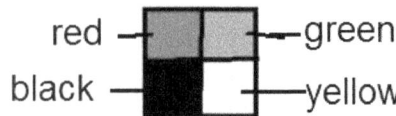
red — green
black — yellow

Use a line to make a one to one connection by the same number.

Connecting the number to the matching number of dots.

Match and connection

Connecting the matching pair of shoes

Connecting the matching shapes

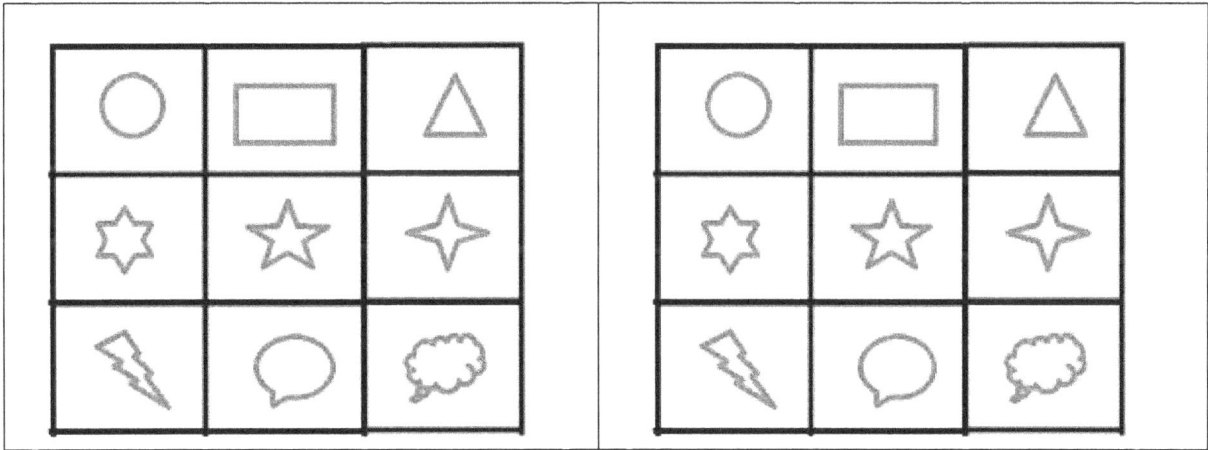

Matching

Connecting the number of the left column to its matching number on the right column

Connecting the total number of shaded squares on the left column to its matching number on the right column

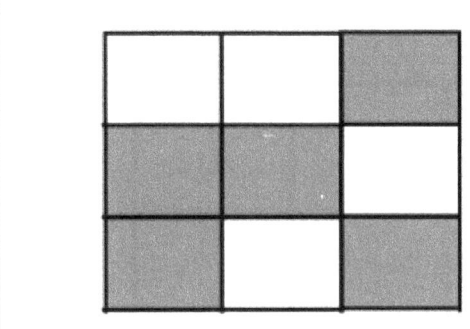

5

Connecting the number of the left column to its matching number on the right column

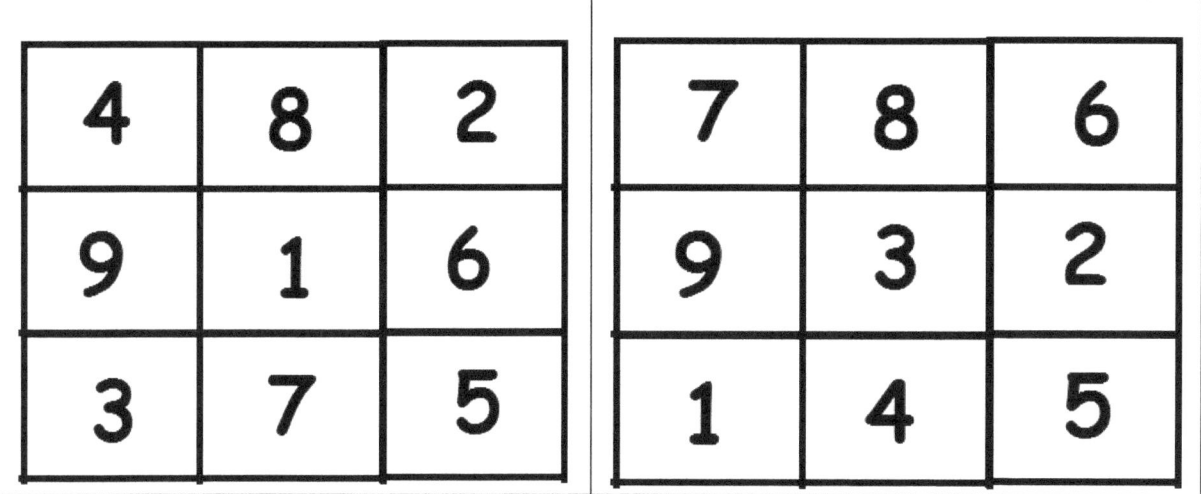

Connecting the total number of shaded squares on the left column to its matching number on the right column

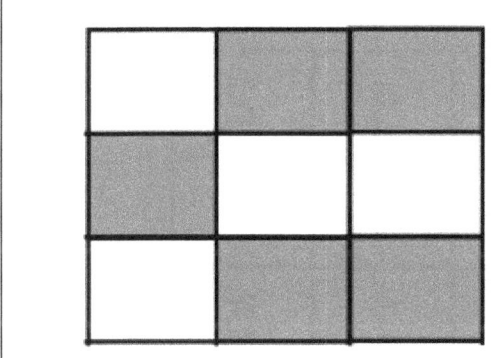

Connecting the number of the left column to its matching number on the right column

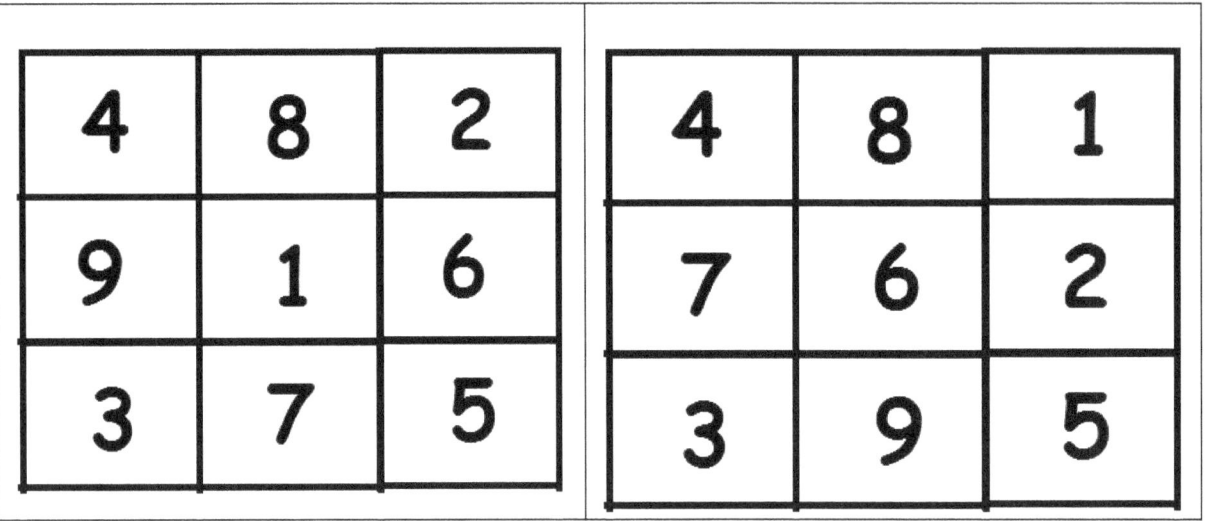

Connecting the total number of shaded squares on the left column to its matching number on the right column

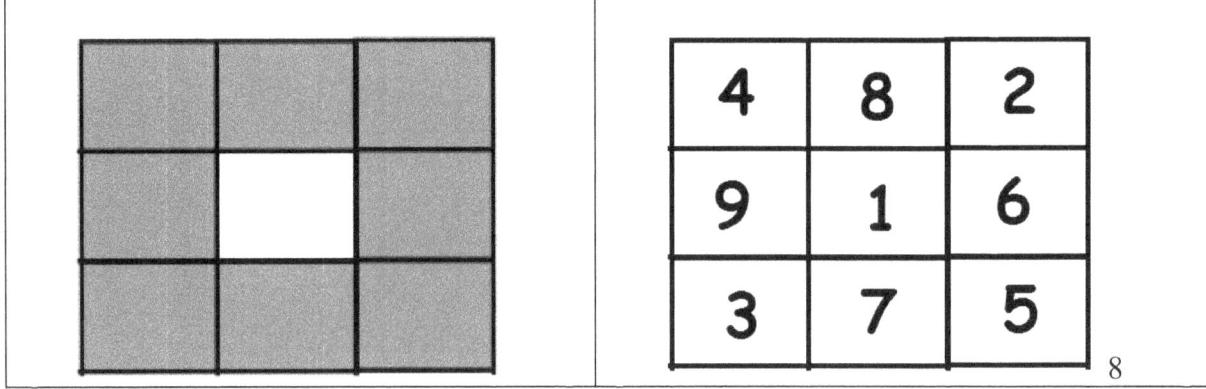

Connecting the number of the left column to its matching number on the right column

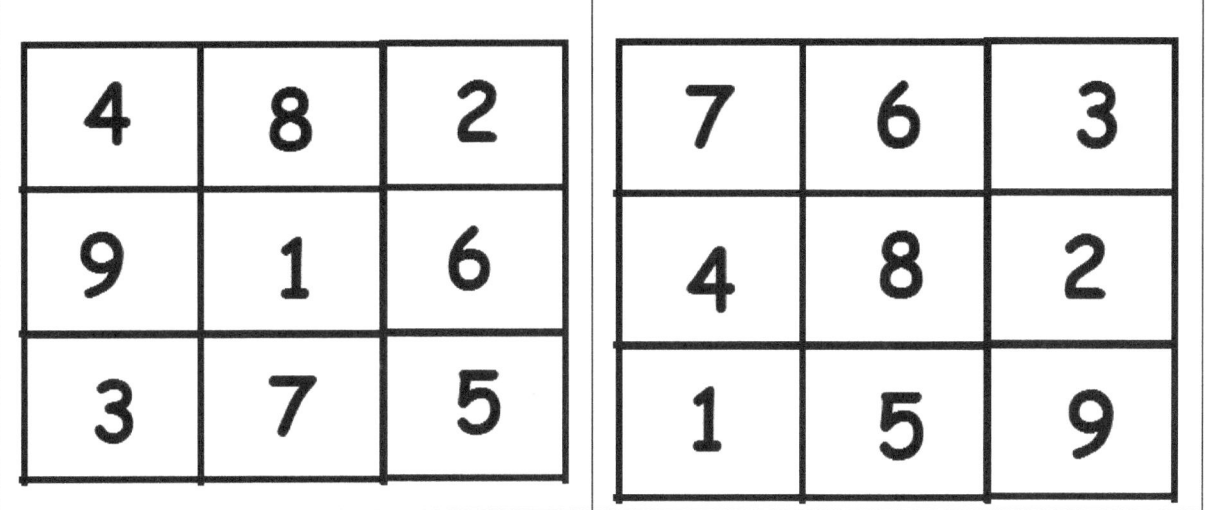

Connecting the total number of shaded squares on the left column to its matching number on the right column

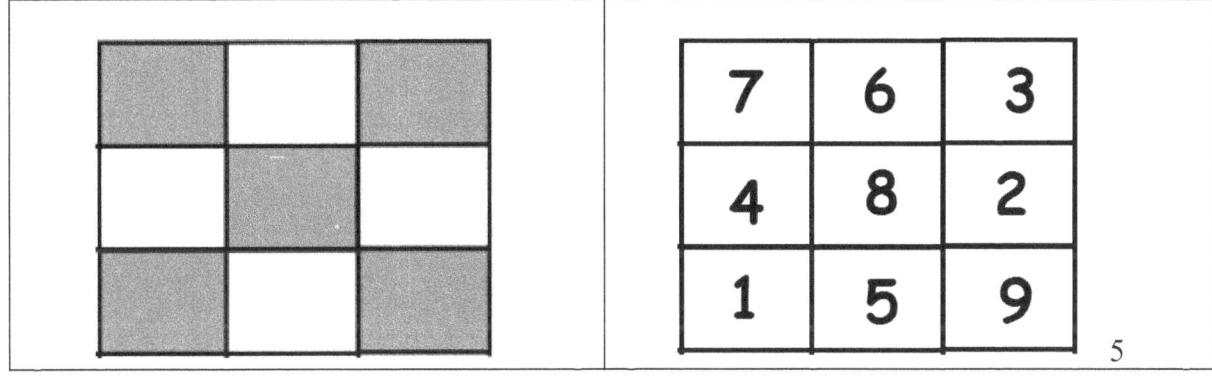

Connecting the number of the left column to its matching number on the right column

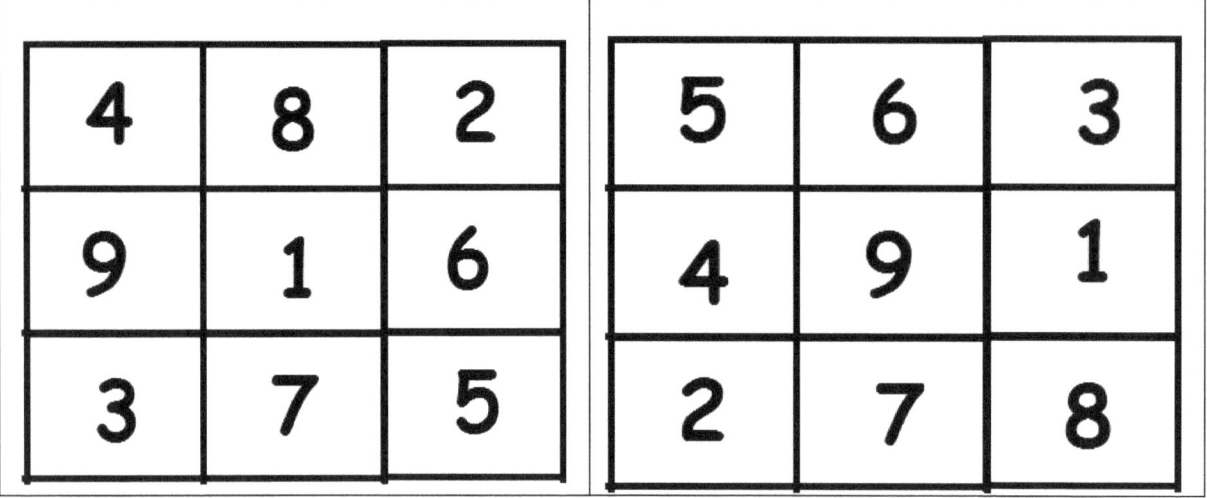

Connecting the total number of shaded squares on the left column to its matching number on the right column

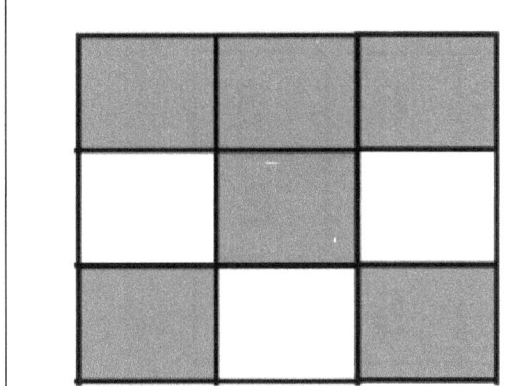

Connecting the number of the left column to its matching number on the right column

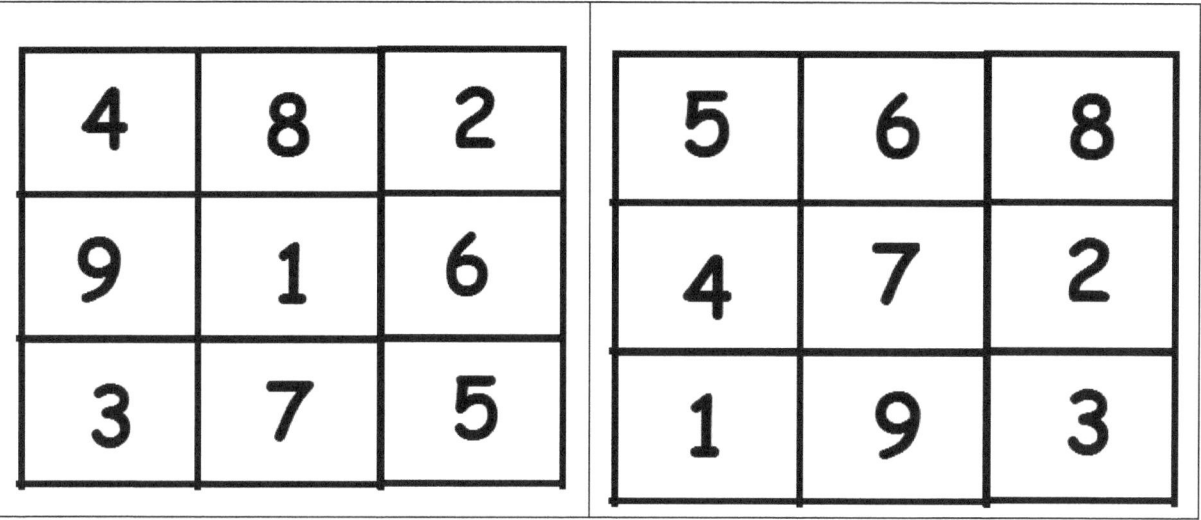

Connecting the total number of shaded squares on the left column to its matching number on the right column

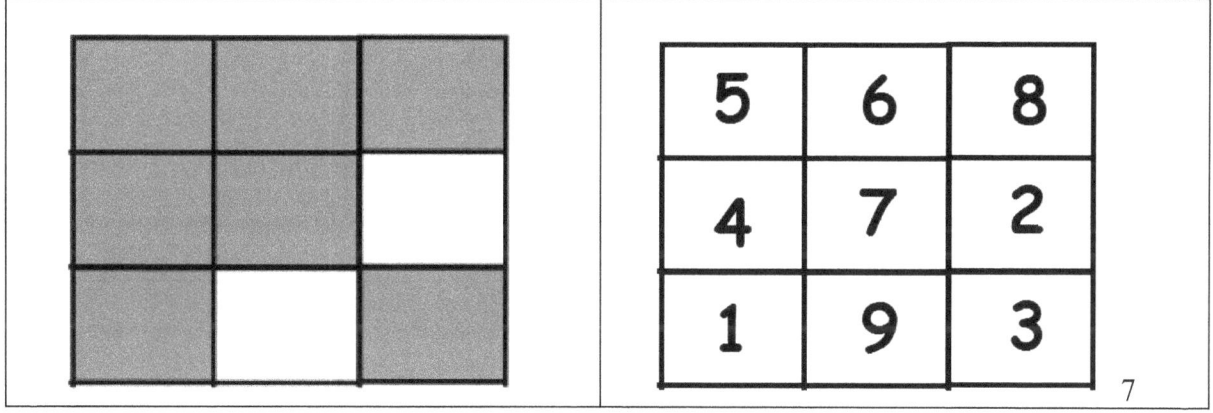

Ho Math Chess — Pre-K and Kindergarten Math

何数棋谜　棋谜式幼儿健脑思维趣味数学

© 2012 – 2021 Frank Ho, Amanda Ho, Canada copyright 1095661, Trademark 771400

Counting and writing numbers

The dotted number showing its value

Count the red dots on the dotted number, including the smaller number written on the numbers 7 and 9 and then write its total on the right side empty box beside each number.

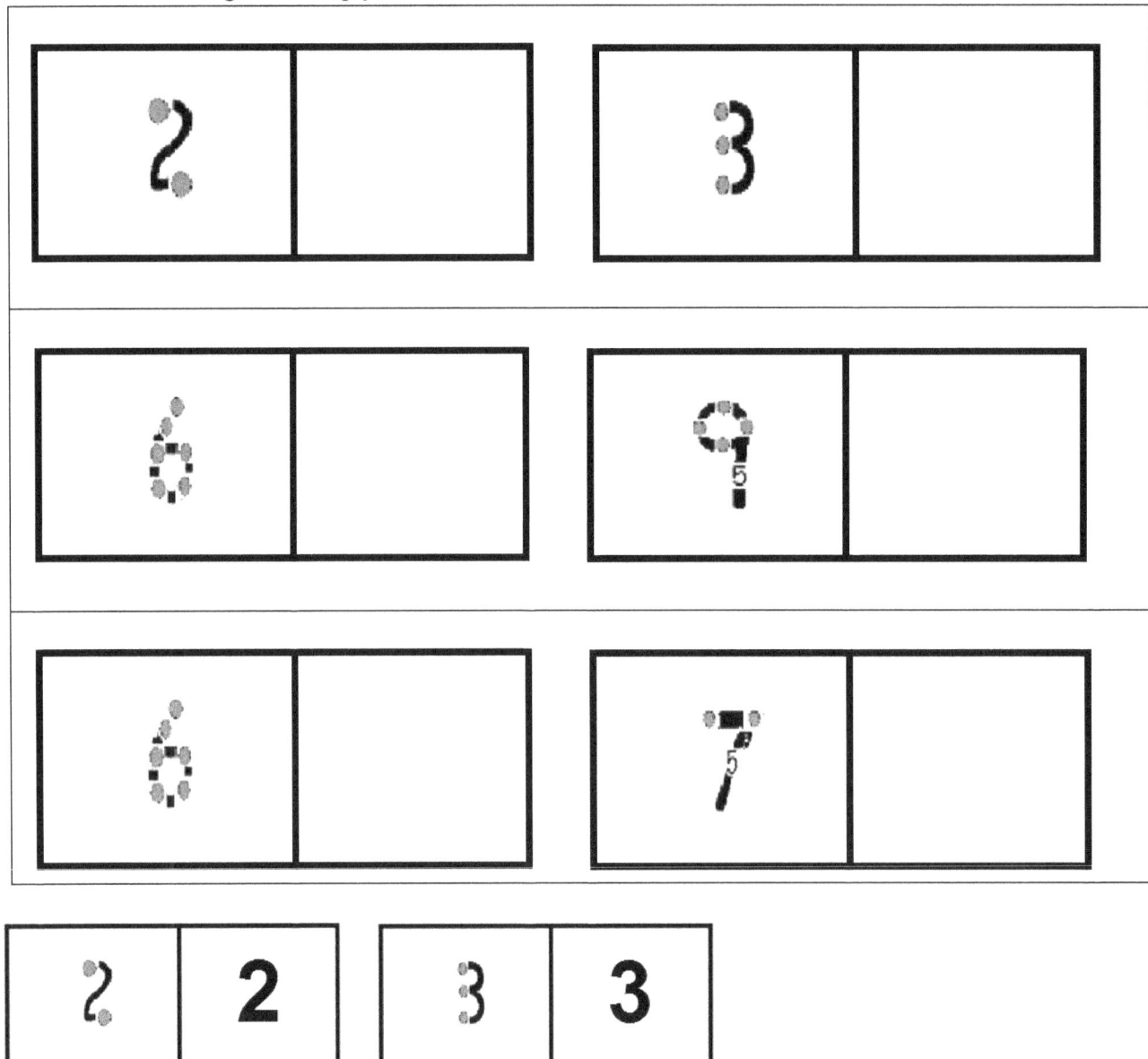

answer

Ho Math Chess — Pre-K and Kindergarten Math

何数棋谜　棋谜式幼儿健脑思维趣味数学

© 2012 – 2021 Frank Ho, Amanda Ho, Canada copyright 1095661, Trademark 771400

The dotted number showing its value

Find the following red matching square to the above-numbered square, then write the matching number beside each square.

9 6
5 1
3 7 .

Ho Math Chess — Pre-K and Kindergarten Math

何数棋谜　棋谜式幼儿健脑思维趣味数学

© 2012 – 2021 Frank Ho, Amanda Ho, Canada copyright 1095661, Trademark 771400

The dotted number showing its value

Find the following red matching square to the above-numbered square, then write the matching number beside each square.

48
21
37

Linking numbers to dots

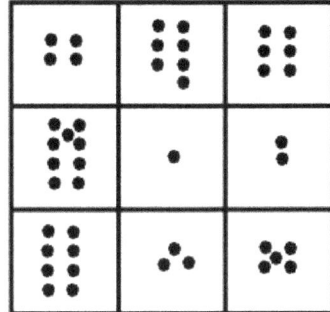

Count the dots to match the shaded square, then write its total on the right side of each 3 by 3 square.

The number is _____.	The number is _____.	The number is _____.
The number is _____.	The number is _____.	The number is _____.
The number is _____.	The number is _____.	The number is _____.

831
594
627

Linking numbers to dots

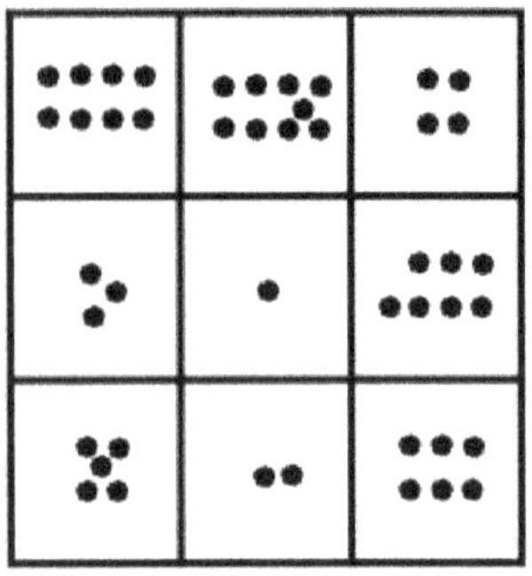

Count the dots to match the shaded square, then write its total on the right side of each 3 by 3 square.

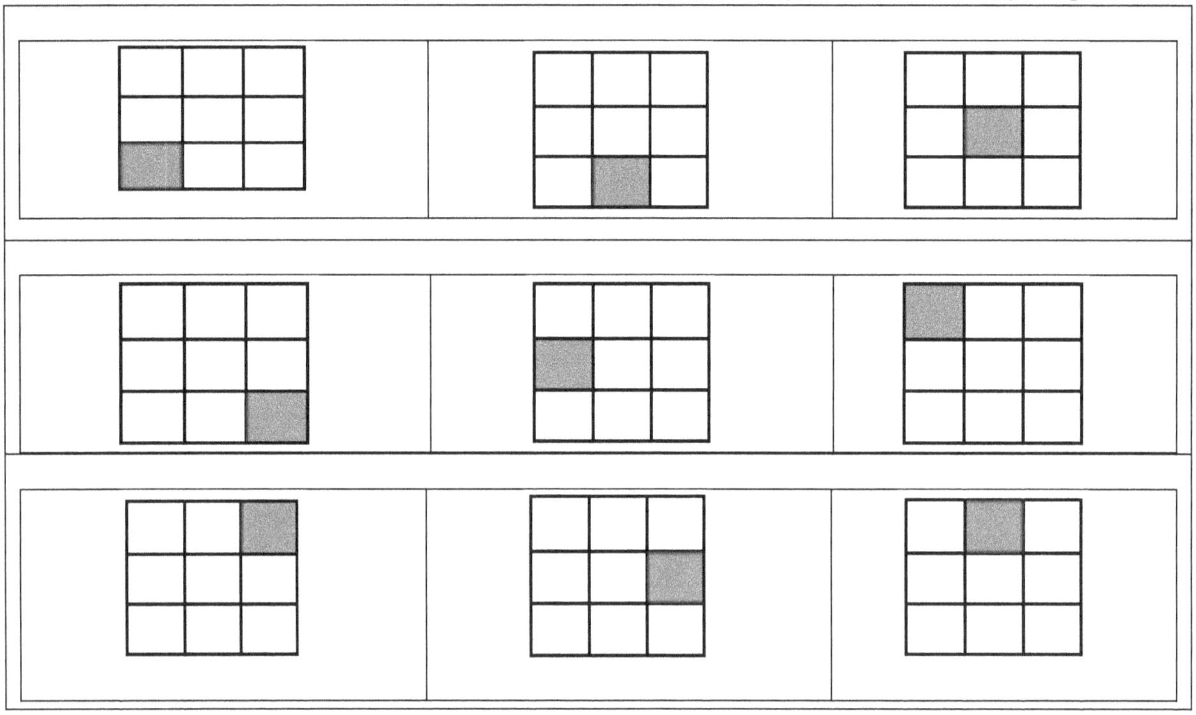

521
638
479

Ho Math Chess   Pre-K and Kindergarten Math

何数棋谜　棋谜式幼儿健脑思维趣味数学

© 2012 – 2021 Frank Ho, Amanda Ho, Canada copyright 1095661, Trademark 771400

Linking numbers to dots

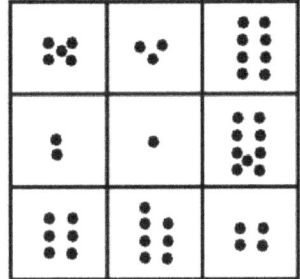

Count the dots to match the shaded square, then write its total on the right side of each 3 by 3 square.

671
425
893

Ho Math Chess — Pre-K and Kindergarten Math

何数棋谜　棋谜式幼儿健脑思维趣味数学

© 2012 – 2021 Frank Ho, Amanda Ho, Canada copyright 1095661, Trademark 771400

Linking numbers to dots

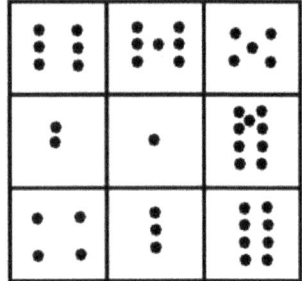

Count the dots to match the shaded square, then write its total on the right side of each 3 by 3 square.

431
826
597

Ho Math Chess — Pre-K and Kindergarten Math

何数棋谜　棋谜式幼儿健脑思维趣味数学

© 2012 – 2021 Frank Ho, Amanda Ho, Canada copyright 1095661, Trademark 771400

Linking numbers to dots

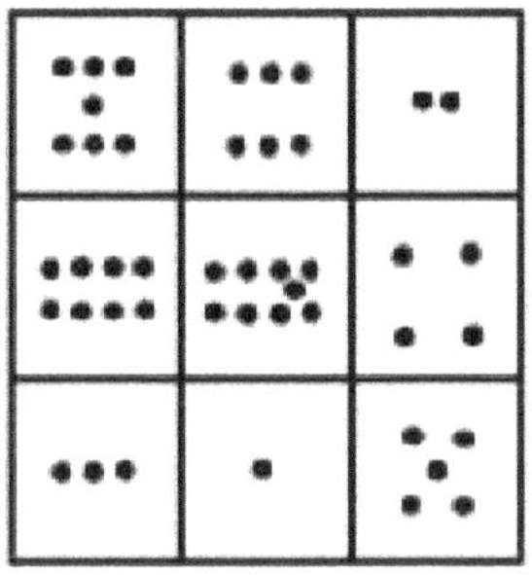

Count the dots to match the shaded square, then write its total on the right side of each 3 by 3 square.

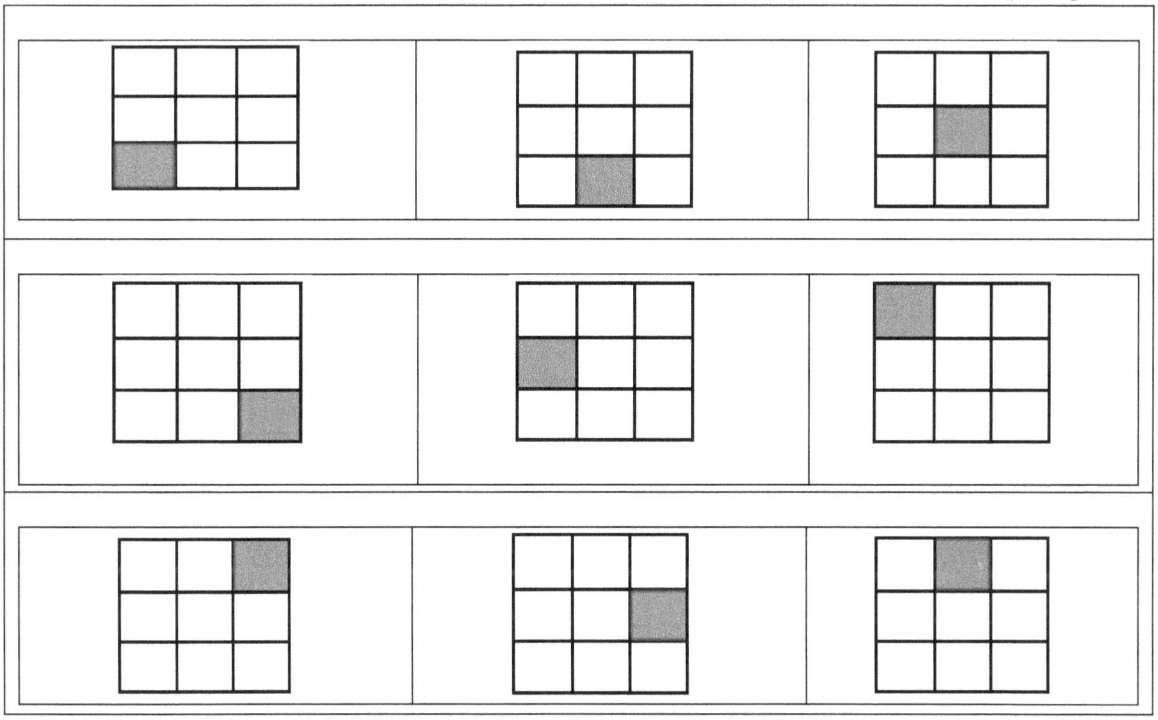

319
587
246

Linking numbers to dots

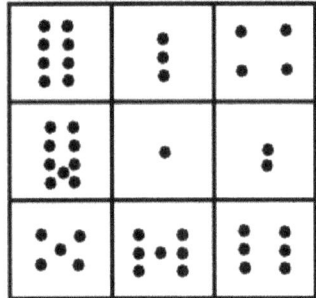

Count the dots to match the shaded square, then write its total on the right side of each 3 by 3 square.

571
698
423

Ho Math Chess — Pre-K and Kindergarten Math

何数棋谜　棋谜式幼儿健脑思维趣味数学

© 2012 – 2021 Frank Ho, Amanda Ho, Canada copyright 1095661, Trademark 771400

Linking numbers to dots

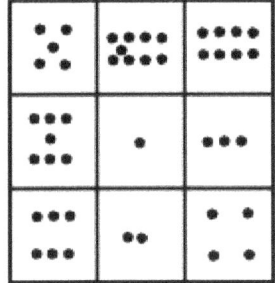

Count the dots to match the shaded square, then write its total on the right side of each 3 by 3 square.

621
475
839

www.homathchess.com 111

Linking numbers to dots

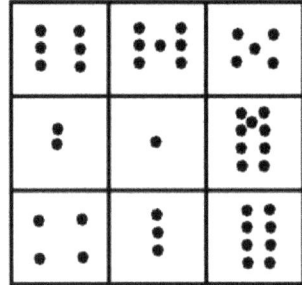

Count the dots to match the shaded square, then write its total on the right side of each 3 by 3 square.

431
826
597

Linking numbers to dots

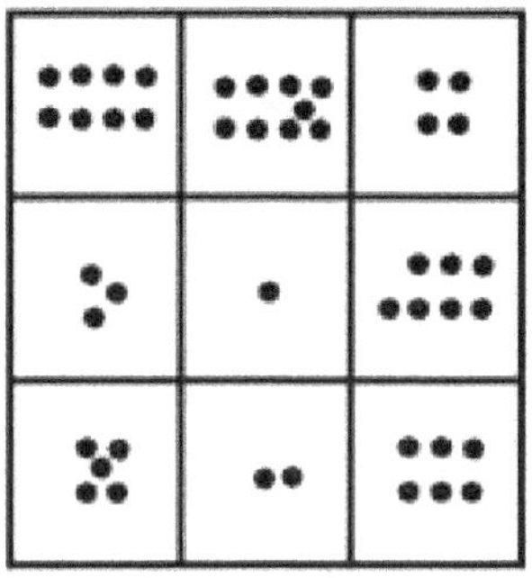

Count the dots to match the shaded square, then write its total on the right side of each 3 by 3 square.

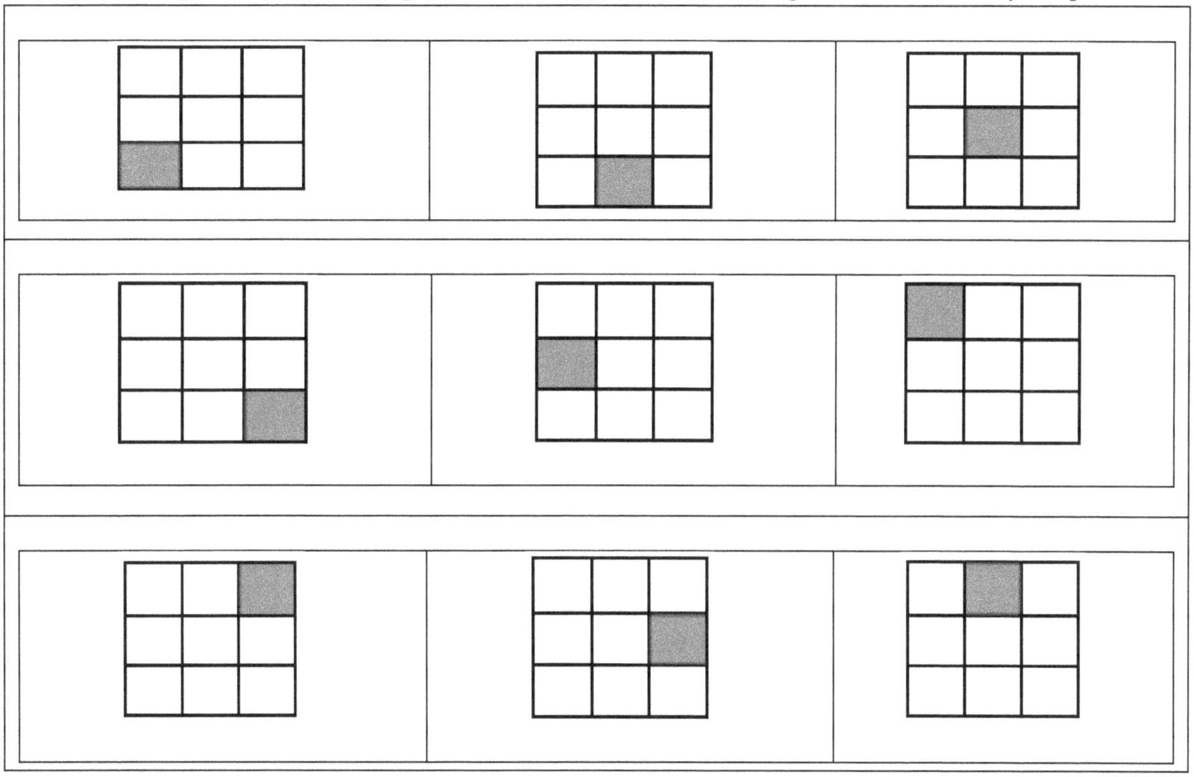

521
638
479

Ho Math Chess — Pre-K and Kindergarten Math

何数棋谜　棋谜式幼儿健脑思维趣味数学

© 2012 – 2021 Frank Ho, Amanda Ho, Canada copyright 1095661, Trademark 771400

Linking numbers to dots

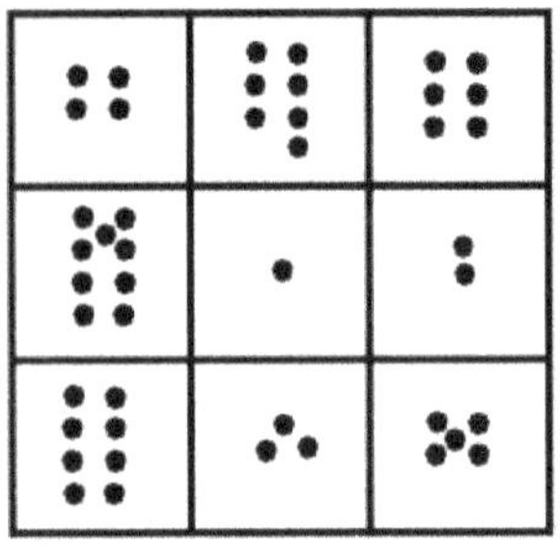

Count the dots to match the shaded square, then write its total on the right side of each 3 by 3 square.

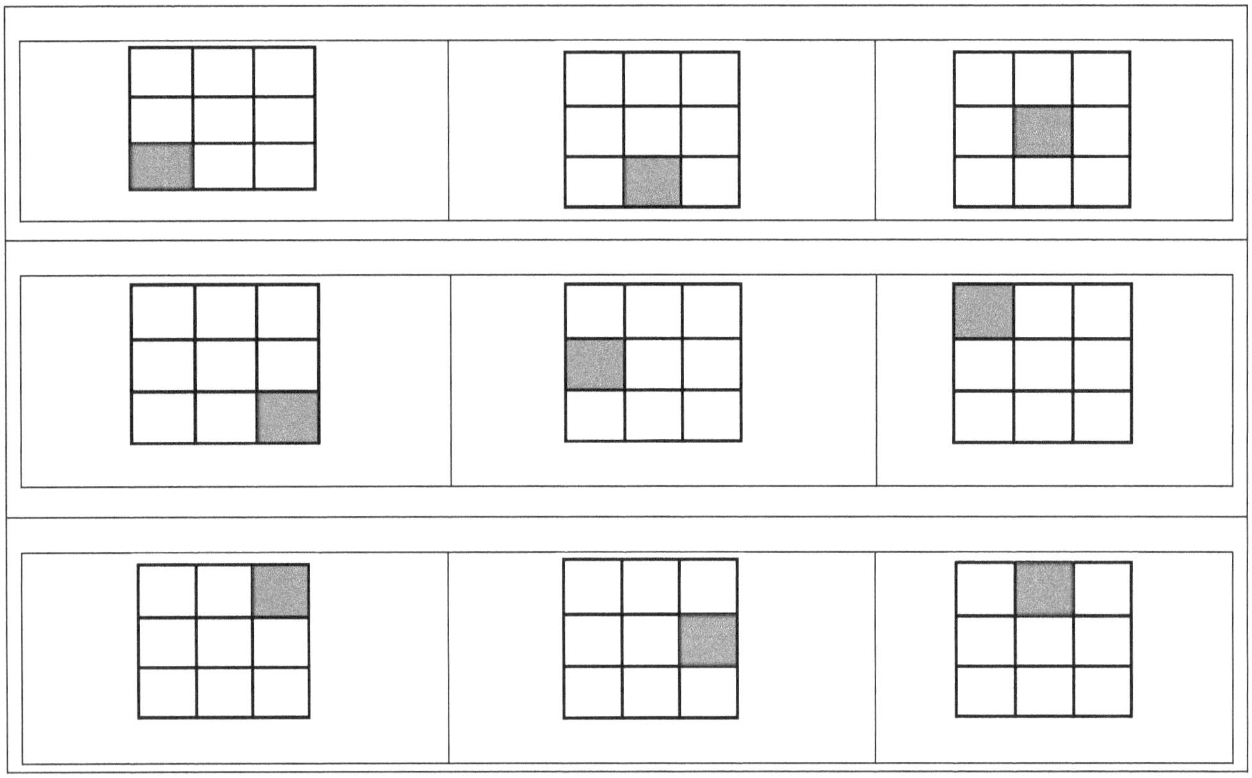

831
594
627

Unmatched numbers

Circle the right-hand side numbers, which are different from their left-hand-side corresponding numbers.

2 on the top and right only.

Ho Math Chess   Pre-K and Kindergarten Math

何数棋谜   棋谜式幼儿健脑思维趣味数学

© 2012 – 2021 Frank Ho, Amanda Ho, Canada copyright 1095661, Trademark 771400

Comparing

The right-hand side picture has 2 changes, which are different from the left-hand side picture. Circle the 2 different changes in the right-hand side picture.

1 6 4 4 6 2	1 6 4 4 6 2
3 1 2 6 5 3	3 1 2 6 5 3
4 3 3 2 6 1	4 3 3 2 6 4
5 2 1 3 3 5	5 2 3 3 3 5
6 4 5 1 2 6	6 4 5 1 2 6
2 5 6 5 1 6	2 5 6 5 1 6

The right-hand side numbers have 5 changes, which are different from the left-hand side numbers. Circle the 5 different changes on the right-hand side.

```
1 2 3 4 5 6 7 8 9 10        1 2 3 4 5 6 7 8 9 10
2 3 4 5 6 7 8 9 10 2        2 3 4 5 6 7 8 9 10 2
3 4 5 6 7 8 9 10 2 3        3 4 5 5 7 8 9 10 2 3
4 5 6 7 8 9 10 2 3 4        4 5 6 7 8 9 9 1 2 3 4
5 6 7 8 9 10 2 3 4 5        5 6 7 8 9 10 2 3 4 5
6 7 8 9 10 2 3 4 5 6        6 7 8 9 10 2 3 4 5 6
7 8 9 0 1 2 3 4 5 6 7       7 8 9 10   2 3 4 5 6 7
```

| nothing | 3, 4 |
| nothing | 5, 9, 1, 10 |

Length, weight, height
長度，重量，高度

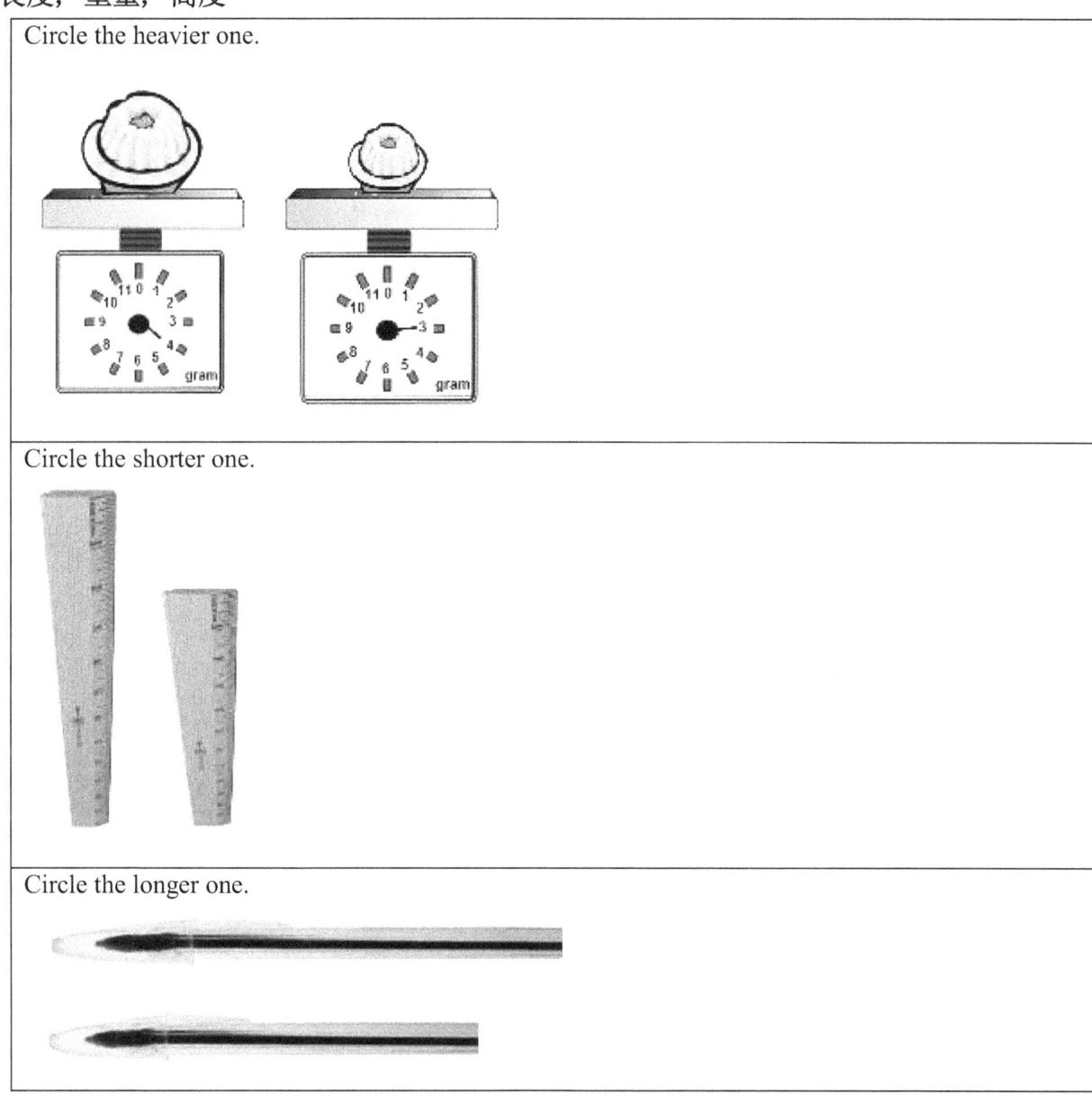

Left
Right
Top

Ho Math Chess — Pre-K and Kindergarten Math

何数棋谜　棋谜式幼儿健脑思维趣味数学

© 2012 – 2021 Frank Ho, Amanda Ho, Canada copyright 1095661, Trademark 771400

Direction

方向

Circle all arrows pointing to the right-hand side.

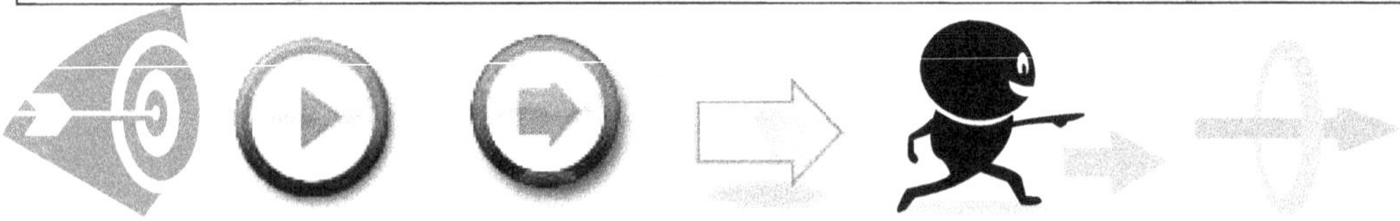

Direction

Circle all arrows pointing to the left-hand side.

answer

Ho Math Chess — Pre-K and Kindergarten Math

何数棋谜　棋谜式幼儿健脑思维趣味数学

© 2012 – 2021 Frank Ho, Amanda Ho, Canada copyright 1095661, Trademark 771400

Direction

Circle all arrows pointing up.

Direction

Circle all arrows pointing down.

Total 6

Ho Math Chess — Pre-K and Kindergarten Math

何数棋谜 棋谜式幼儿健脑思维趣味数学

© 2012 – 2021 Frank Ho, Amanda Ho, Canada copyright 1095661, Trademark 771400

Clockwise and counter clockwise

顺时针，反时针

Circle the motions of the following objects turning in a clockwise rotation.

Total 7

www.homathchess.com   122

Ho Math Chess   Pre-K and Kindergarten Math

何数棋谜   棋谜式幼儿健脑思维趣味数学

© 2012 – 2021 Frank Ho, Amanda Ho, Canada copyright 1095661, Trademark 771400

Clockwise and counter clockwise

Circle the motions of the following objects turning in the counter clockwise rotation.

Total 5

| Ho Math Chess | Pre-K and Kindergarten Math |

何数棋谜　棋谜式幼儿健脑思维趣味数学

© 2012 – 2021 Frank Ho, Amanda Ho, Canada copyright 1095661, Trademark 771400

Traditional chess setup and Ho Math Chess setup
传统棋及何数棋谜教学棋

Frank Ho said, "Chess is a hands-on brain work using directions."
Frank Ho said, "Chess move is decided by visualizing how lines intersect each other."

Ho Math Chess Training Set is designed to train children to play chess using geometry concepts of lines and line segments. For details, please see www.mathandchess.com.

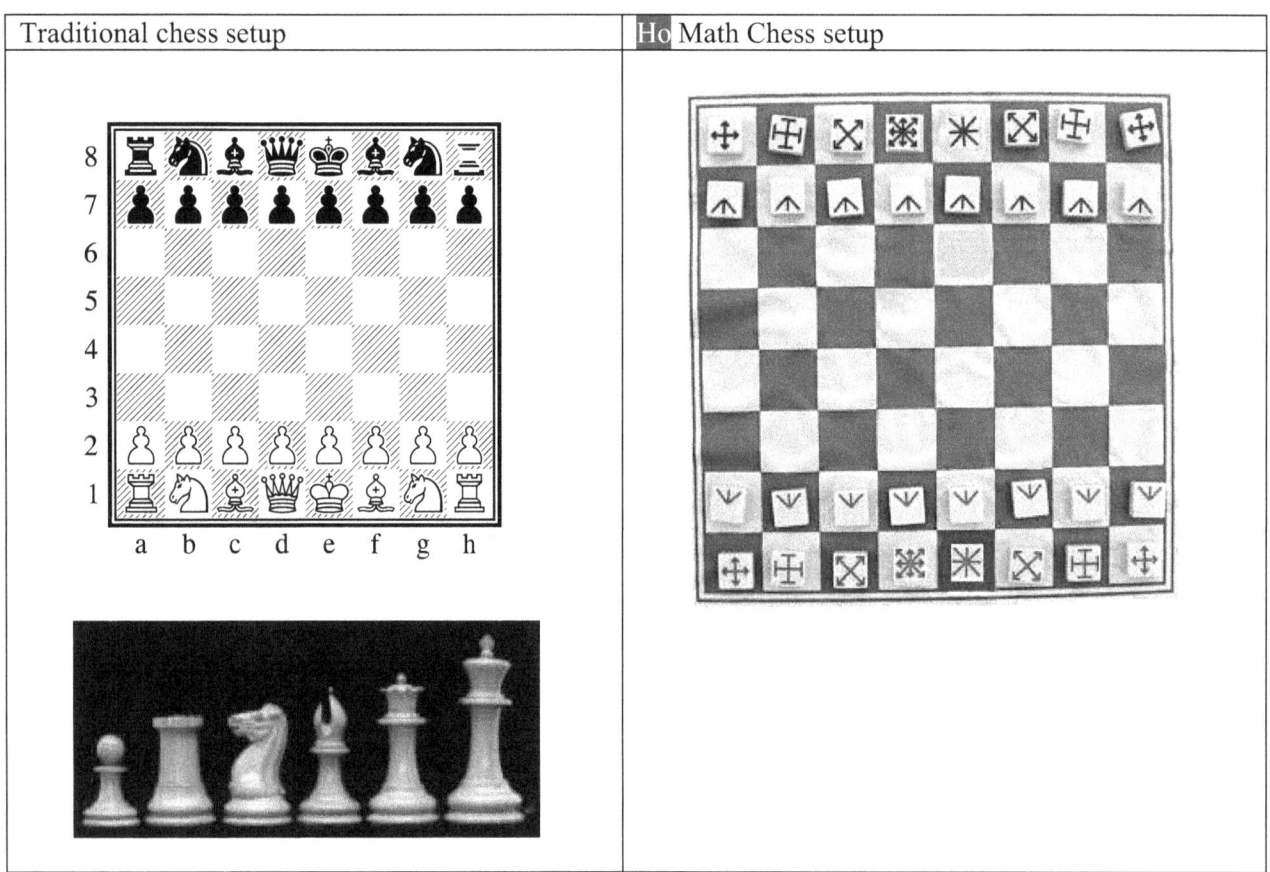

Ho Math Chess   Pre-K and Kindergarten Math

何数棋谜   棋谜式幼儿健脑思维趣味数学

© 2012 – 2021 Frank Ho, Amanda Ho, Canada copyright 1095661, Trademark 771400

Ho Math Chess pieces' moves

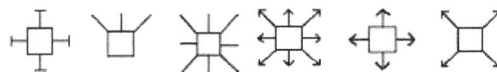

(Canada Trademark TMA771400, Copyright 1095661)

Ho Math Chess Teaching Set	Traditional chess set	English name
		Pawn
		Knight
		Bishop
		Rook
		Queen
		King

www.homathchess.com

Ho Math Chess   Pre-K and Kindergarten Math

何数棋谜   棋谜式幼儿健脑思维趣味数学

© 2012 – 2021 Frank Ho, Amanda Ho, Canada copyright 1095661, Trademark 771400

Ho Math Chess chessboard setup

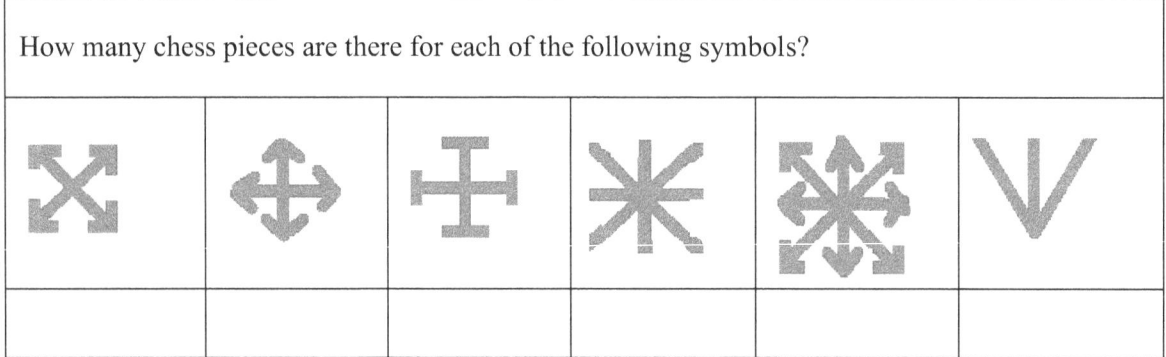

1 1 2 2 2 8

1 1 2 2 2 8

Ho Math Chess — Pre-K and Kindergarten Math

Chess moves

Knight moves	Rook moves
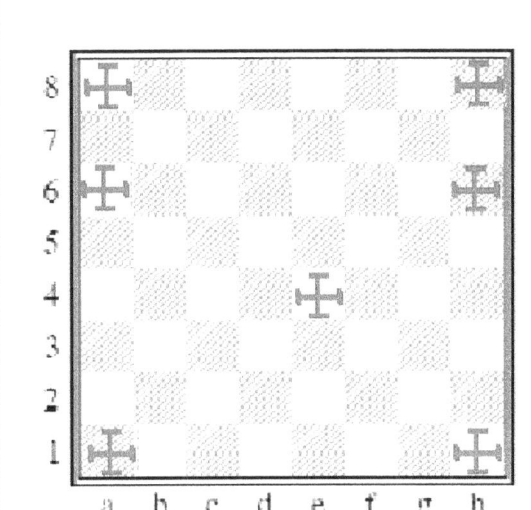	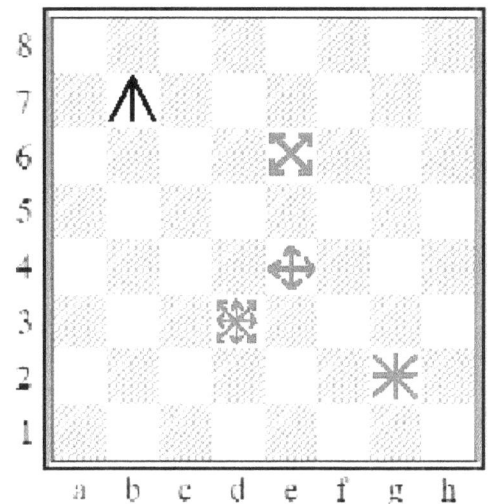

Use "X" to mark each of all knights' possible moves.

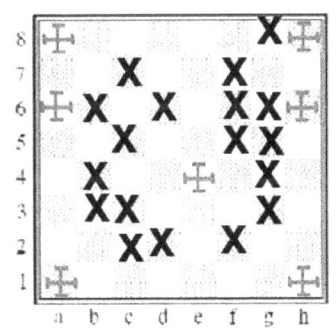   answer

Write the final location notation for each of the following chess pieces' moves. For example, the king moves up and right. ____ h3

The queen moves up 3 and right diagonal up 2. ____ f8

The rook moves up 1 and left 2. ____ c5

The bishop moves left diagonal down 3 and left diagonal right 2. _____ a6

The black pawn moves up 2. _____. b5

Rook's moves
車的走法

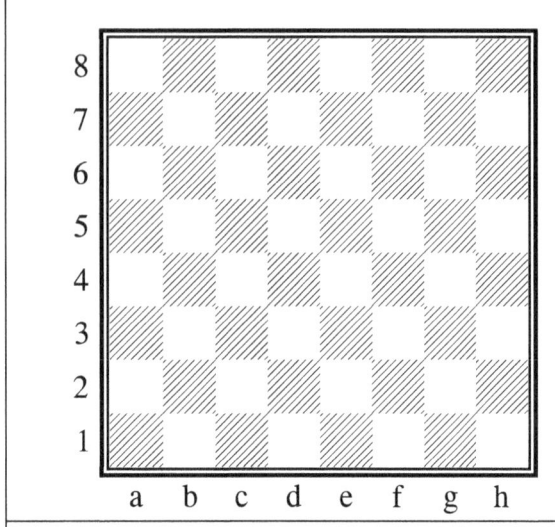

1. I am a rook and am located at e5.

My moves are as follows.

Right 2, left 5, right 6, left 4. Where am I? Mark an "X" on the square where I move to.   d5

2. I am a rook, and I am located at d6.

My moves are as follows;

Right 3, left 4, right 2, left 1. Where am I? Mark an "X" on the square where I move to.   d6

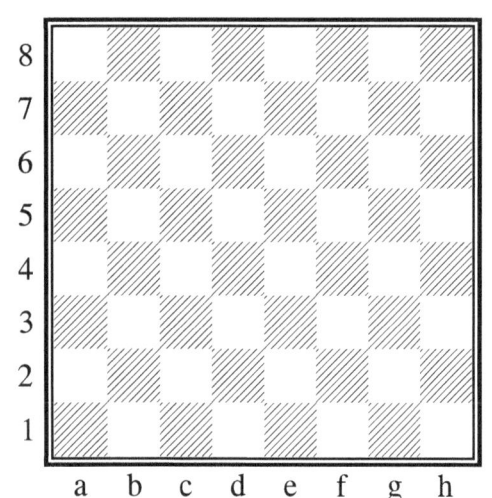

1. I am a rook, and I am located at d4.

My moves are as follows.

Right 3, left 3, right 4, left 2. Where am I? Mark an "X" on the square where I move to. f4

2. I am a rook and located at d3.

My moves are as follows.

Right 2, left 3, right 1, left 2. Where am I? Mark an "X" on the square where I move to.   b3

Ho Math Chess   Pre-K and Kindergarten Math

Directions and chess moves

棋谜式数学

Circle the answer.

Dog
Cat
Chicken
Hoarse

Ho Math Chess — Pre-K and Kindergarten Math

Direction

You are a chess piece located at c3.

● = 1

Circle the answer.

Chair
Dog
Elephant
Bird

Direction

You are a chess piece located at c3.

⬢ = 1

Circle the answer.

Cat
Chicken
Dog
Hoarse

Ho Math Chess — Pre-K and Kindergarten Math

何数棋谜　棋谜式幼儿健脑思维趣味数学

© 2012 – 2021 Frank Ho, Amanda Ho, Canada copyright 1095661, Trademark 771400

Direction

You are a chess piece located at c3.
● = 1

Circle the answer.

Bird
Chair
Elephant
Dog

www.homathchess.com

Ho Math Chess — Pre-K and Kindergarten Math

何数棋谜　棋谜式幼儿健脑思维趣味数学

© 2012 – 2021 Frank Ho, Amanda Ho, Canada copyright 1095661, Trademark 771400

Directions and Frankho ChessMaze

何数棋谜迷宫

Frankho ChessMaze

Trace the path from ✻ to ♚ through the non-shaded squares only.
Movement direction is shown by a darker line segment on a chess piece.

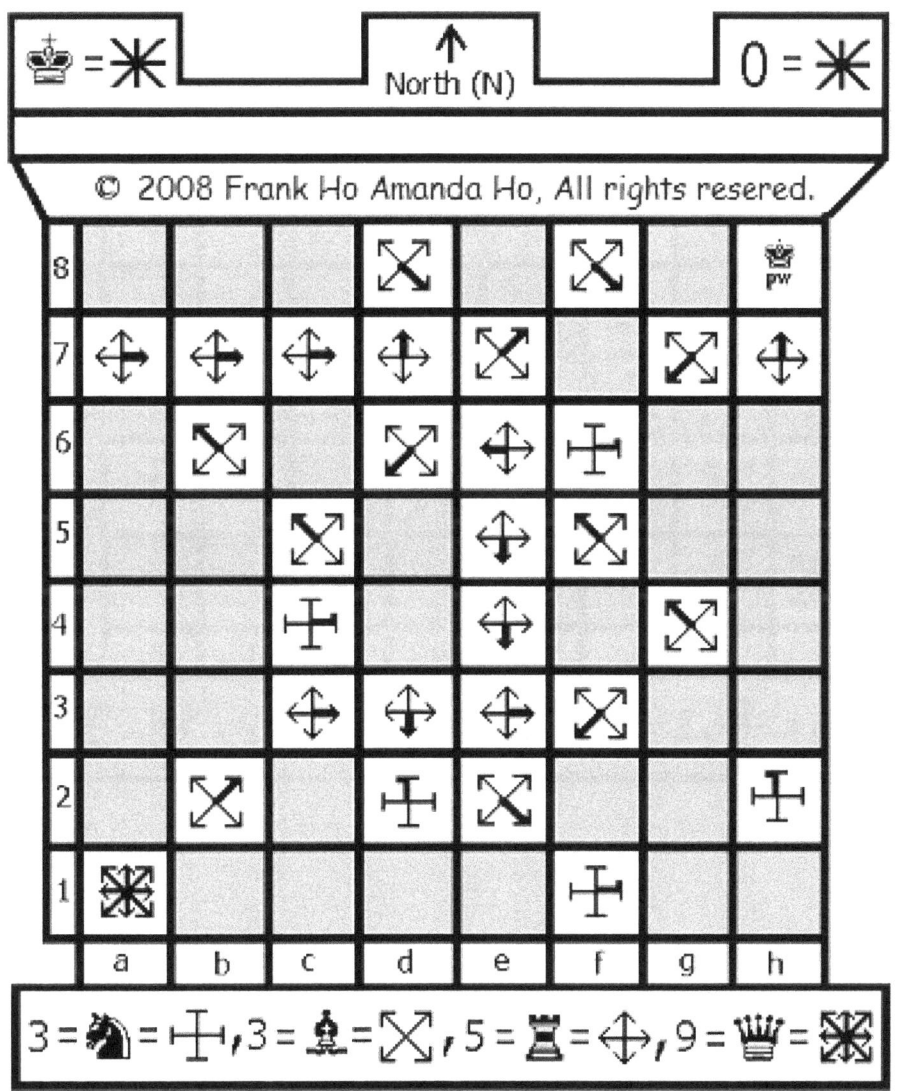

Ho Math Chess   Pre-K and Kindergarten Math

© 2012 – 2021 Frank Ho, Amanda Ho, Canada copyright 1095661, Trademark 771400

Answer

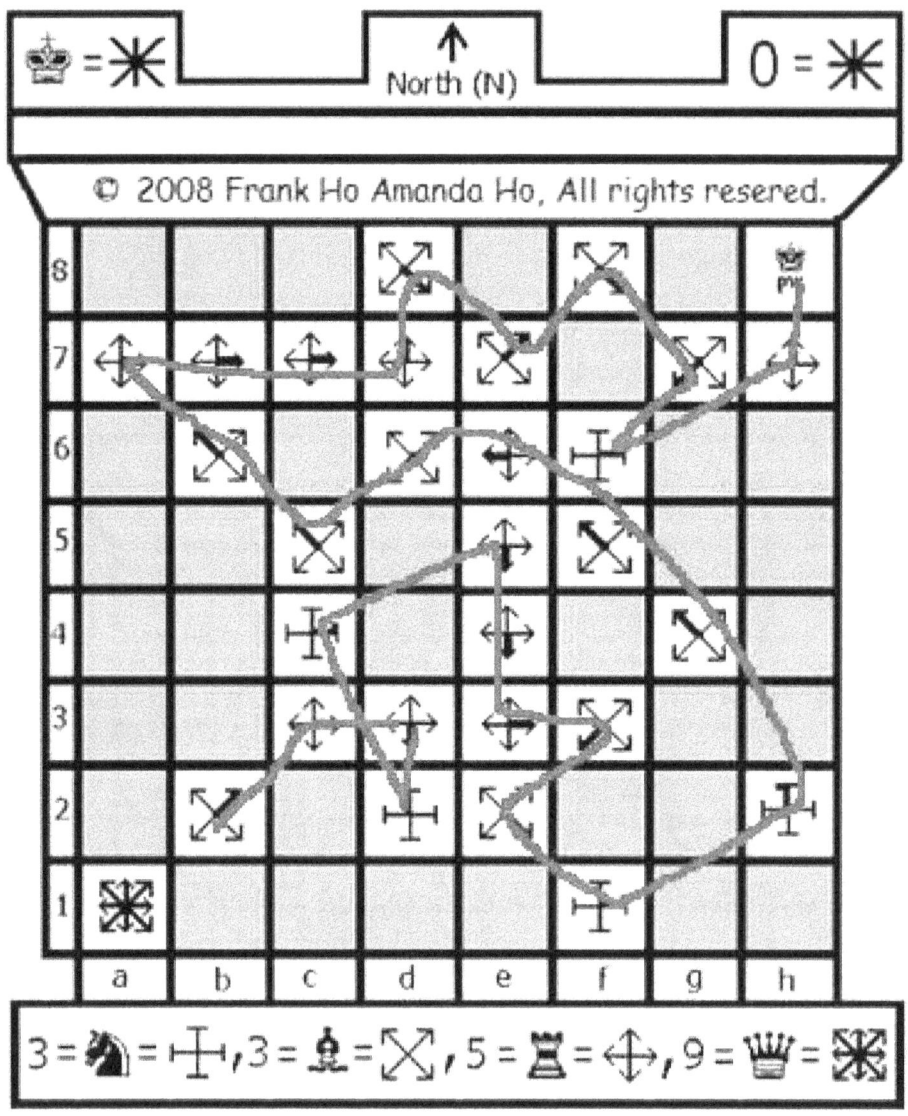

answer

© 2012 – 2021 Frank Ho, Amanda Ho, Canada copyright 1095661, Trademark 771400

Directions and chess mazes
何数棋谜迷宫

Frankho Chess Mazes and Castle Math

Trace the path from 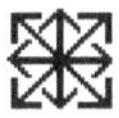 to .

Movement direction is shown by a darker line segment on a chess piece.

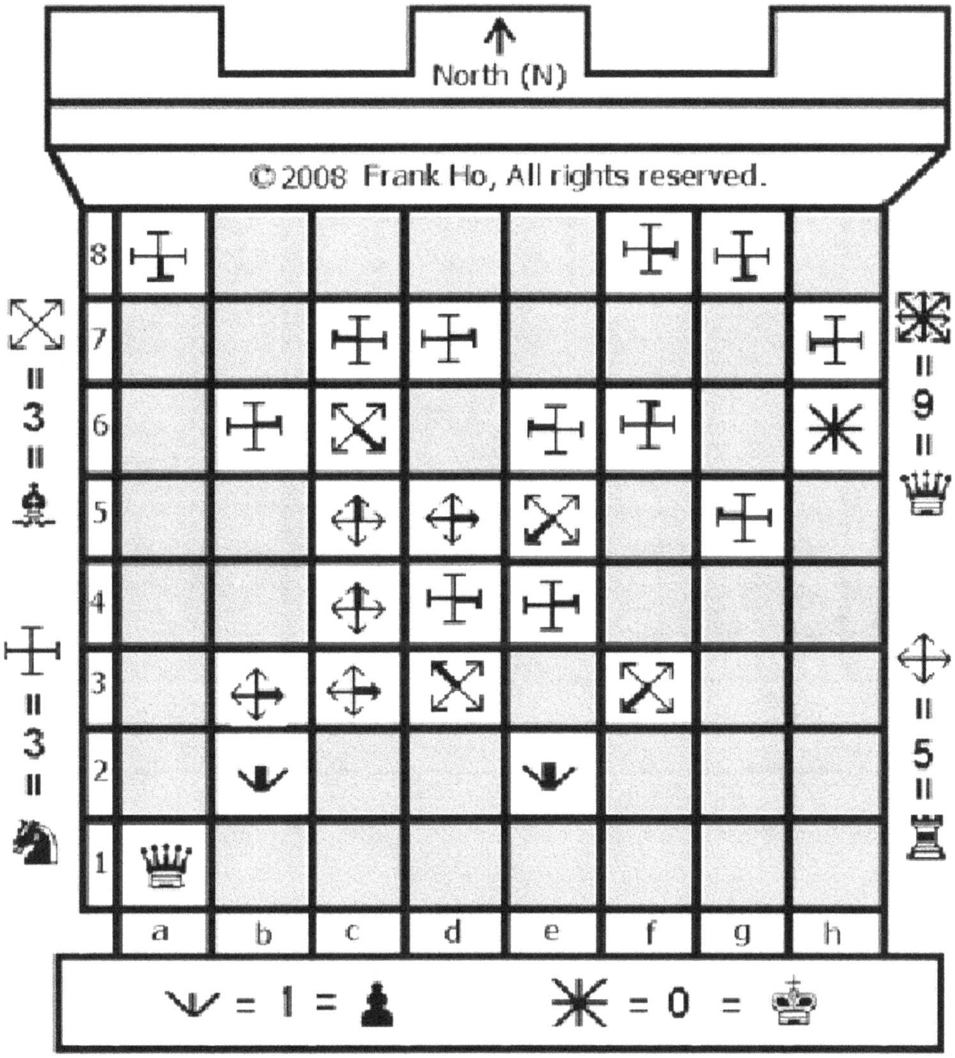

Answer

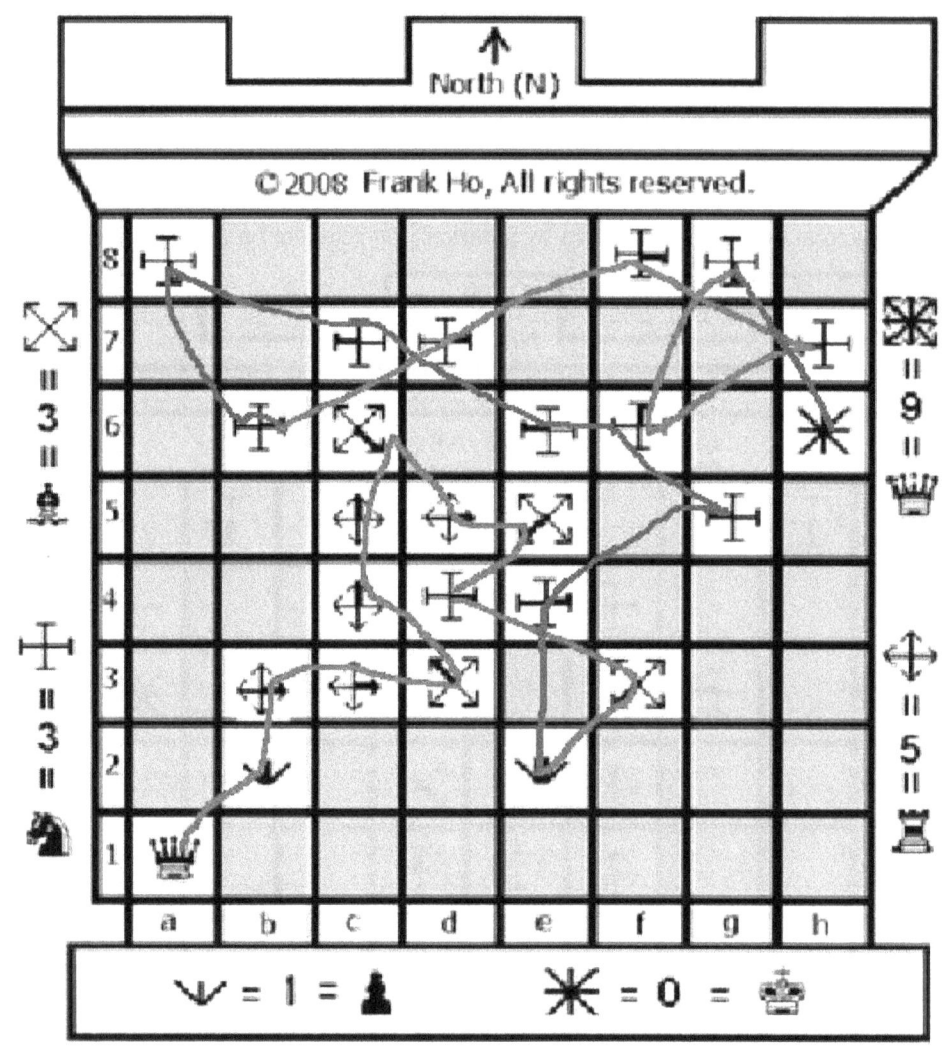

Ho Math Chess — Pre-K and Kindergarten Math

何数棋谜　棋谜式幼儿健脑思维趣味数学

© 2012 – 2021 Frank Ho, Amanda Ho, Canada copyright 1095661, Trademark 771400

Matrix reasoning

矩阵规律思考

Find an answer to each question mark.

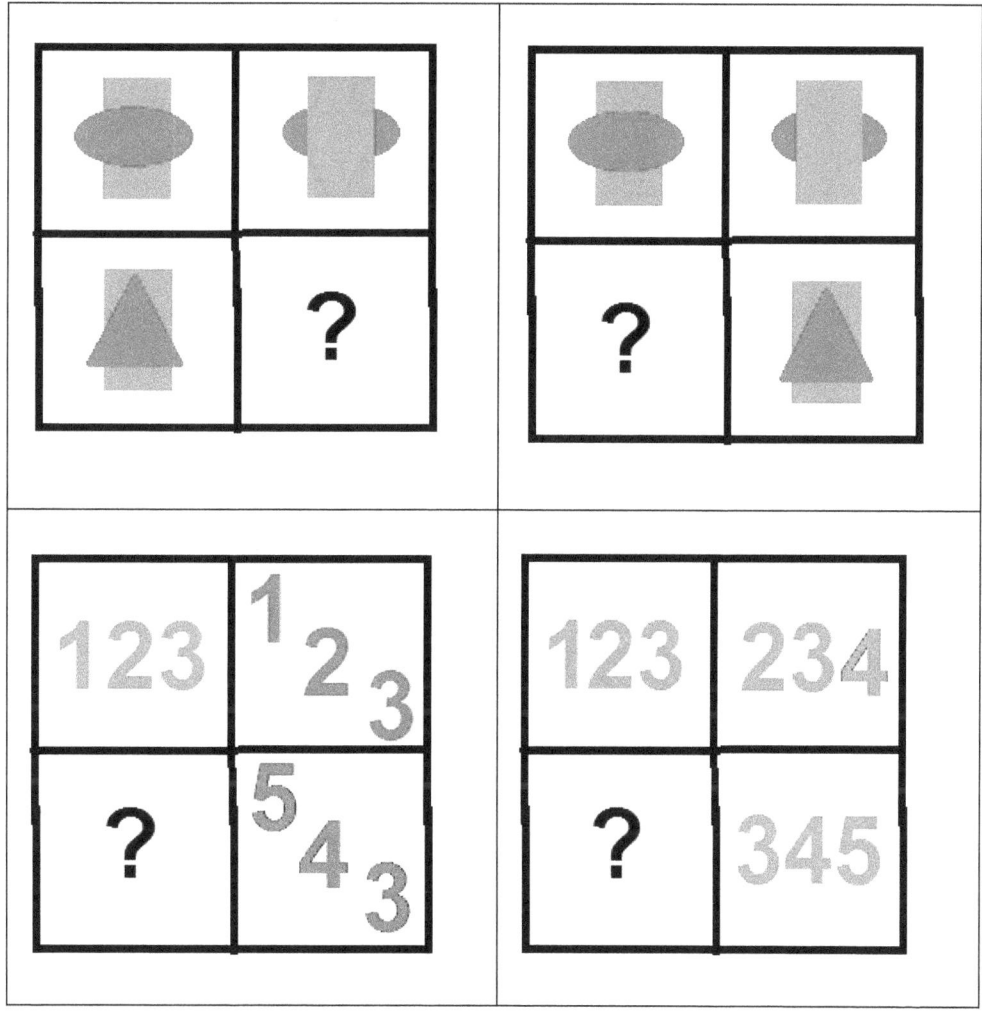

Answer

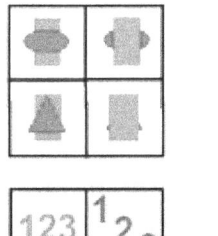

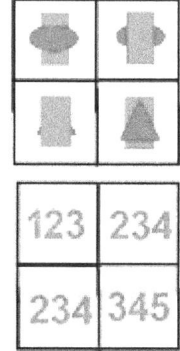

Matrix reasoning

Find an answer to each question mark.

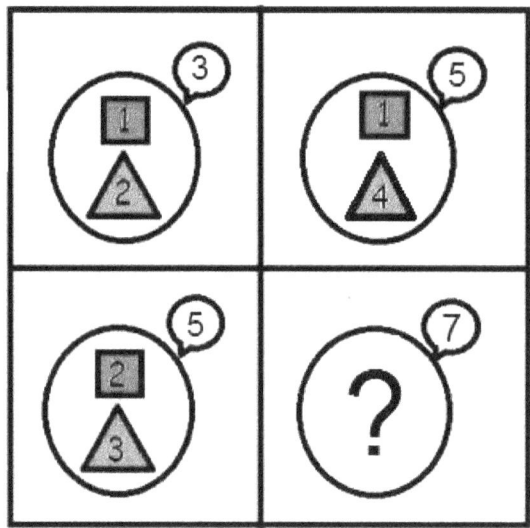

answer

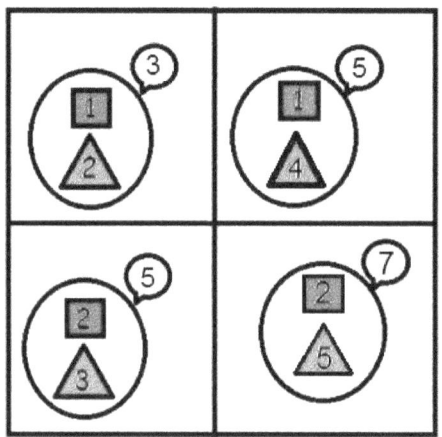

Matrix reasoning

Find an answer to each question mark.

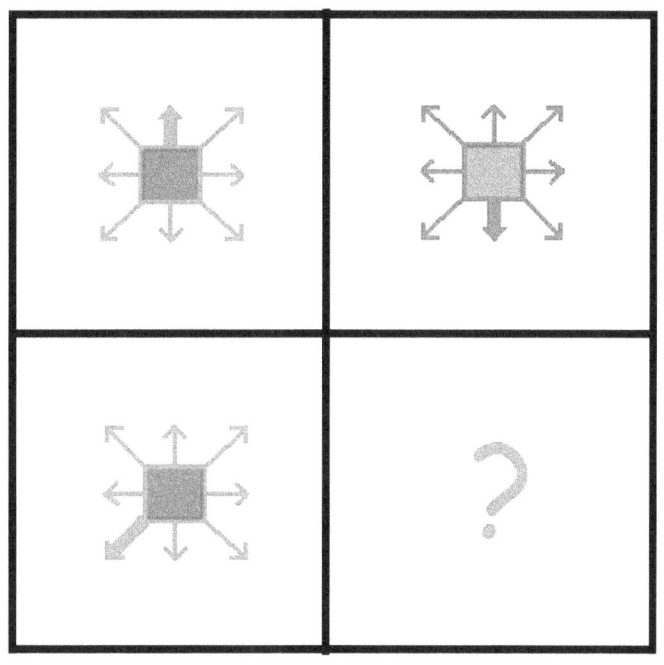

answer

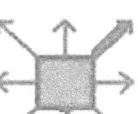

Matrix reasoning

Find an answer to each question mark.

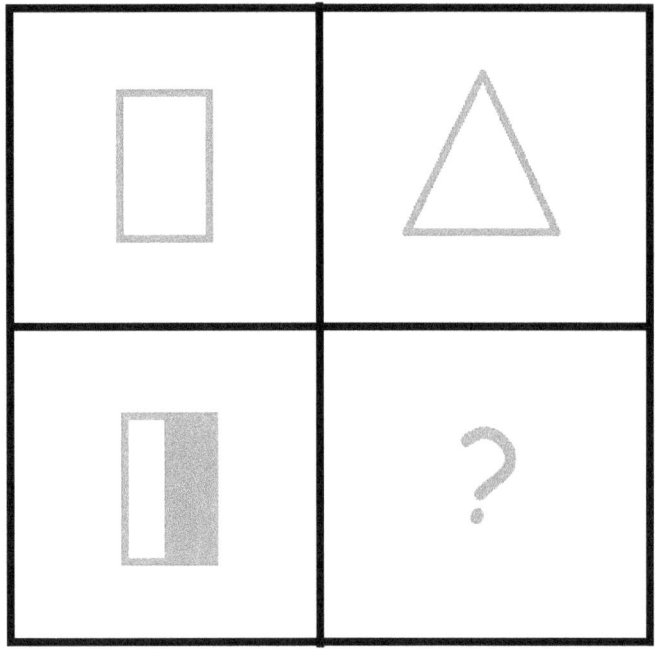

answer

Matrix reasoning

Find an answer to each question mark.

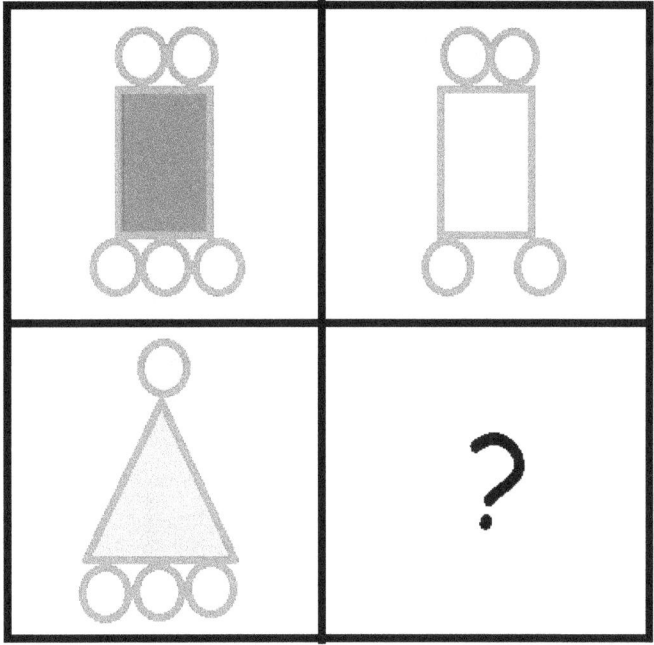

 answer

Matrix reasoning

Find an answer to each question mark.

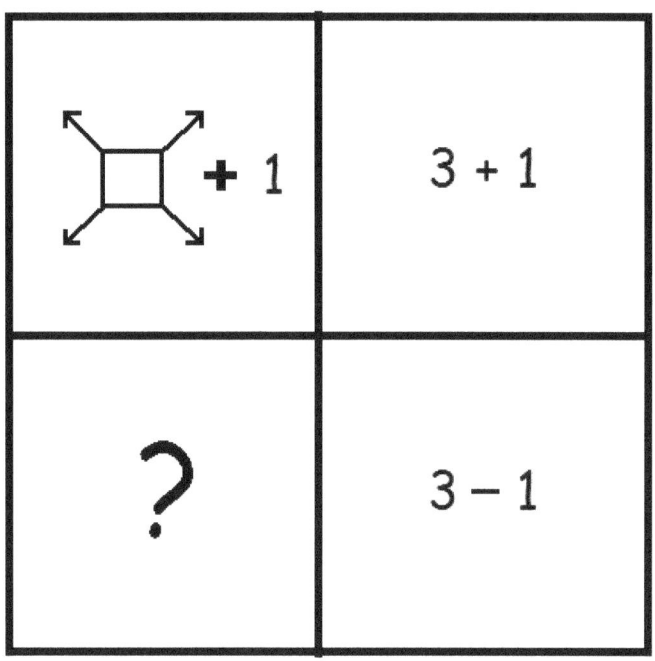

Answer

□ − 1

Matrix reasoning

Find an answer to each question mark.

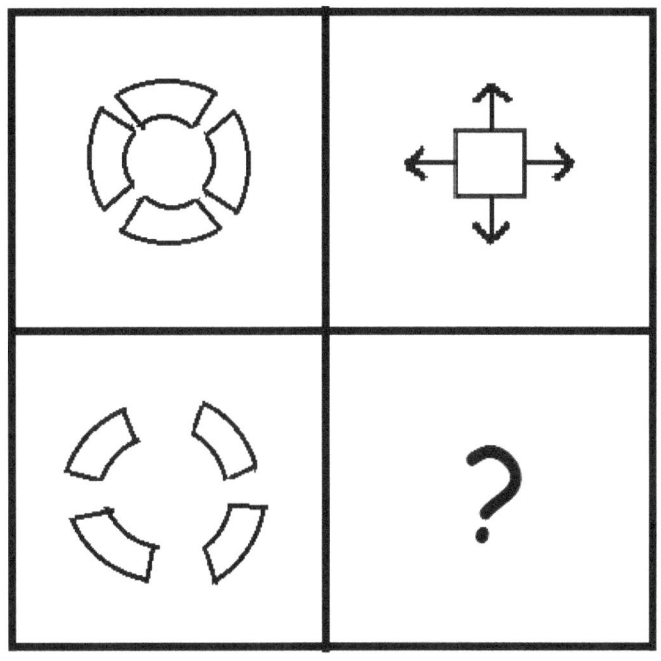

answer

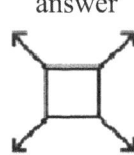

A curve or straight line
曲线或直线

Total 2

Total 11

A curve or straight line

Circle all objects having part or whole circle shapes.

Circle the left 2.

Circle the right 2.

Circle all.

Ho Math Chess — Pre-K and Kindergarten Math

何数棋谜　棋谜式幼儿健脑思维趣味数学

© 2012 – 2021 Frank Ho, Amanda Ho, Canada copyright 1095661, Trademark 771400

Handwriting of number

Circle the correct handwriting of 1, 2, and 3.

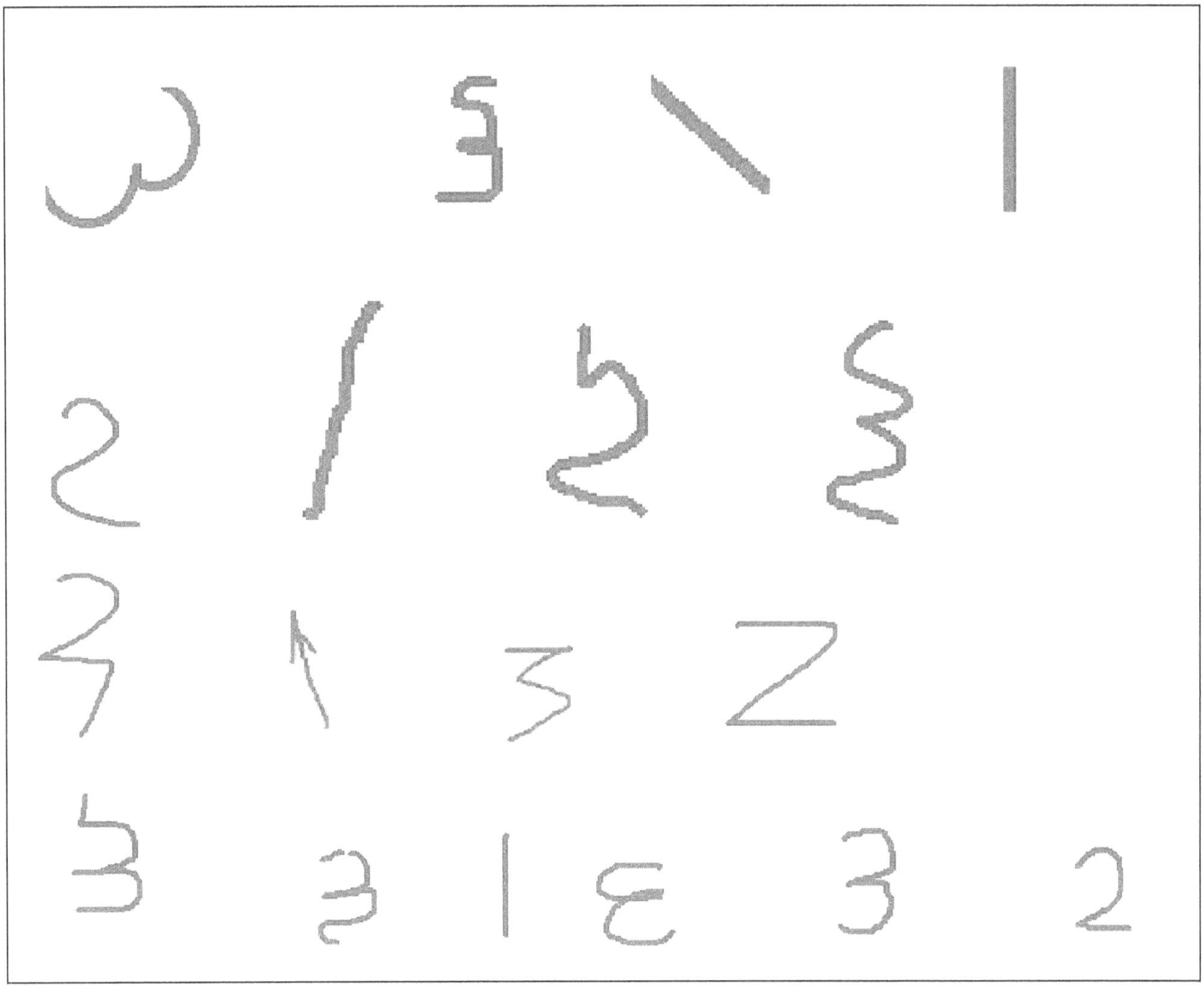

Ho Math Chess   Pre-K and Kindergarten Math

何数棋谜   棋谜式幼儿健脑思维趣味数学

© 2012 – 2021 Frank Ho, Amanda Ho, Canada copyright 1095661, Trademark 771400

Figures made by connecting numbers.

Duck

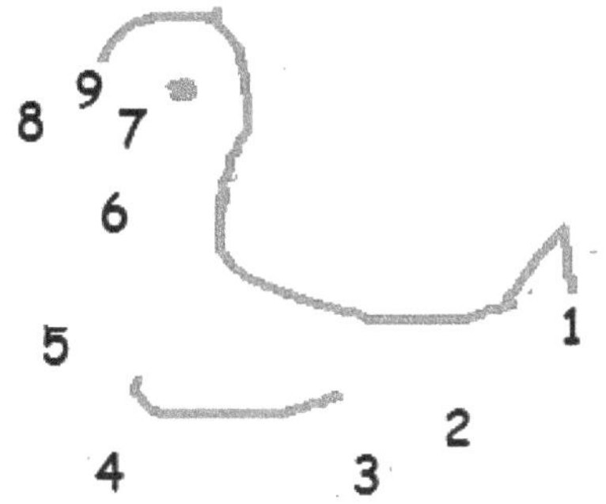

Swan

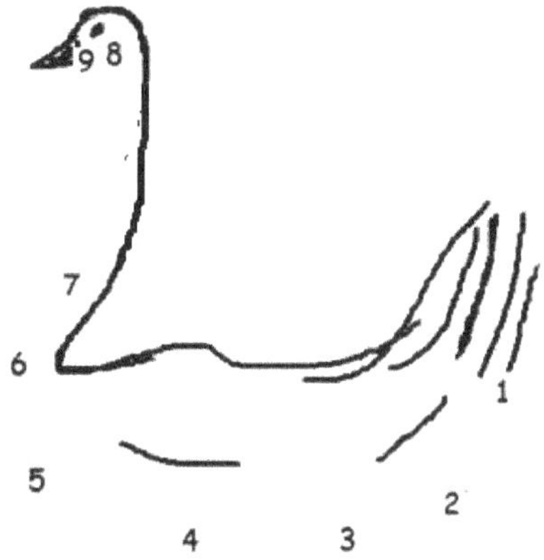

Find numbers (6, 5, 2, 3, and 7) shown in the following figure.

Find numbers (9, 1, 3, 3, 9, and 7) shown in the following figure.

answer

www.homathchess.com   147

The number and its writing

Circle all 2's or every figure which has the image of 2.

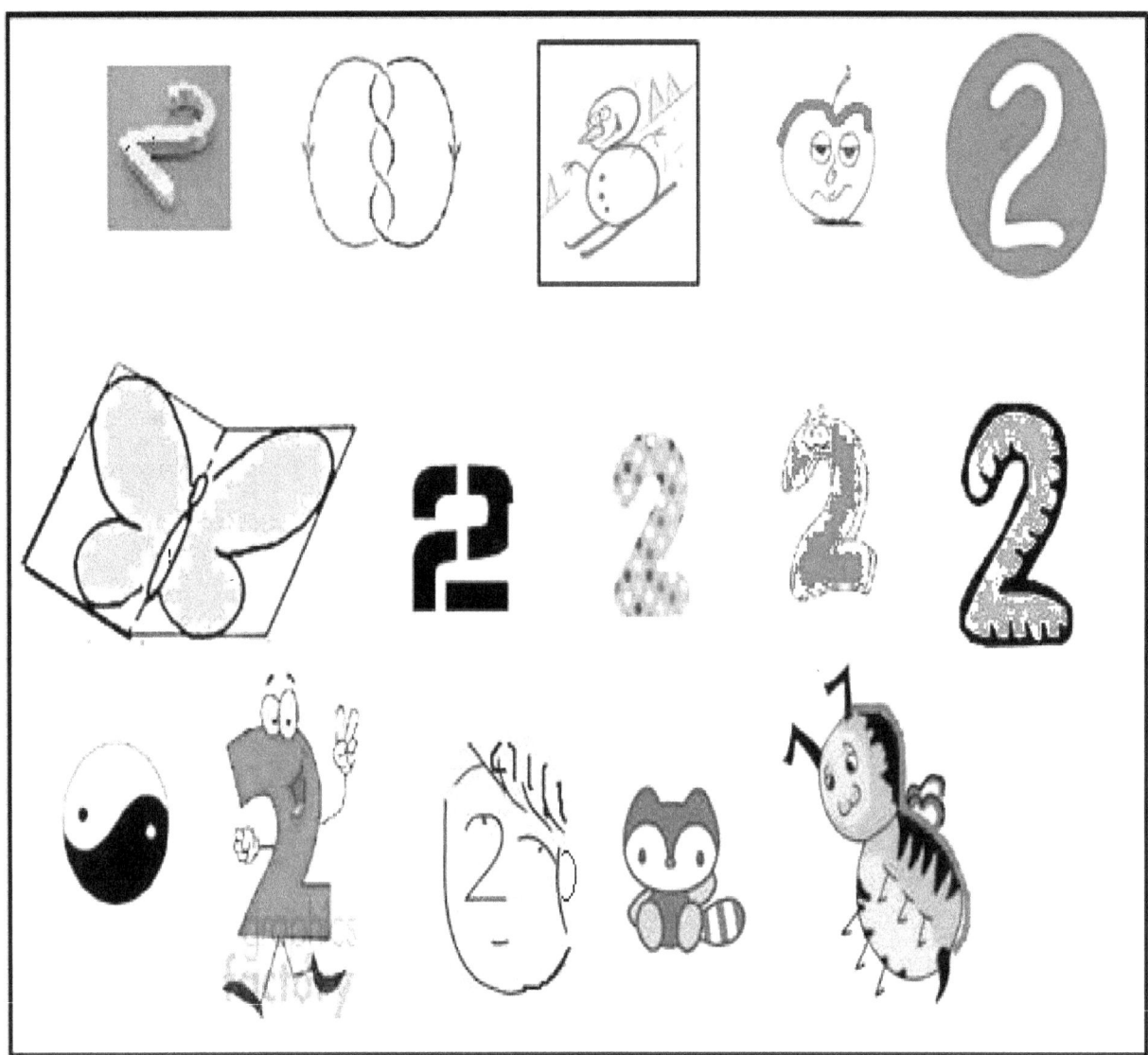

Total 8.

Circle all 3's or every figure which has the image of 3.

Total 6

Ho Math Chess — Pre-K and Kindergarten Math

何数棋谜　棋谜式幼儿健脑思维趣味数学

© 2012 – 2021 Frank Ho, Amanda Ho, Canada copyright 1095661, Trademark 771400

More than (>), larger than (>) or less than (<)
多於，大於，少於

Write numbers starting from 1 all the way to 9 on each step of the ladder when climbing up.

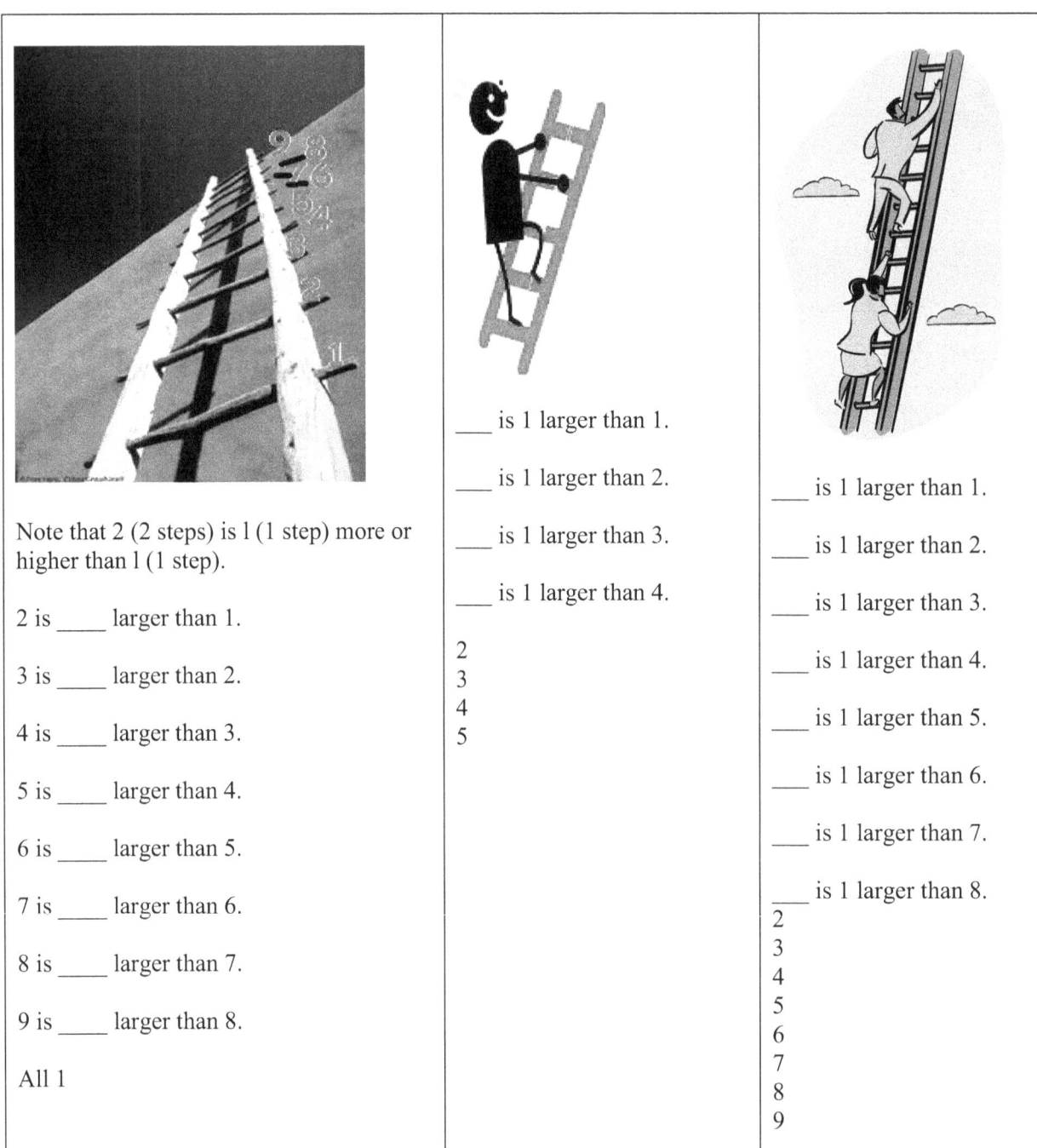

Note that 2 (2 steps) is 1 (1 step) more or higher than 1 (1 step).

2 is ____ larger than 1.

3 is ____ larger than 2.

4 is ____ larger than 3.

5 is ____ larger than 4.

6 is ____ larger than 5.

7 is ____ larger than 6.

8 is ____ larger than 7.

9 is ____ larger than 8.

All 1

___ is 1 larger than 1.
___ is 1 larger than 2.
___ is 1 larger than 3.
___ is 1 larger than 4.
2
3
4
5

___ is 1 larger than 1.
___ is 1 larger than 2.
___ is 1 larger than 3.
___ is 1 larger than 4.
___ is 1 larger than 5.
___ is 1 larger than 6.
___ is 1 larger than 7.
___ is 1 larger than 8.
2
3
4
5
6
7
8
9

More than (>) or less than (<)

Note that 2 (2 steps) is 1 (1 step) more or higher than 1 (1 step).

2 is ____ larger than (>) 1.

3 is ____ larger than (>) 2.

4 is ____ larger than (>) 3.

5 is ____ larger than (>) 4.

6 is ____ larger than (>) 5.

7 is ____ larger than (>) 6.

8 is ____ larger than (>) 7.

9 is ____ larger than (>) 8.

All 1

____ is 1 larger than 1.

____ is 1 larger than 2.

____ is 1 larger than 3.

____ is 1 larger than 4.

____ is 1 larger than 5.

____ is 1 > 1.

____ is 1 > 2.

____ is 1 > 3.

____ is 1 > 4.

____ is 1 > 5.

2 3 4 5 6

____ is 1 > 1.

____ is 1 > 2.

____ is 1 > 3.

____ is 1 > 4.

____ is 1 > 5.

____ is 1 > 6.

____ is 1 > 7.

____ is 1 > 8.

2
3
4
5
6
7
8
9

More than or less than

1 is ____ less than (<) 2.

2 is ____ less than (<) 3.

3 is ____ less than (<) 4.

4 is ____ less than (<) 5.

5 is ____ less than (<) 6.

6 is ____ less than (<) 7.

7 is ____ less than (<) 8.

8 is ____ less than (<) 9.

All 1

1 is ____ less than (<) 2.

2 is ____ less than (<) 3.

3 is ____ less than (<) 4.

4 is ____ less than (<) 5.

5 is ____ less than (<) 6.

6 is ____ less than (<) 7.

7 is ____ less than (<) 8.

8 is ____ less than (<) 9.

All 1

___ is 1 < 2.

___ is 1 < 3.

___ is 1 < 4.

___ is 1 < 5.

___ is 1 < 6.

___ is 1 < 7.

___ is 1 < 8.

___ is 1 < 9.

1 2 3 4 5 6 7 8

More than or less than

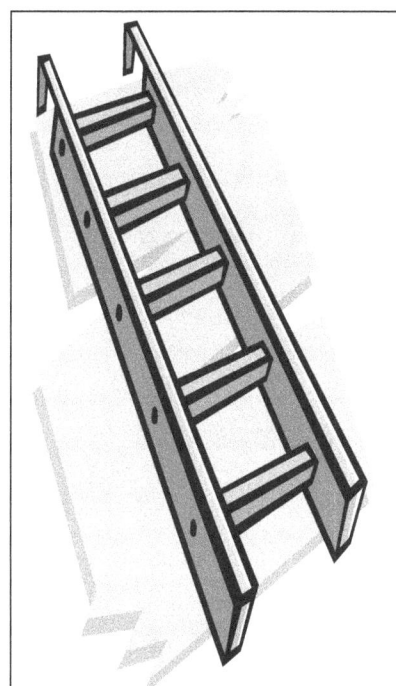

___ is 1 < 2.
___ is 1 > 3.
___ is 1 < 4.
___ is 1 > 5.
___ is 1 < 6.
___ is 1 > 7.
___ is 1 > 8.
___ is 1 < 9.

1 2 3 4 5 6 7 8

___ is 1 > 2.
___ is 1 < 3.
___ is 1 > 4.
___ is 1 < 5.
___ is 1 > 6.
___ is 1 < 7.
___ is 1 > 8.
___ is 1 < 9.

3 2 5 4 7 8 9 8

___ is 1 < 2.
___ is 1 > 3.
___ is 1 < 4.
___ is 1 < 5.
___ is 1 > 6.
___ is 1 > 7.
___ is 1 < 8.
___ is 1 < 9.

1 4 3 4 5 8 7 8

Depth of shapes

Place a number 1, 2, 3, or 4 on each set of shapes such that it shows the lowest number is on the top, and the largest number is at the bottom.

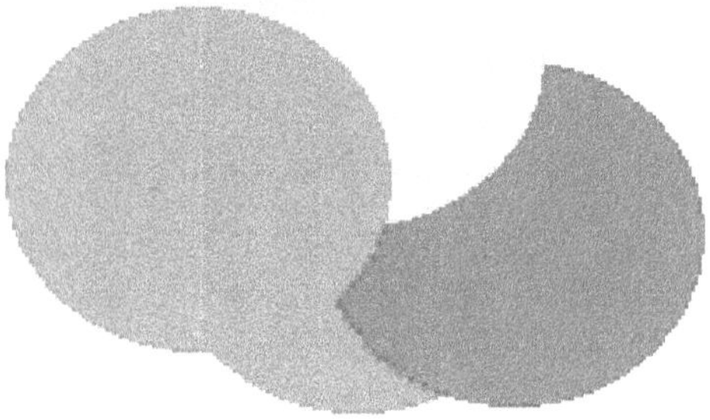

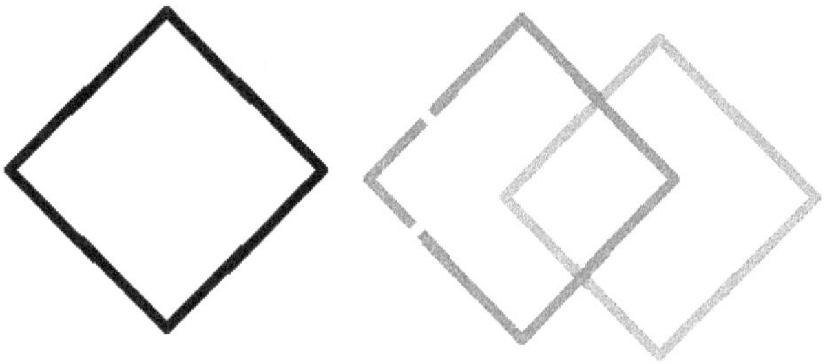

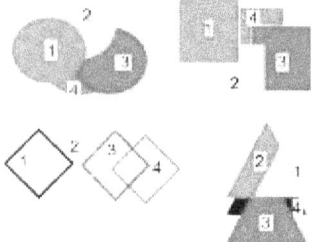
answer

Ho Math Chess   Pre-K and Kindergarten Math

何数棋谜   棋谜式幼儿健脑思维趣味数学

© 2012 – 2021 Frank Ho, Amanda Ho, Canada copyright 1095661, Trademark 771400

Checking sets matching the number of objects

Each set is indicated by a circle.

Cheque (✓) the sets with only 1 object in each set (within the circle).	Cheque (✓) the sets with only 2 objects in each set (within the circle).	Cheque (✓) the sets with 3 or more objects in each set (within the circle).

Circling identical parts

Circle all three identical parts	Circle all two identical parts.	Circle all two identical parts.
answer	answer	answer
Circle all sides with the same length.	Circle all two same parts.	Circle all one unique parts.
answer	answer	Answer

Ho Math Chess — Pre-K and Kindergarten Math

Circle the following figures, which cannot be traced by one stroke without lifting the pencil.

answer

	Ho Math Chess Pre-K and Kindergarten Math
	何数棋谜 棋谜式幼儿健脑思维趣味数学
	© 2012 – 2021 Frank Ho, Amanda Ho, Canada copyright 1095661, Trademark 771400

Counting 1, 2 or 3

Image or figure	How many adults are there?	How many children are there?	How many family members are there?
(family silhouette)	2	1	3

Image or figure	How many lions are there?	How many rabbits are there?	How many pandas are there?
(lions, rabbits, panda)	3	2	1

Image or figure	How many snakes are there	How many pigs are there?	How many roosters are there?
(roosters, pigs, snake)	1	2	3

www.homathchess.com   158

Ho Math Chess   Pre-K and Kindergarten Math
何数棋谜   棋谜式幼儿健脑思维趣味数学
© 2012 – 2021 Frank Ho, Amanda Ho, Canada copyright 1095661, Trademark 771400

Comparing objects

Circle all unique one objects.	Circle all two objects which are identical in size and shape.	Circle all 3 objects which are identical in size and shape.

No pigs, pandas, chickens, snakes	circle snake, chicken, lion goat	circle pig

www.homathchess.com

Ho Math Chess — Pre-K and Kindergarten Math

何数棋谜　棋谜式幼儿健脑思维趣味数学

© 2012 – 2021 Frank Ho, Amanda Ho, Canada copyright 1095661, Trademark 771400

Matching objects

Draw a line from this column to the right-most column to match the same objects with the same matching number of objects. Draw each line by connecting only one object to the matching object.		
(left column of objects)		(right column of objects)

No pig, chicken, goat

Connecting numbers

Count the number of each animal and write its answer below.

Pig – 2
Chicken – 6
Panda – 6
Rabbit – 6
Snake – 6
Lion – 7
Goat – 6

Ho Math Chess — Pre-K and Kindergarten Math

Counting objects

Circle every three hearts from row to row from row to tow.

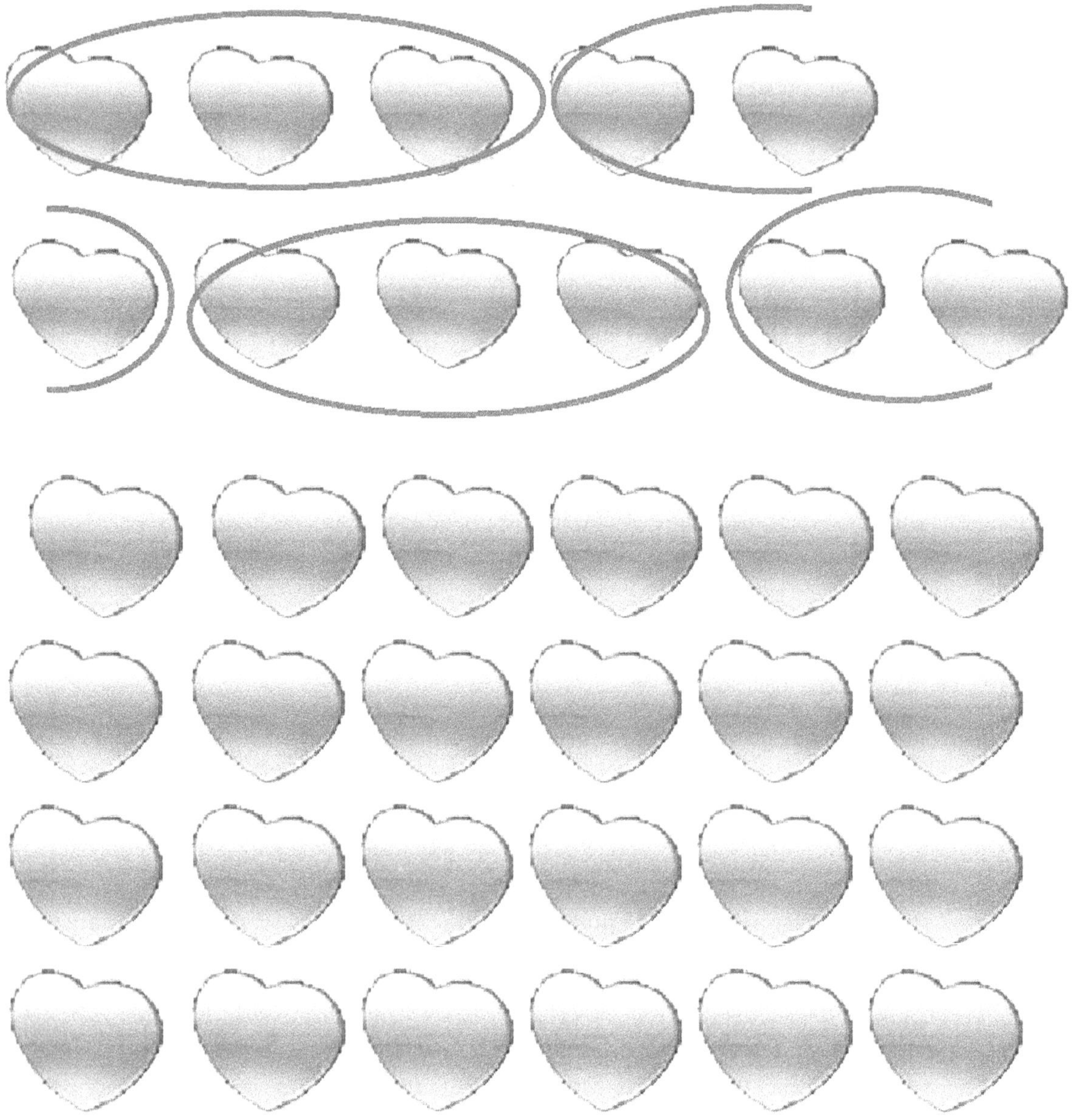

Circle every two balls from row to row.

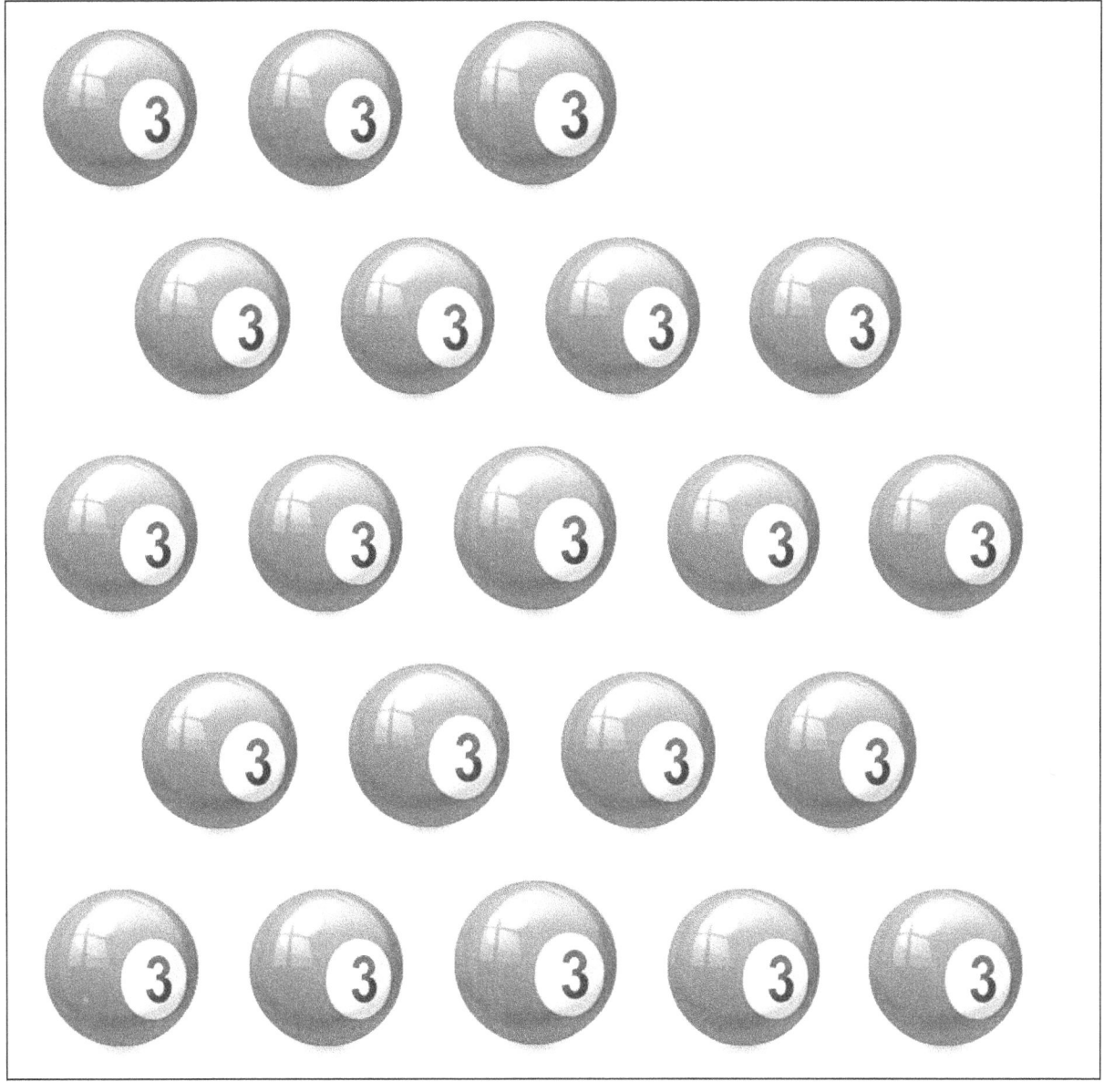

Logic connection

Connect the left column to the right column with similar objects.	

Ho Math Chess — Pre-K and Kindergarten Math

何数棋谜　棋谜式幼儿健脑思维趣味数学

© 2012 – 2021 Frank Ho, Amanda Ho, Canada copyright 1095661, Trademark 771400

Review of numbers writing.

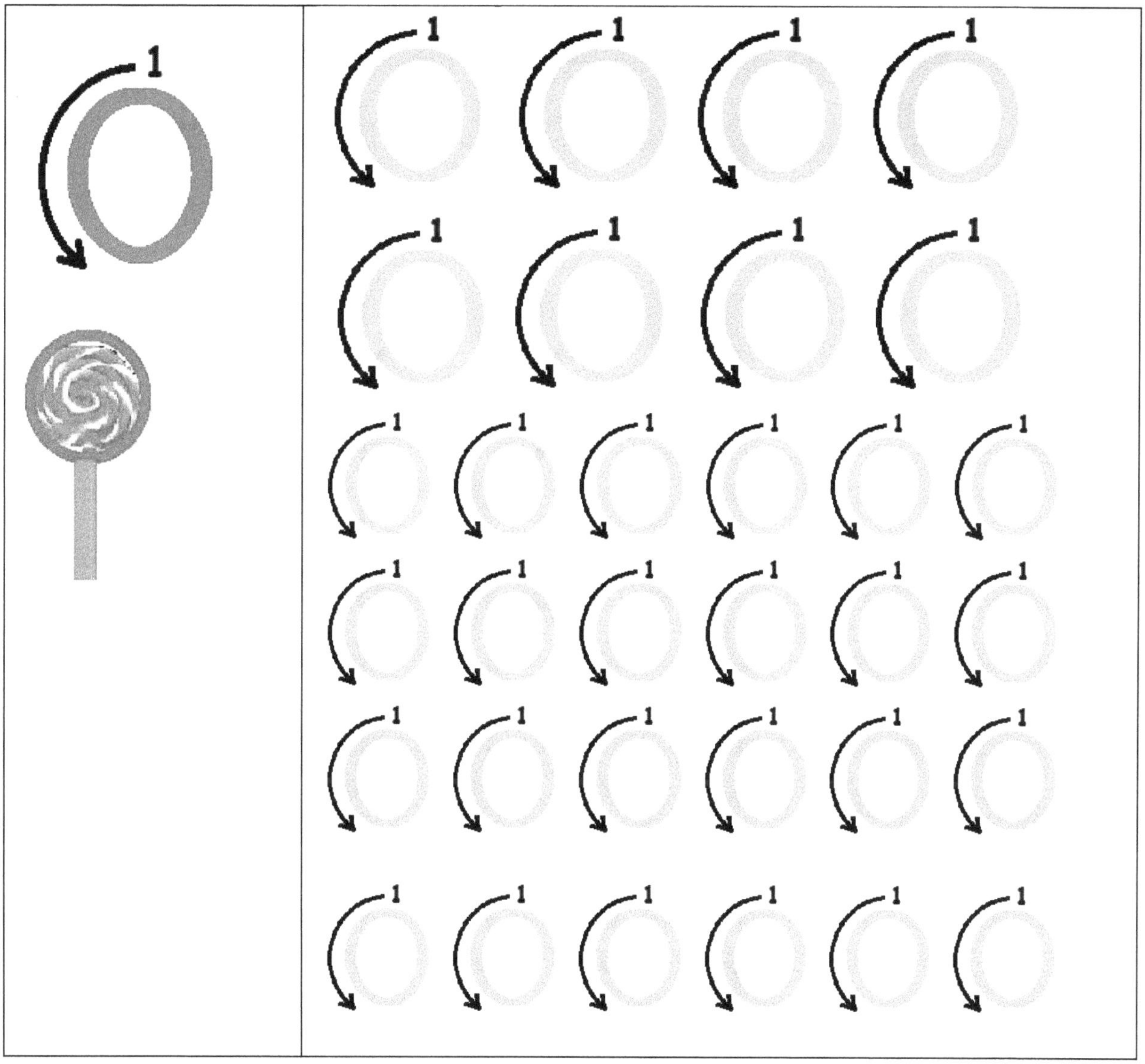

Ho Math Chess Pre-K and Kindergarten Math
何数棋谜 棋谜式幼儿健脑思维趣味数学

© 2012 – 2021 Frank Ho, Amanda Ho, Canada copyright 1095661, Trademark 771400

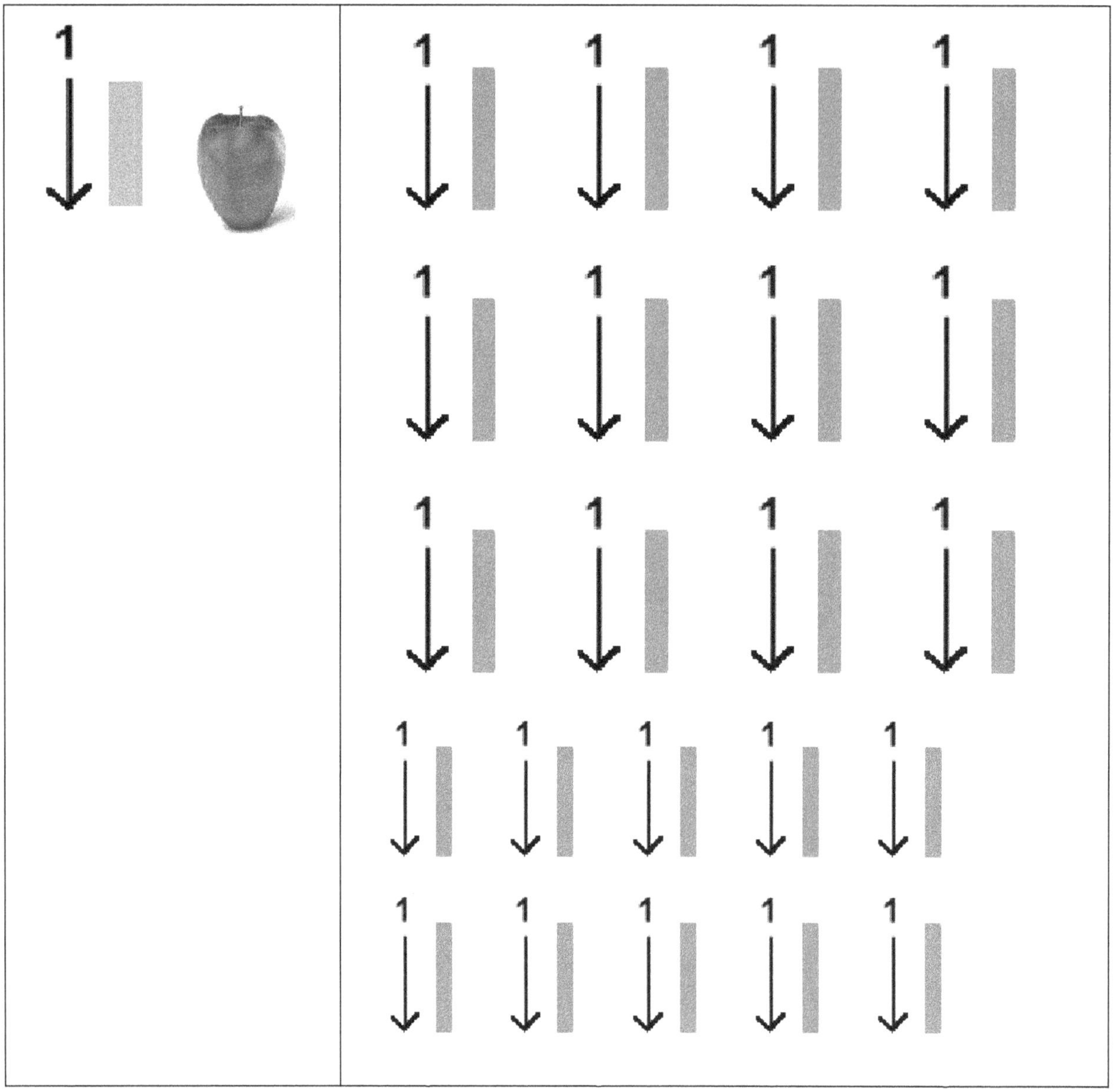

Ho Math Chess — Pre-K and Kindergarten Math

何数棋谜　棋谜式幼儿健脑思维趣味数学

© 2012 – 2021 Frank Ho, Amanda Ho, Canada copyright 1095661, Trademark 771400

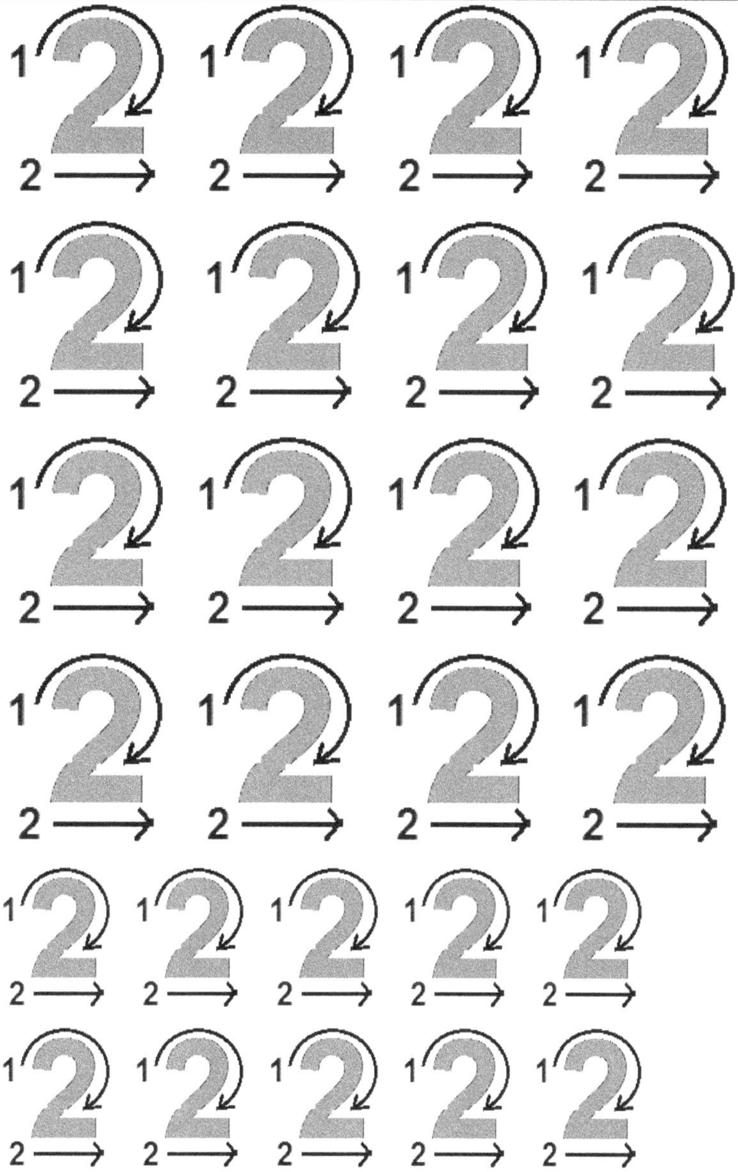

Ho Math Chess — Pre-K and Kindergarten Math

何数棋谜 棋谜式幼儿健脑思维趣味数学

© 2012 – 2021 Frank Ho, Amanda Ho, Canada copyright 1095661, Trademark 771400

Ho Math Chess — Pre-K and Kindergarten Math

何数棋谜　棋谜式幼儿健脑思维趣味数学

© 2012 – 2021 Frank Ho, Amanda Ho, Canada copyright 1095661, Trademark 771400

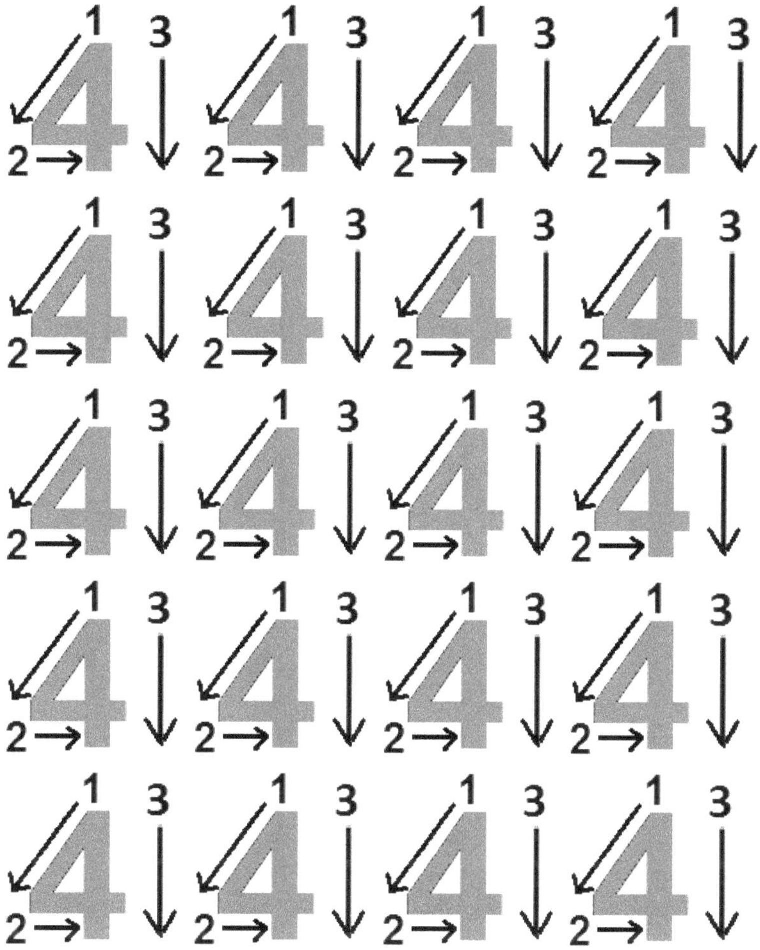

Ho Math Chess Pre-K and Kindergarten Math
何数棋谜 棋谜式幼儿健脑思维趣味数学
© 2012 – 2021 Frank Ho, Amanda Ho, Canada copyright 1095661, Trademark 771400

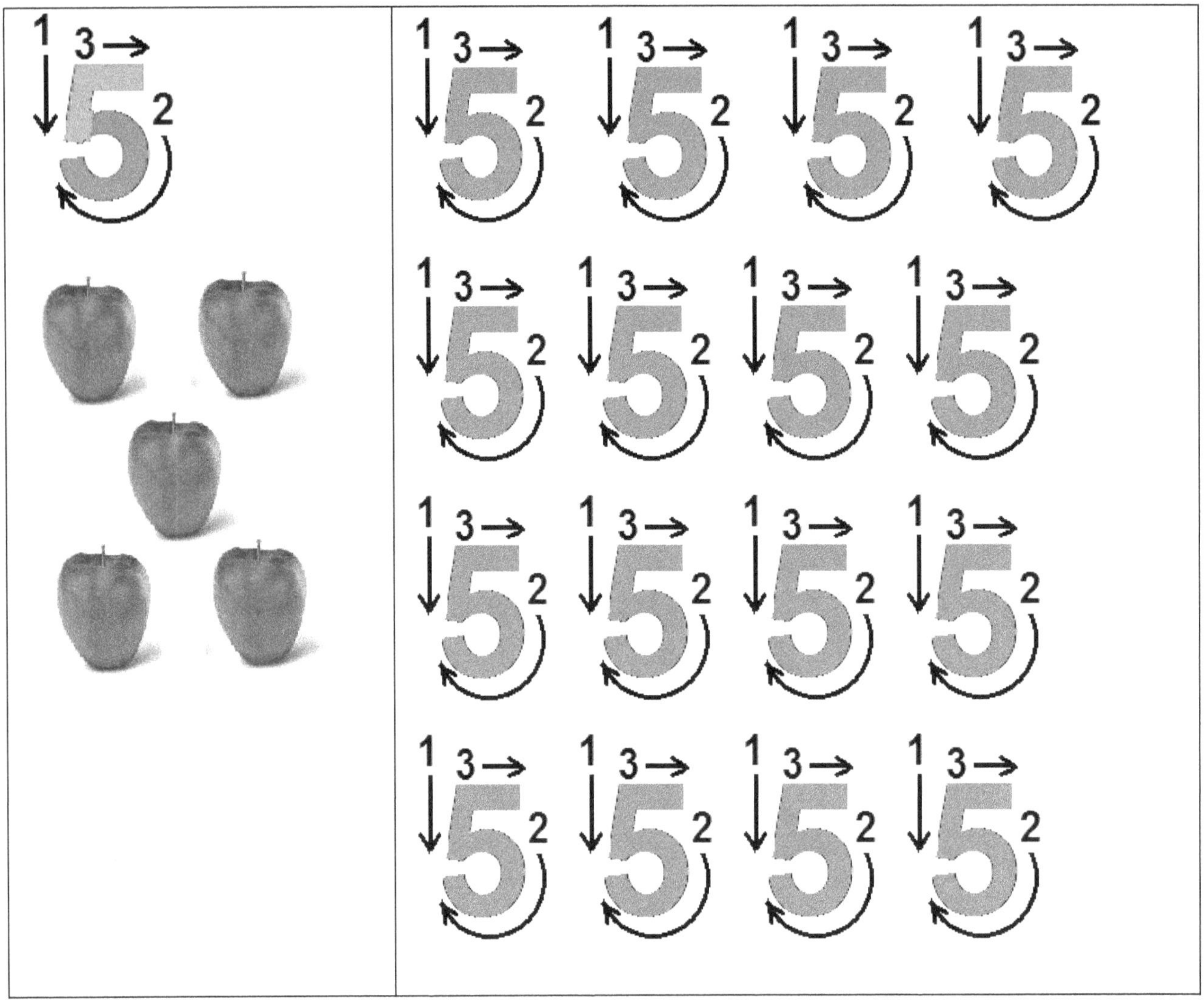

www.homathchess.com

Ho Math Chess   Pre-K and Kindergarten Math
何数棋谜   棋谜式幼儿健脑思维趣味数学
© 2012 – 2021 Frank Ho, Amanda Ho, Canada copyright 1095661, Trademark 771400

Ho Math Chess   Pre-K and Kindergarten Math
何数棋谜　棋谜式幼儿健脑思维趣味数学
© 2012 – 2021 Frank Ho, Amanda Ho, Canada copyright 1095661, Trademark 771400

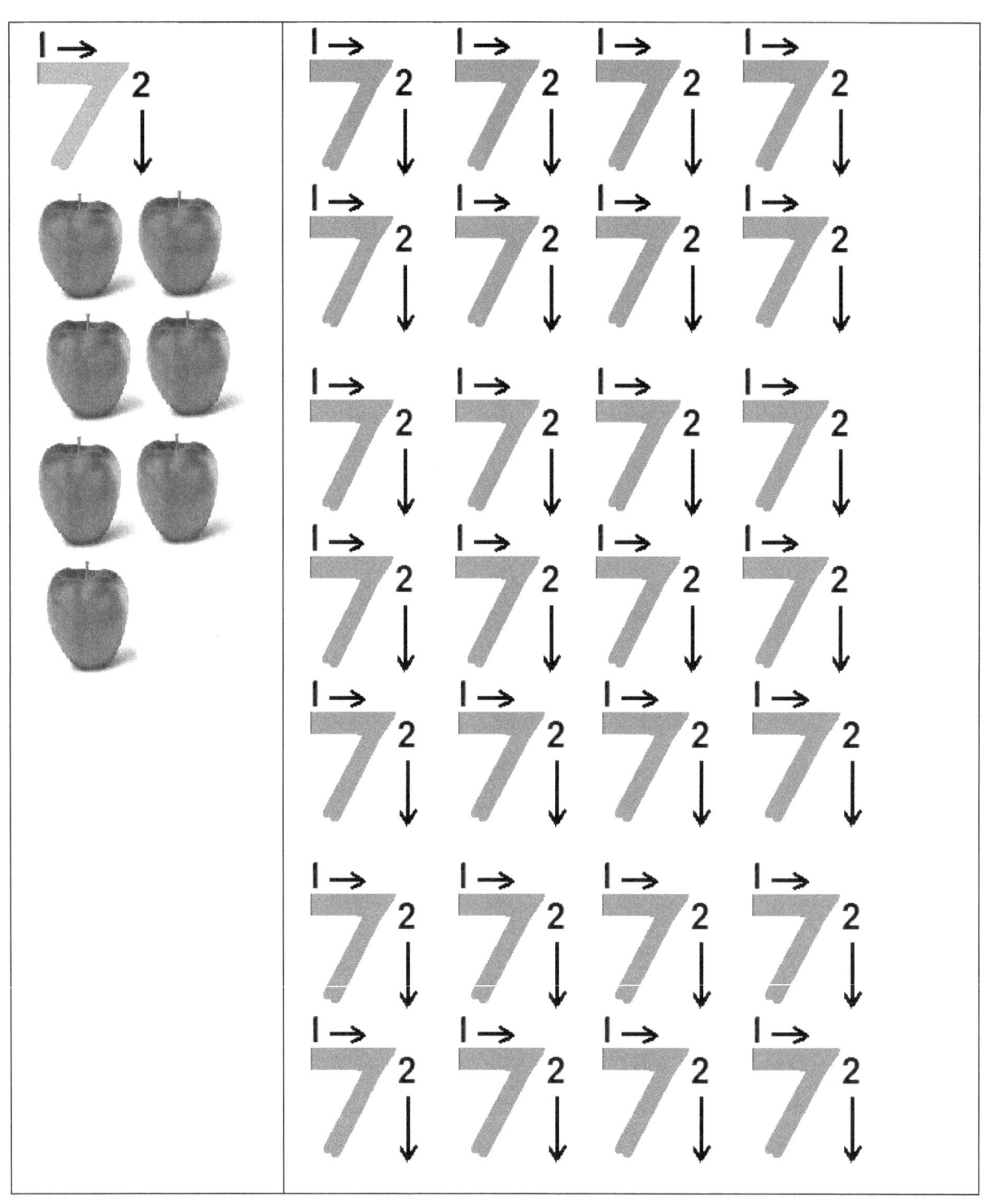

www.homathchess.com   172

Ho Math Chess — Pre-K and Kindergarten Math

何数棋谜　棋谜式幼儿健脑思维趣味数学

© 2012 – 2021 Frank Ho, Amanda Ho, Canada copyright 1095661, Trademark 771400

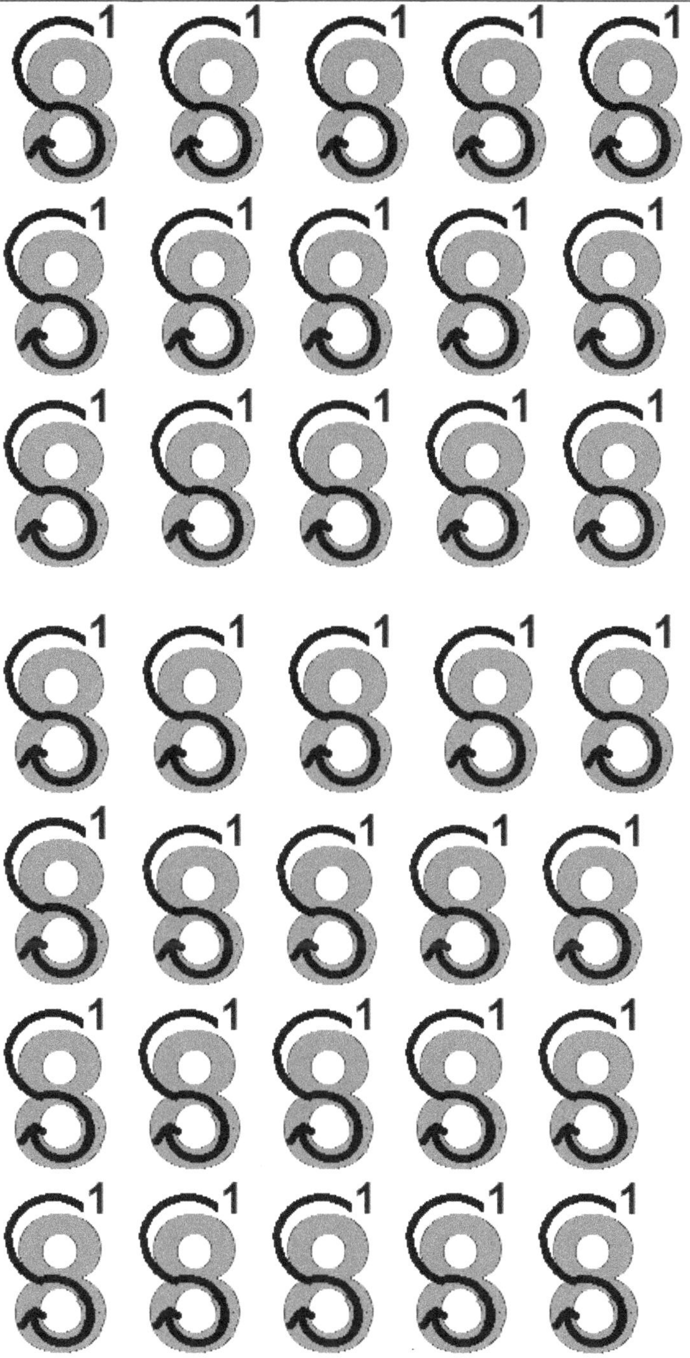

Ho Math Chess   Pre-K and Kindergarten Math

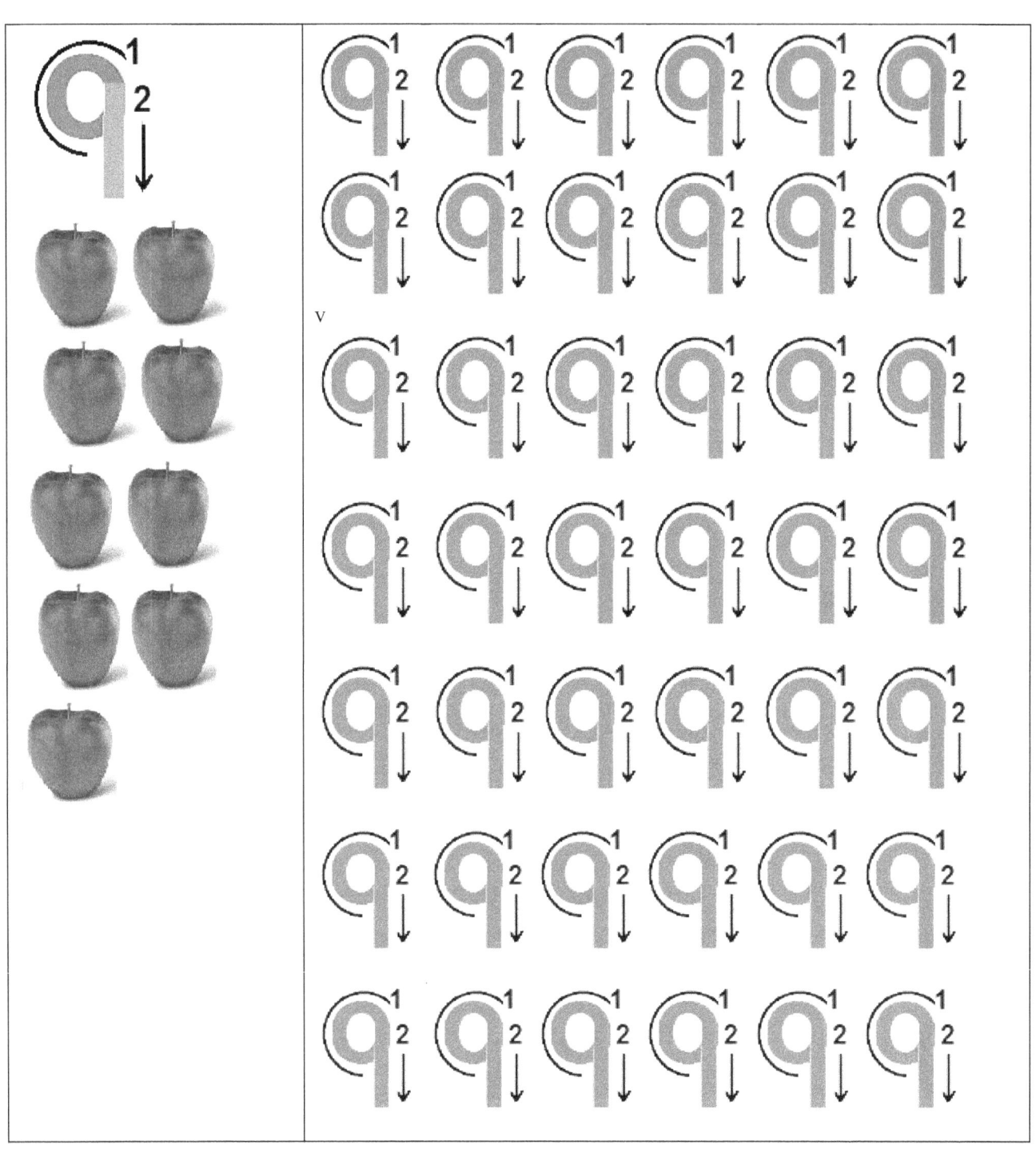

Ho Math Chess — Pre-K and Kindergarten Math

何数棋谜　棋谜式幼儿健脑思维趣味数学

© 2012 – 2021 Frank Ho, Amanda Ho, Canada copyright 1095661, Trademark 771400

Counting and writing numbers

Verbally count the objects in each set and then write the number of objects counted.

(4 soccer balls)	Write the number of objects. _____
(5 mice)	Write the number of objects. _____
(7 kettles)	Write the number of objects. _____
(9 candles)	Write the number of objects. _____
(6 hats)	Write the number of objects. _____

4 5 7 9 6

Ho Math Chess   Pre-K and Kindergarten Math
何数棋谜 棋谜式幼儿健脑思维趣味数学
© 2012 – 2021 Frank Ho, Amanda Ho, Canada copyright 1095661, Trademark 771400

Counting and writing numbers
Verbally count the objects in each set and then write the number of objects counted.

(5 sailboats)	Write the number of objects. _____
(8 snowmen)	Write the number of objects. _____
(9 swans)	Write the number of objects. _____
(8 zebras)	Write the number of objects. _____

5 8 9 8

Counting and writing numbers

Verbally count the objects in each set and then write the number of objects counted.

(5 pandas)	Write the number of objects. _____
(6 cupcakes)	Write the number of objects. _____
(5 squares)	Write the number of objects. _____
(7 boots)	Write the number of objects. _____
(9 sweaters)	Write the number of objects. _____

5 6 5 7 9

Ho Math Chess Pre-K and Kindergarten Math

何数棋谜 棋谜式幼儿健脑思维趣味数学

© 2012 – 2021 Frank Ho, Amanda Ho, Canada copyright 1095661, Trademark 771400

Counting and writing numbers

Verbally count each number in the following 9 by 9 table and then write the number of each counted number.

	0	1	2	3	4	5	6	7	8	9
Total counts										

0 3 1 3 3 3 3 4 3 2

3		1					4	
8		6		3				
			8					
	6		9			3	5	
5			7		1			4
	1	9			8		7	
					2			
				4		7		8
	7					5		6

Counting and writing numbers using the table

Verbally count each number in the following 9 by 9 table and then write the number of each counted number.

	0	1	2	3	4	5	6	7	8	9
Total counts										

0 3 1 6 3 5 2 3 2 5

						8	5		3
4		9			3	8	6		7
	5	3							
					1			9	4
		5				6			
1	2		9						
						9	3		
3	9	4	1			7			5
5		7	3						

Ho Math Chess   Pre-K and Kindergarten Math

何数棋谜   棋谜式幼儿健脑思维趣味数学

© 2012 – 2021 Frank Ho, Amanda Ho, Canada copyright 1095661, Trademark 771400

Counting and writing numbers

Verbally count each number in the following 9 by 9 table and then write the number of each counted number.

	0	1	2	3	4	5	6	7	8	9
Total counts										

0 4 4 3 4 3 3 3 4 0

	1	7	4	8				
5		6	1					
4	8				2		3	
		8						
2			3		4			1
						2		
	7		2				5	3
						6	7		8
				4	5	6	1	

Ho Math Chess Pre-K and Kindergarten Math

何数棋谜 棋谜式幼儿健脑思维趣味数学

© 2012 – 2021 Frank Ho, Amanda Ho, Canada copyright 1095661, Trademark 771400

Counting and writing numbers

Verbally count each number in the following 9 by 9 table and then write the number of each counted number.

	0	1	2	3	4	5	6	7	8	9
Total counts										

0 3 3 3 2 4 3 3 3 5

			6	8	3			2
							3	5
	3		2		7	9		
	4				9	6		
8			4		5			9
		6	7				1	
		9	5		8		6	
1	2							
5			9	7	1			

Ho Math Chess   Pre-K and Kindergarten Math

何数棋谜　棋谜式幼儿健脑思维趣味数学

© 2012 – 2021 Frank Ho, Amanda Ho, Canada copyright 1095661, Trademark 771400

Counting and writing numbers

Verbally count each number in the following 16 by 16 table and then write the number of each counted number.

	0	1	2	3	4	5	6	7	8	9
Total counts										

0 7 6 6 6 7 6 4 6 4

f		2		a	b			5		d			3	1	c
5	a	8			f										
								1	4				a	f	d
7					8		f						5		
6	d						f		3					9	
9	4		a		1		b		8	g	6	d			
	2		b		d		3	4	e					g	6
1	f				c	e			2	d					5
b				2	g			e	f					3	9
	g	a				6	8	b		9		5		c	
			3	e	b	5		d			4		6	g	1
	6					1		c						7	e
		e				5		a							8
4	7	9				2	c								
									3				9	4	g
2	8	6		1		a			9	f		b			7

www.homathchess.com

Ho Math Chess   Pre-K and Kindergarten Math

何数棋谜 棋谜式幼儿健脑思维趣味数学

© 2012 – 2021 Frank Ho, Amanda Ho, Canada copyright 1095661, Trademark 771400

Counting and writing numbers

Verbally count each number in the following 16 by 16 table and then write the number of each counted number.

	0	1	2	3	4	5	6	7	8	9
Total counts										

0 3 2 9 7 6 7 6 6 9

d	9		6		c	e				2			b		
		3				g		d	e	4			5	9	8
	e		b		4		3			5		d	a		
		2				5	a	6		7	b				c
3														6	f
	7	1		d	f					2	6			a	5
6			c	3				9			1	g			
5			4	a					c		7	9			
		9	1		8						f	3			a
		8	4			5					d	e			7
4	d			g		e				9	3		c	f	
1	3														b
a				8	7		6	3	e				d		
		g	3		9				f		d		b		4
8		9	d			f	4	b		6			5		
	c				5				9	a		7		8	3

www.homathchess.com

Counting and writing numbers

Verbally count the objects in each set and then write the number of objects counted.

◯ = red,   ◯ = blue,   ◯ = black

(set 1 of circles)	Write the number of red objects. _____ Write the number of black objects. _____ Write the number of blue objects. _____
(set 2 of circles)	Write the number of red objects. _____ Write the number of black objects. _____ Write the number of blue objects. _____
(set 3 of circles)	Write the number of red objects. _____ Write the number of black objects. _____ Write the number of blue objects. _____
(set 4 of circles)	Write the number of red objects. _____ Write the number of black objects. _____ Write the number of blue objects. _____

Counting and writing numbers

Verbally count the objects and then write the number of objects counted.

	Triangles	Circles	Squares
Red	How many?	How many?	How many?
Blue	How many?	How many?	How many?
Black	How many?	How many?	How many?

2 3 3
3 4 2
2 3 4

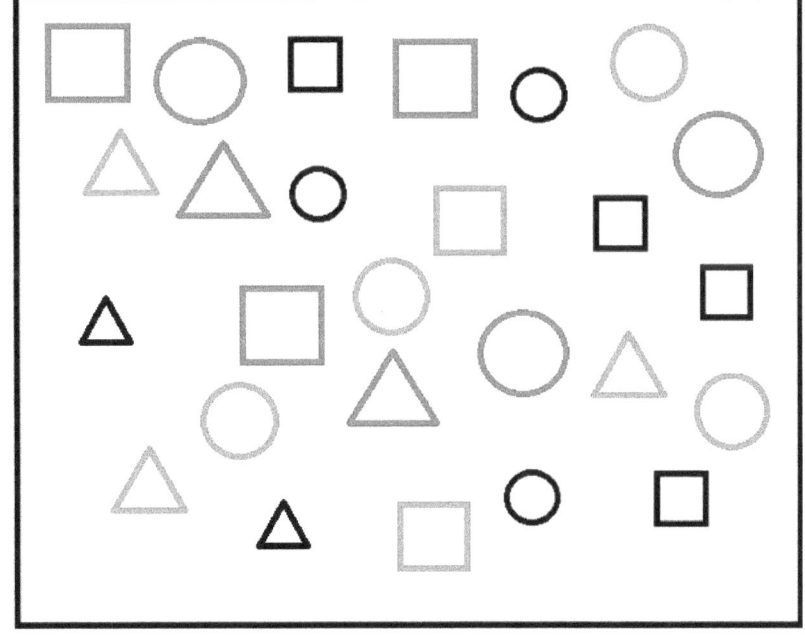

Ho Math Chess — Pre-K and Kindergarten Math

何数棋谜　棋谜式幼儿健脑思维趣味数学

© 2012 – 2021 Frank Ho, Amanda Ho, Canada copyright 1095661, Trademark 771400

Counting and writing numbers

Verbally count the objects and then write the number of objects counted.

Write the number of **red** objects. ____ 21

Write the number of **black** objects. ____ 15

Write the number of **blue** objects. ____ 10

Ho Math Chess — Pre-K and Kindergarten Math

何数棋谜　棋谜式幼儿健脑思维趣味数学

© 2012 – 2021 Frank Ho, Amanda Ho, Canada copyright 1095661, Trademark 771400

Counting and writing numbers

Verbally count the objects and then write the number of objects counted.

red objects. =

black objects. =

blue objects. =

Count every three red circles as one set and then write the number of red sets. _____ 7

Count every four blue circles as one set and then write the number of blue sets. _____ 5

Count every five black circles as one set and then write the number of black sets. _____ 3

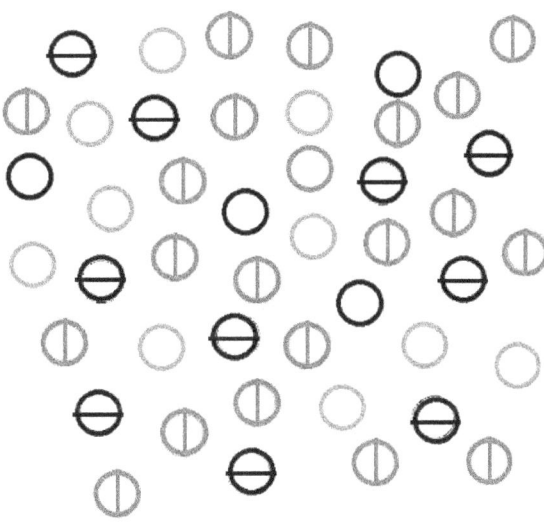

Subitizing
知数感，一见即知数

Knowing at a glance exactly how many objects without actually counting.

How many 1's are there? _____   7

How many 2's are there? _____   4

How many 3's are there? _____   5

Ho Math Chess   Pre-K and Kindergarten Math

何数棋谜   棋谜式幼儿健脑思维趣味数学

© 2012 – 2021 Frank Ho, Amanda Ho, Canada copyright 1095661, Trademark 771400

Subitizing

How many objects without actually counting?

How many 1's are there? How many 2's are there? How many 3's are there? ‖ ‖ 2 2 3 3 3	How many 1's are there? How many 2's are there? How many 3's are there? 3 ‖ ‖ 3 3 3 2 2
How many 1's are there? How many 2's are there? How many 3's are there? 3 3 2 ‖ 3 3 2 ‖	How many 1's are there? How many 2's are there? How many 3's are there? 3 3 3 3 2 2 ‖ ‖

423 224
214 224

Subitizing

How many objects without actually counting?

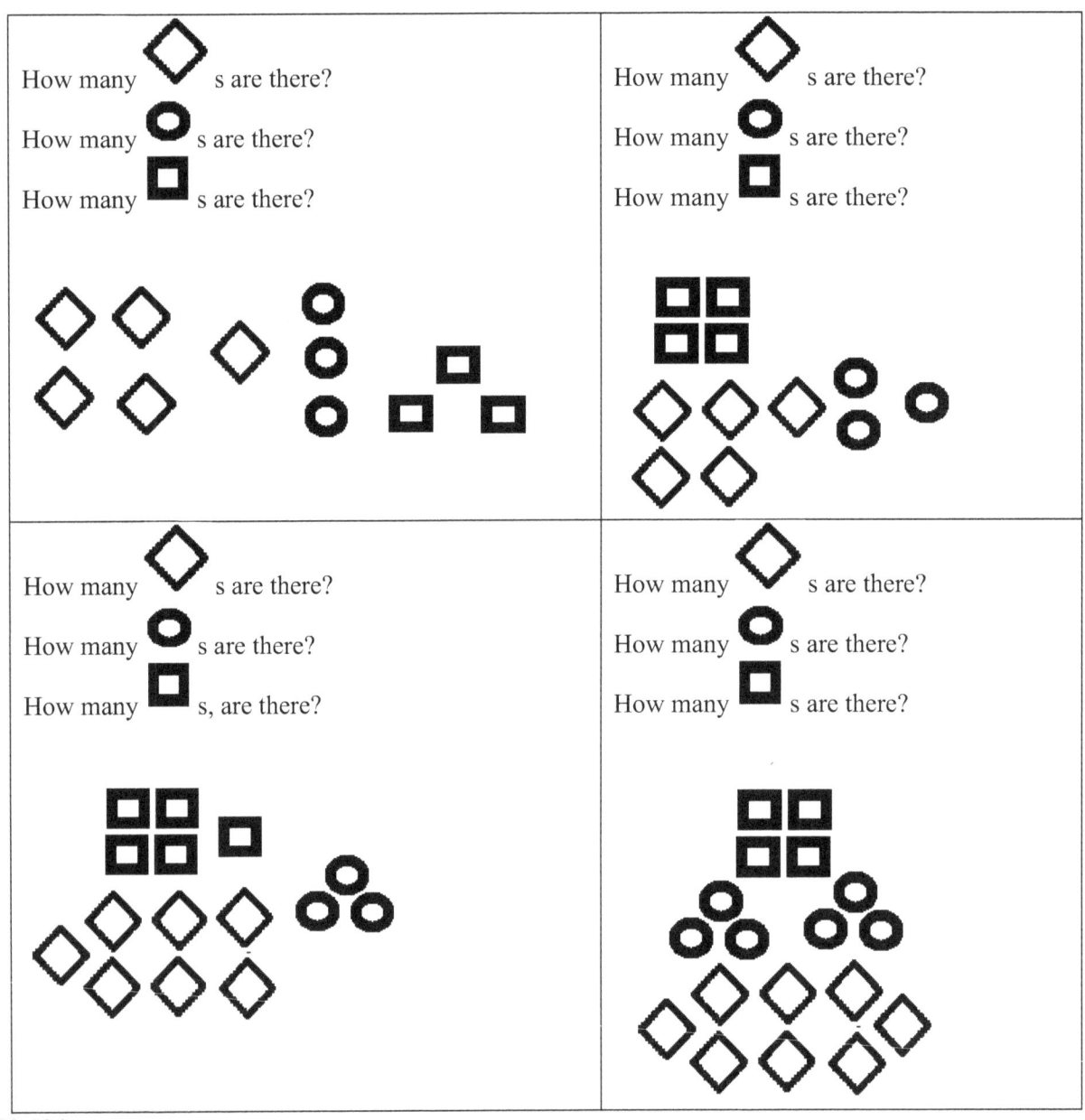

533 534
735 864

Subitizing (知数感, 一見物即知和.)

This section trains a child to know the sum of a group of objects without counting them.

Circle the number of dots on the surface face of each die when its value is 1. 6

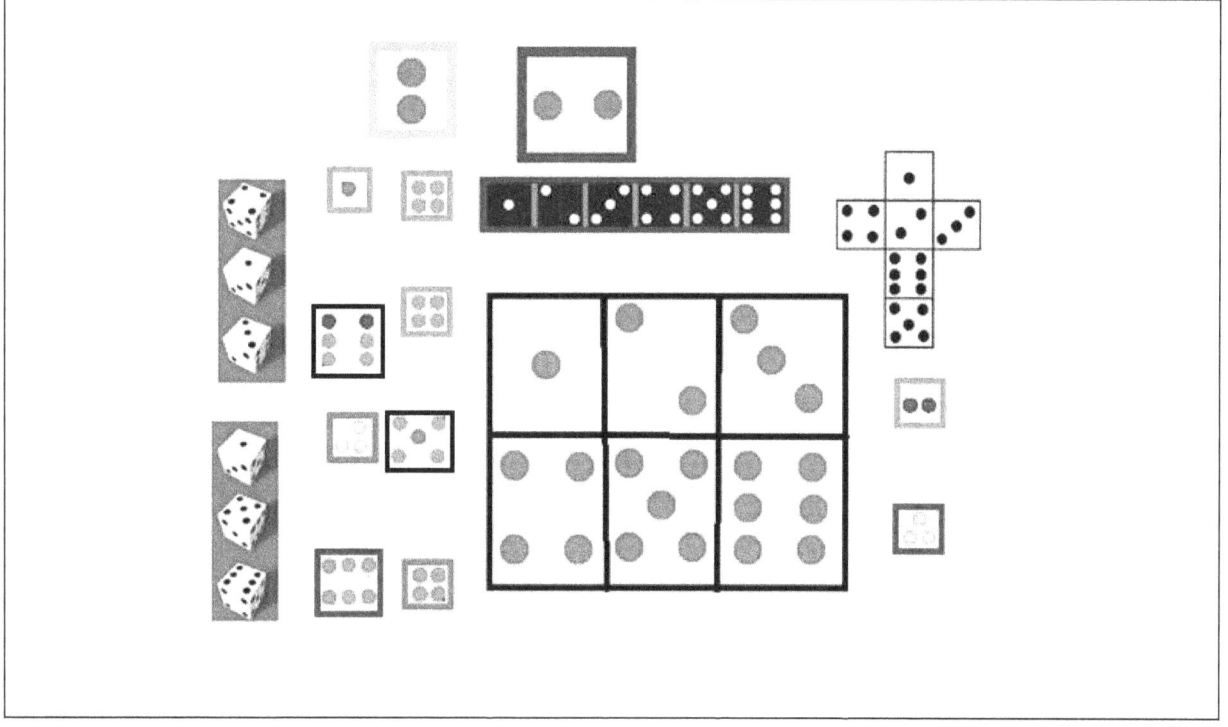

Ho Math Chess — Pre-K and Kindergarten Math

何数棋谜　棋谜式幼儿健脑思维趣味数学

© 2012 – 2021 Frank Ho, Amanda Ho, Canada copyright 1095661, Trademark 771400

Subitizing

Circle the number of dots on the surface face of each die when its value is 2. 6

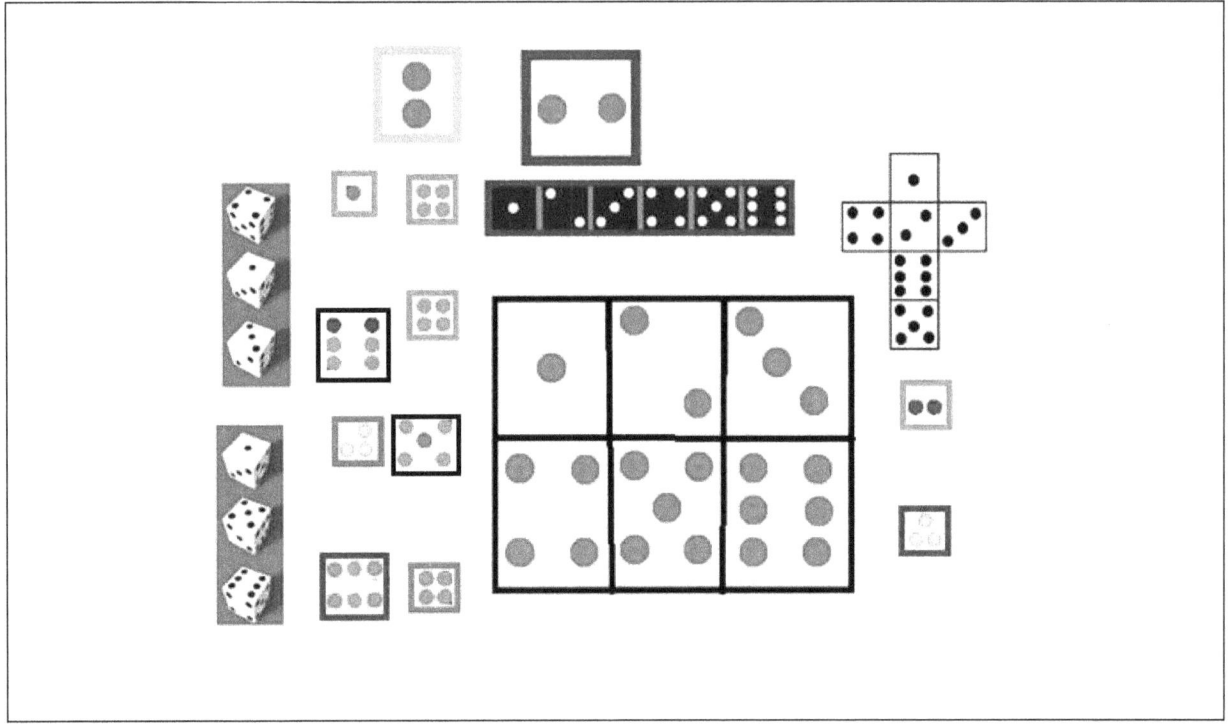

Subitizing

Circle the number of dots on the surface face of each die when its value is 3. 6

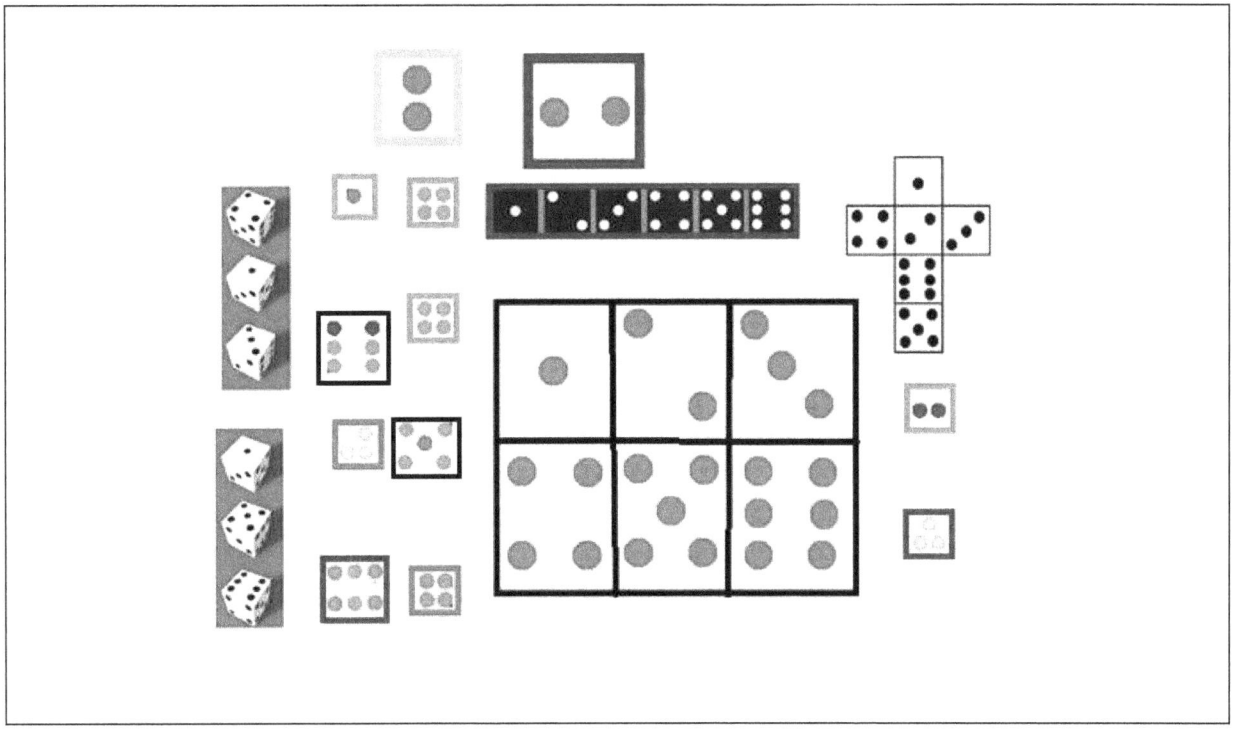

Ho Math Chess — Pre-K and Kindergarten Math

Subitizing

Circle the number of dots on the surface face of each die when its value is 4. 7

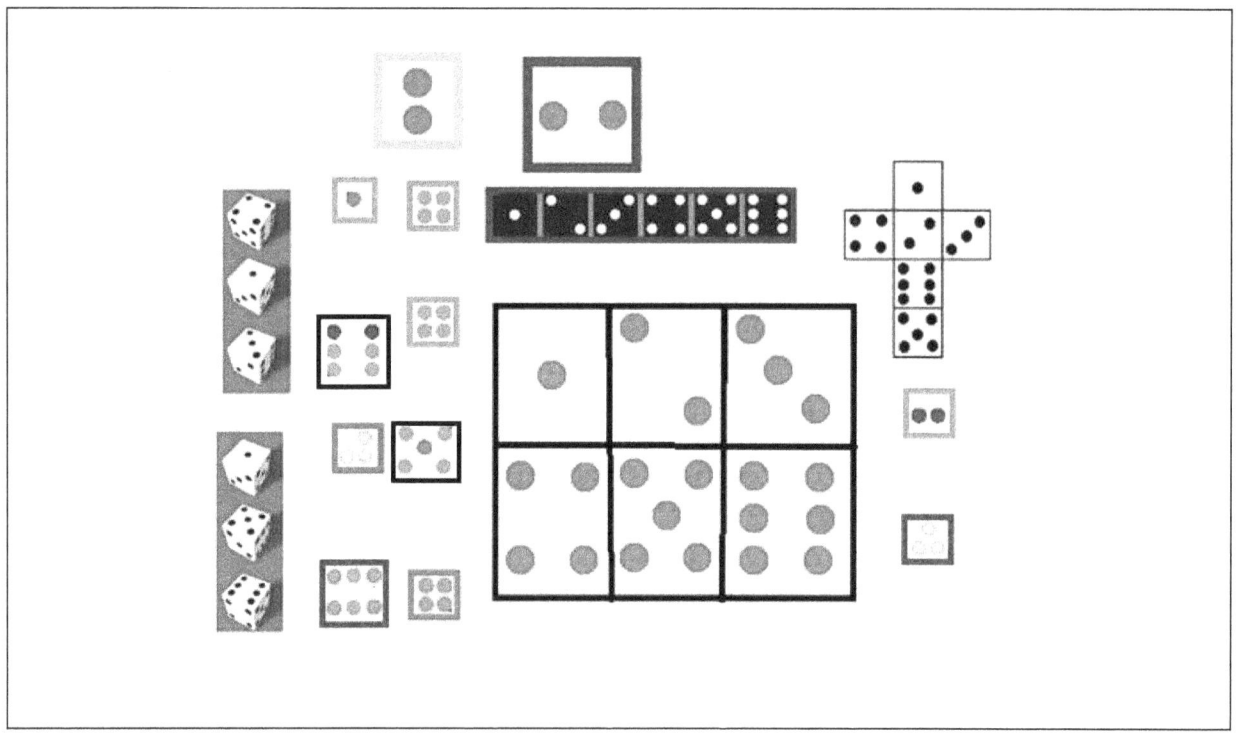

Subitizing

Circle the number of dots on the surface face of each die when its value is 5. 8

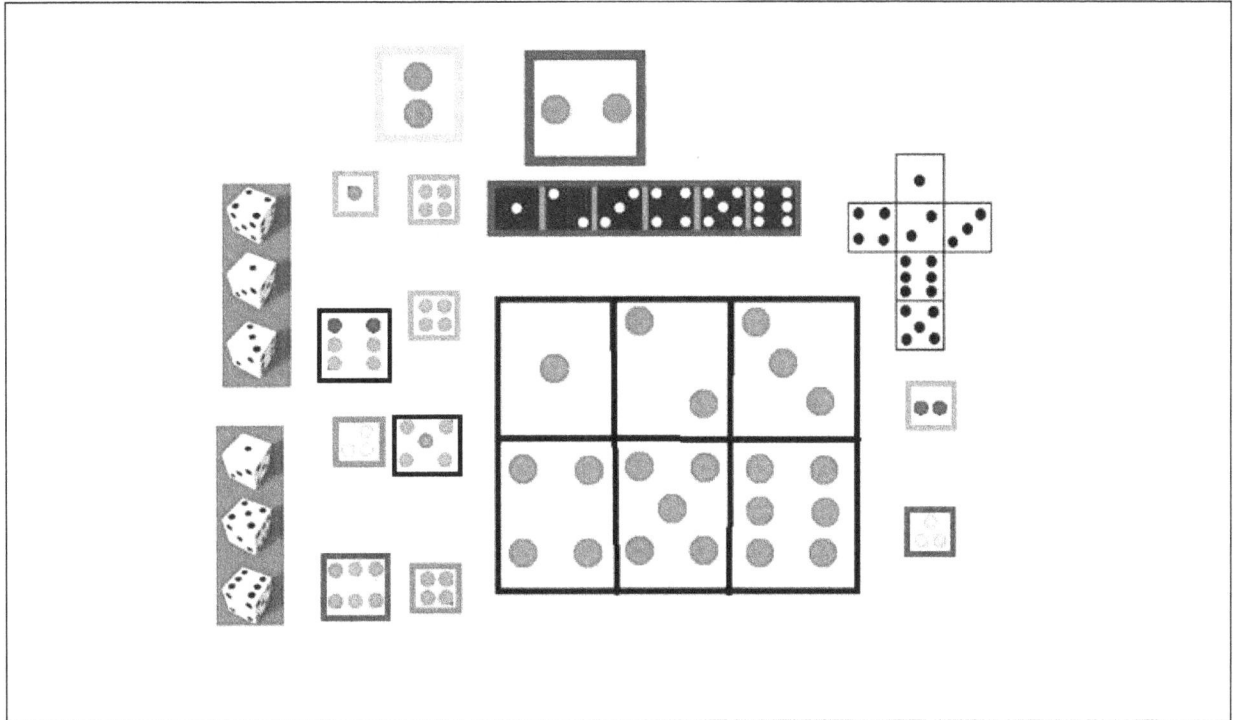

Subitizing

Circle the number of dots on the surface face of each die when its value is 6. 3

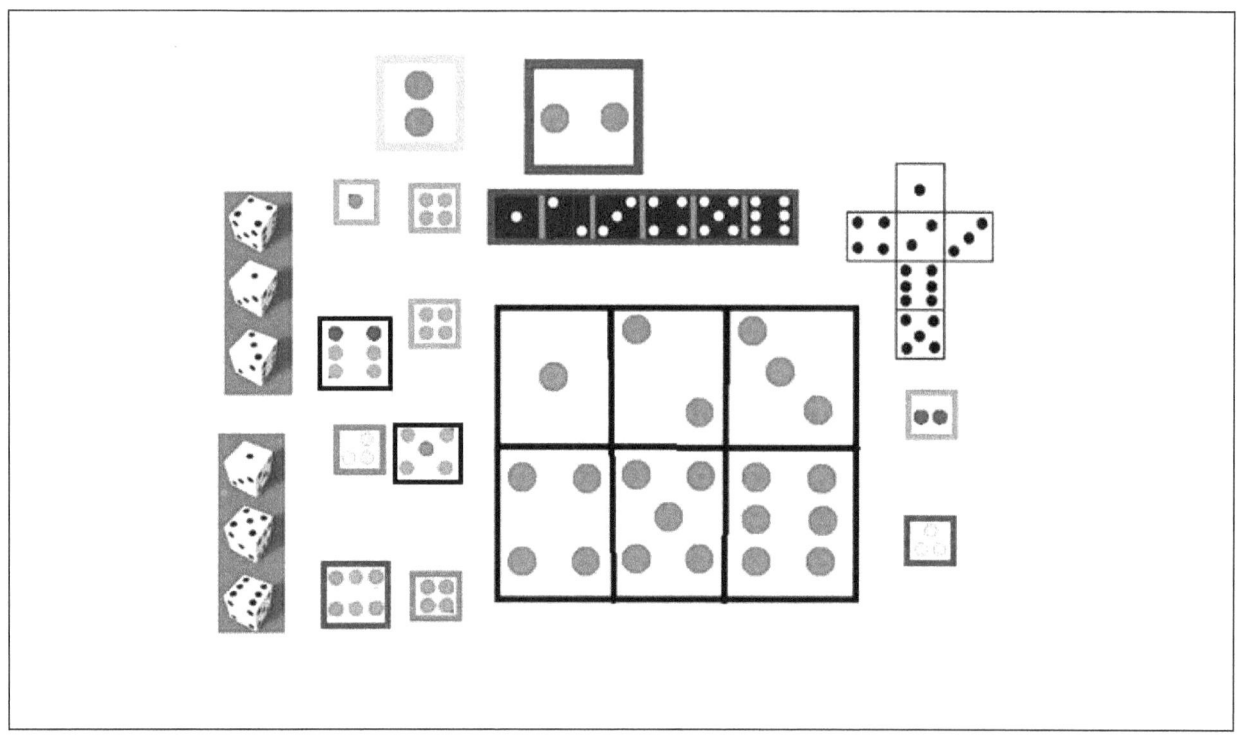

Subitizing

Circle the number of dots on each of the domino square or rectangular surface face when its value is 1.　　3

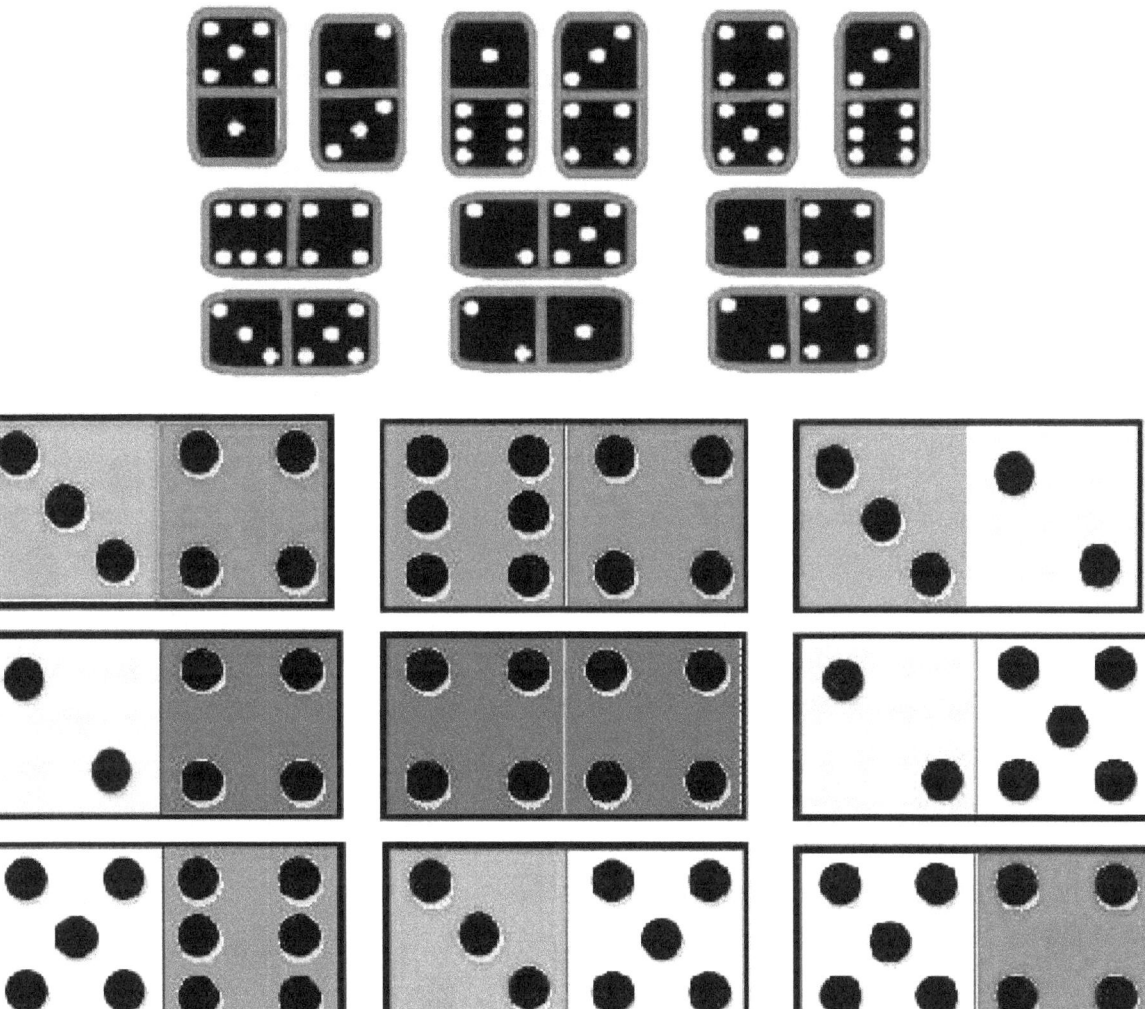

Ho Math Chess — Pre-K and Kindergarten Math

Subitizing

Circle the number of dots on each of the domino square or rectangular surface face when its value is 2. 6

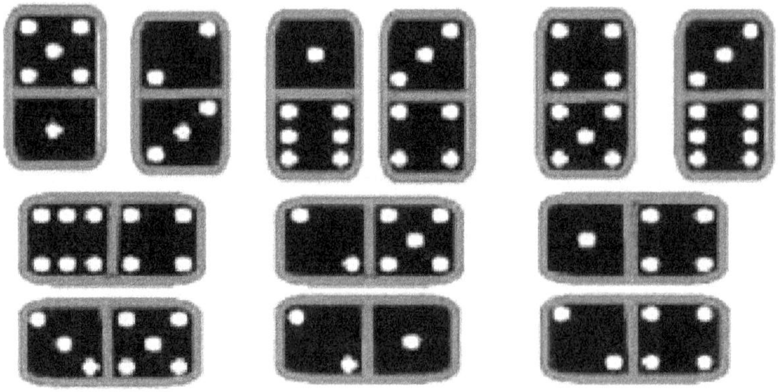

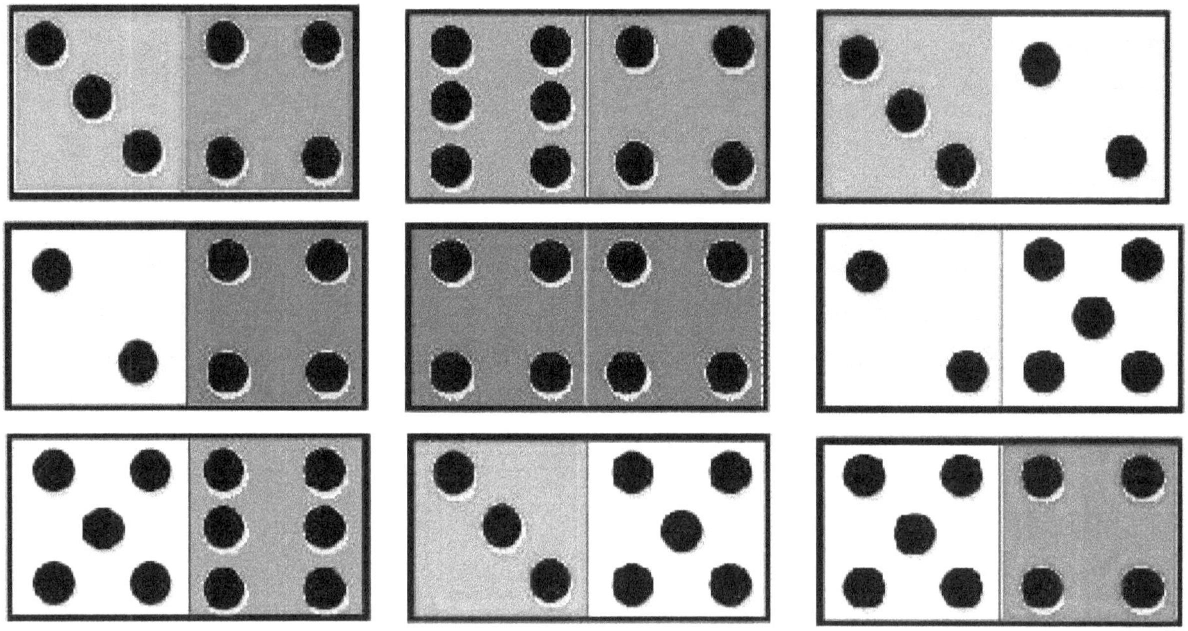

Subitizing

Circle the number of dots on each of the domino square or rectangular surface face when its value is 3.   7

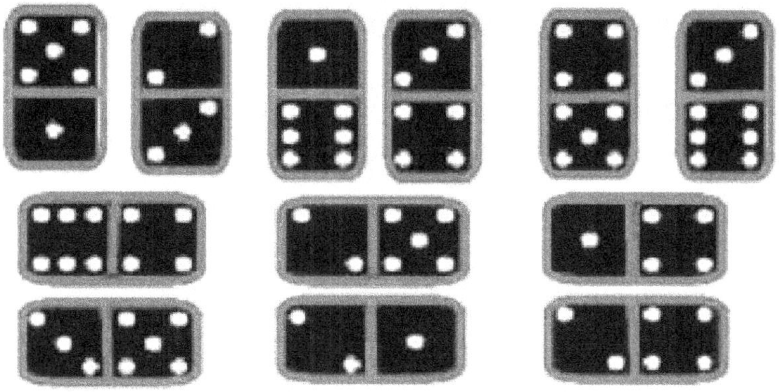

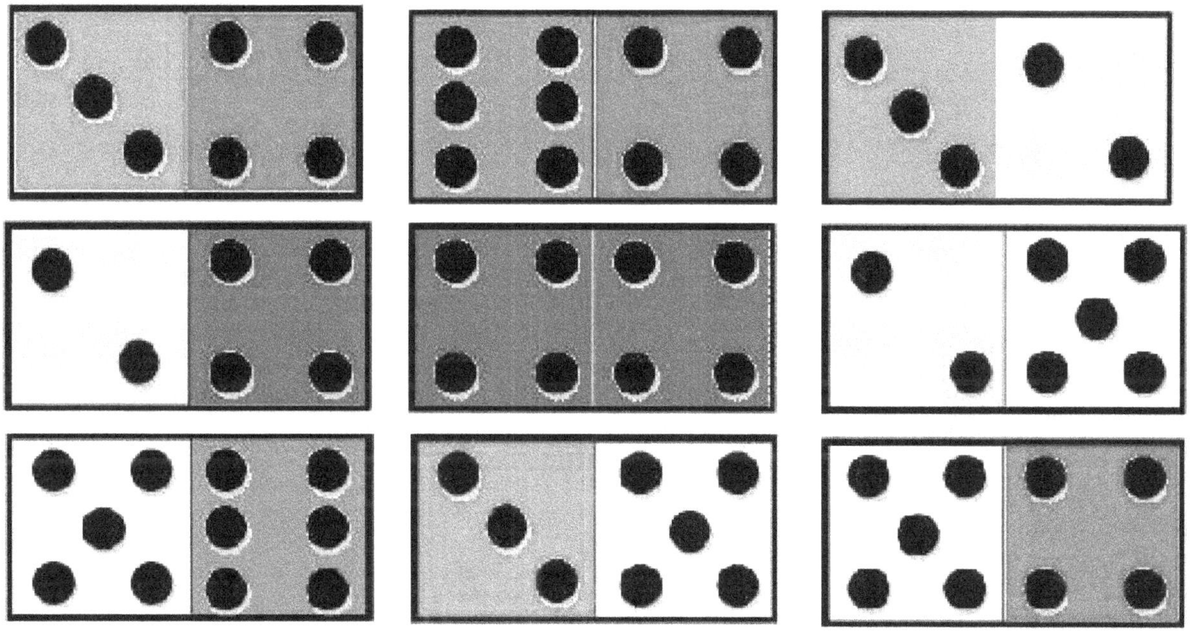

Ho Math Chess — Pre-K and Kindergarten Math

Subitizing

Circle the number of dots on each of the domino square or rectangular surface face when its value is 4. 11

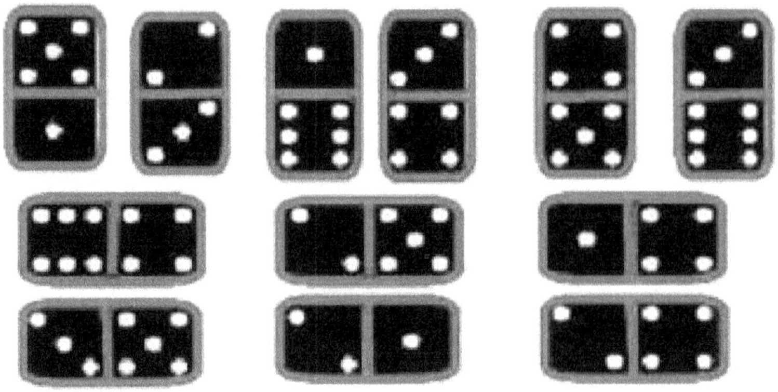

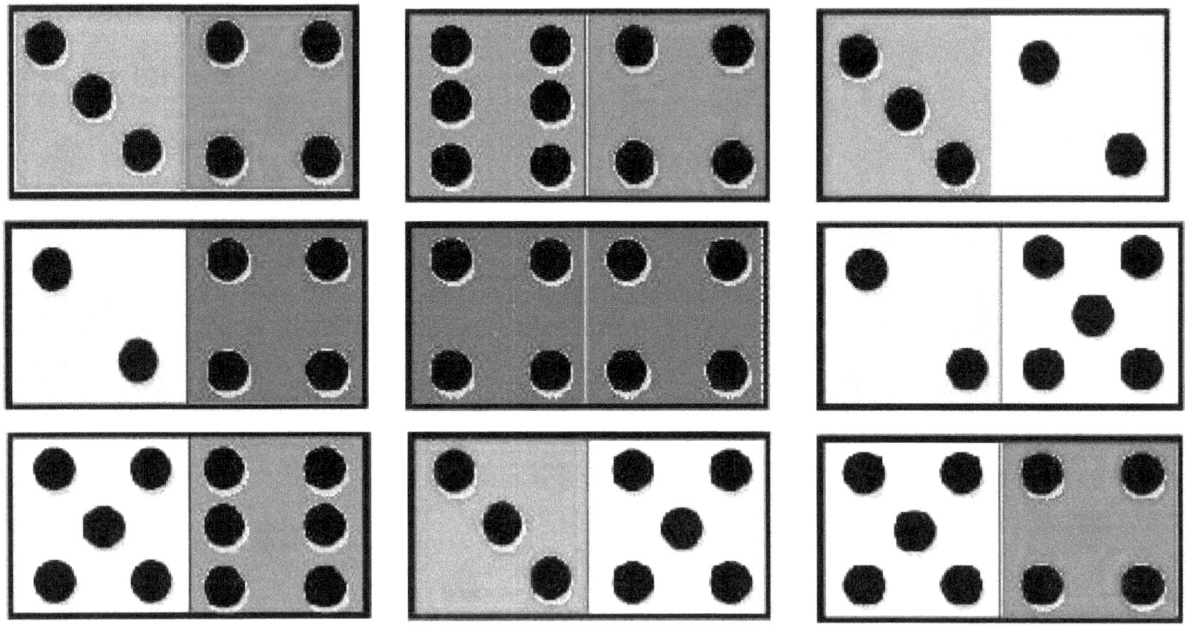

Subitizing

Circle the number of dots on each of the domino square or rectangular surface face when its value is 5. 7

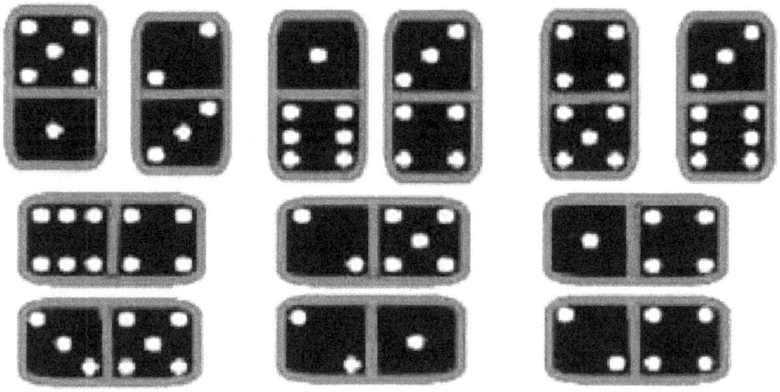

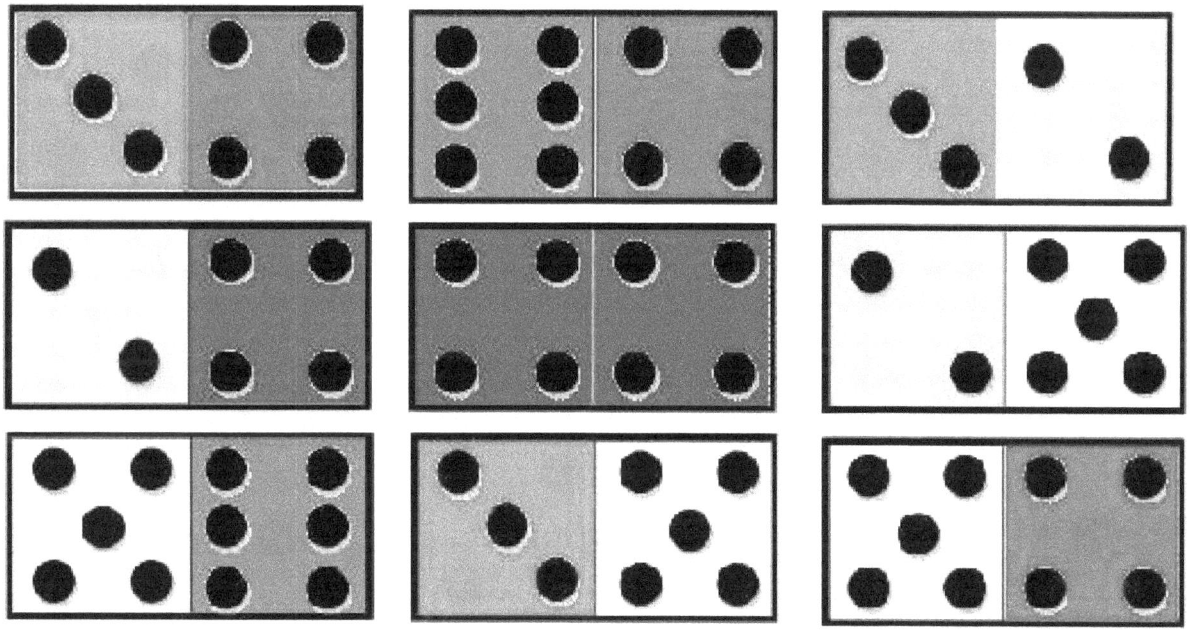

Subitizing

Circle the number of dots on each of the domino square or rectangular surface face when its value is 6. 8

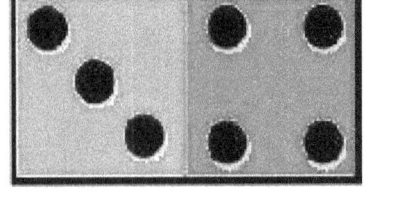 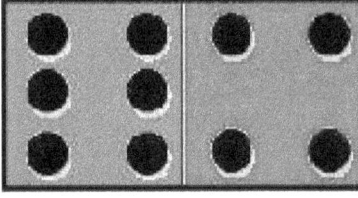 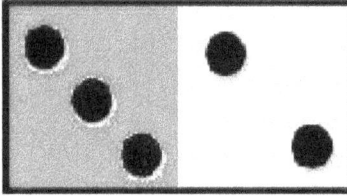

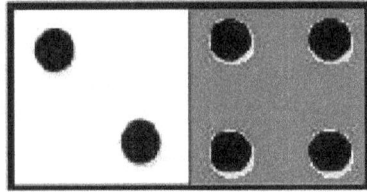 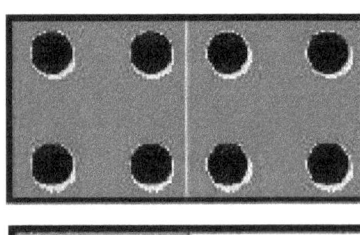 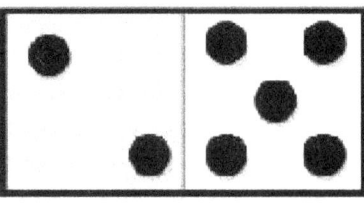

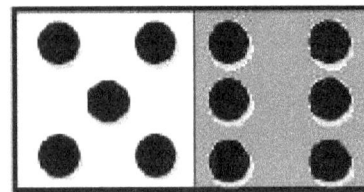 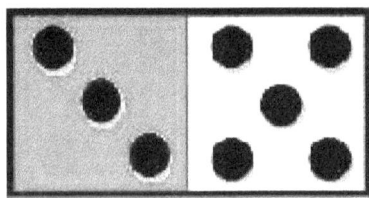 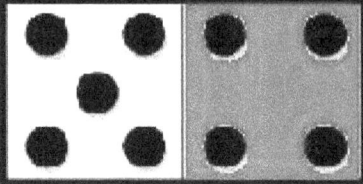

Subitizing

Circle the number of dots on each of the domino square surface faces when its value is 1.   8

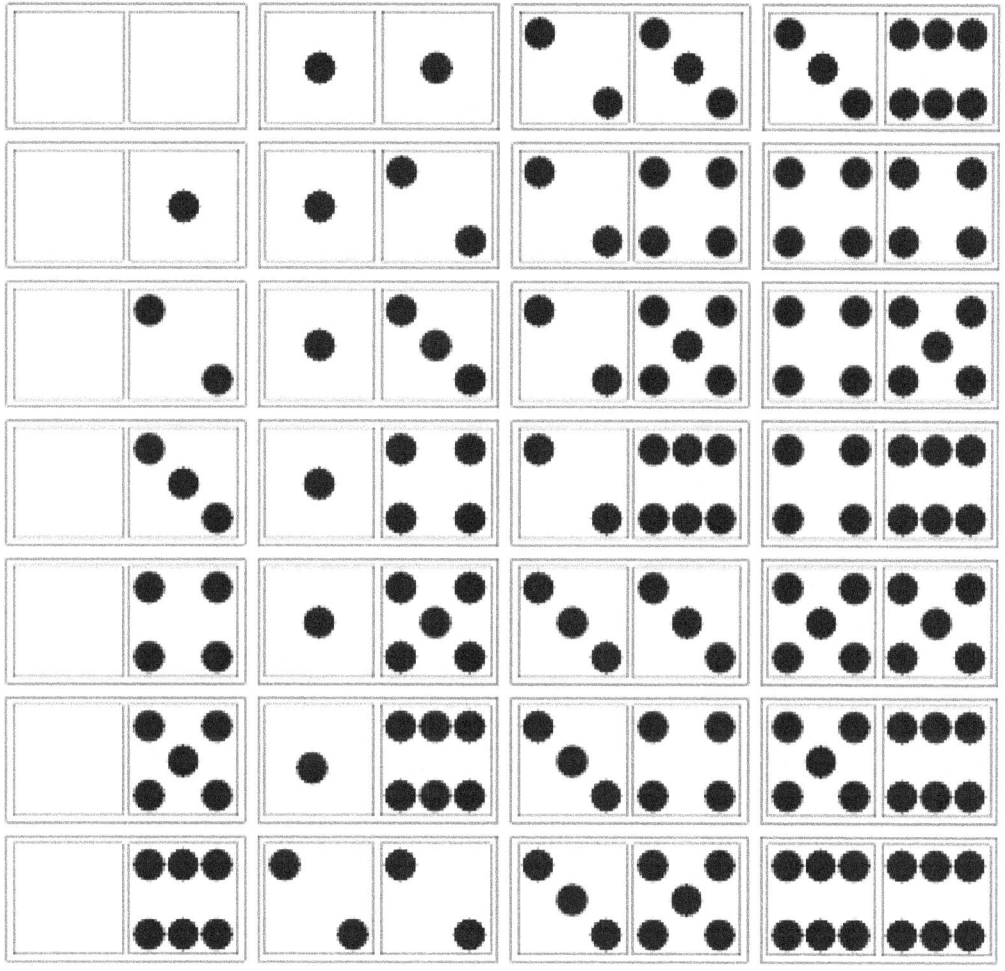

Subitizing

Circle the number of dots on each of the domino square surface faces when its value is 3.   11

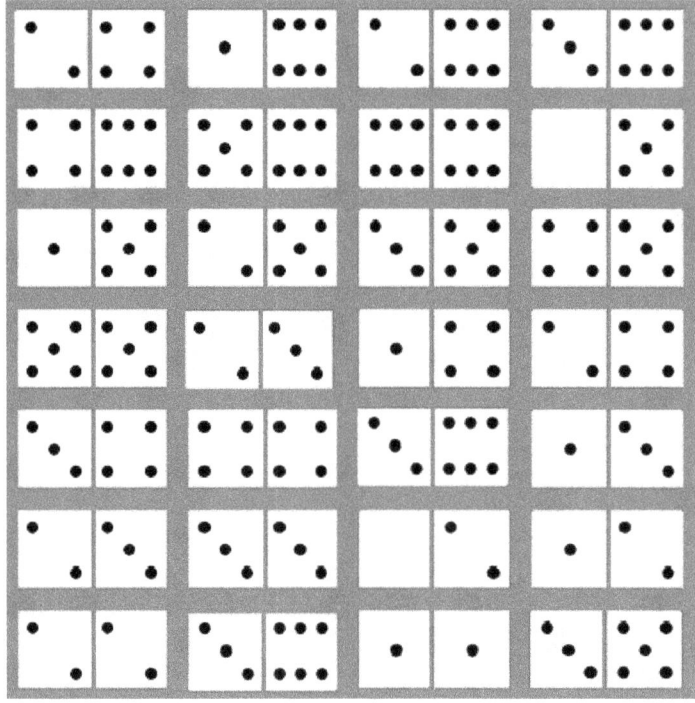

Subitizing

Circle the number of dots on each of the Lego® surface faces when its value is 2. 4

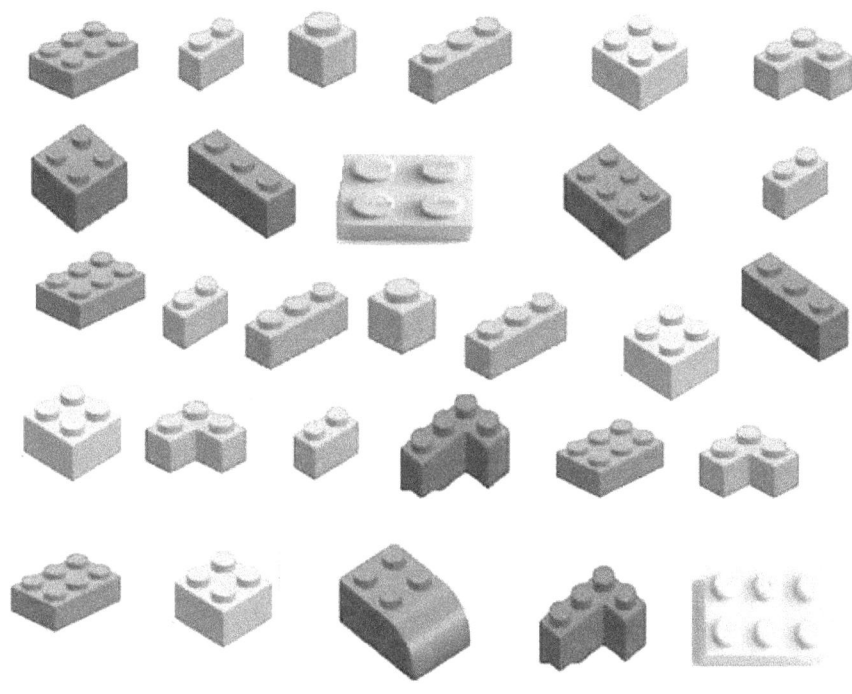

Subitizing

Circle the number of dots on each of the Lego® surface faces when its value is 3. 8

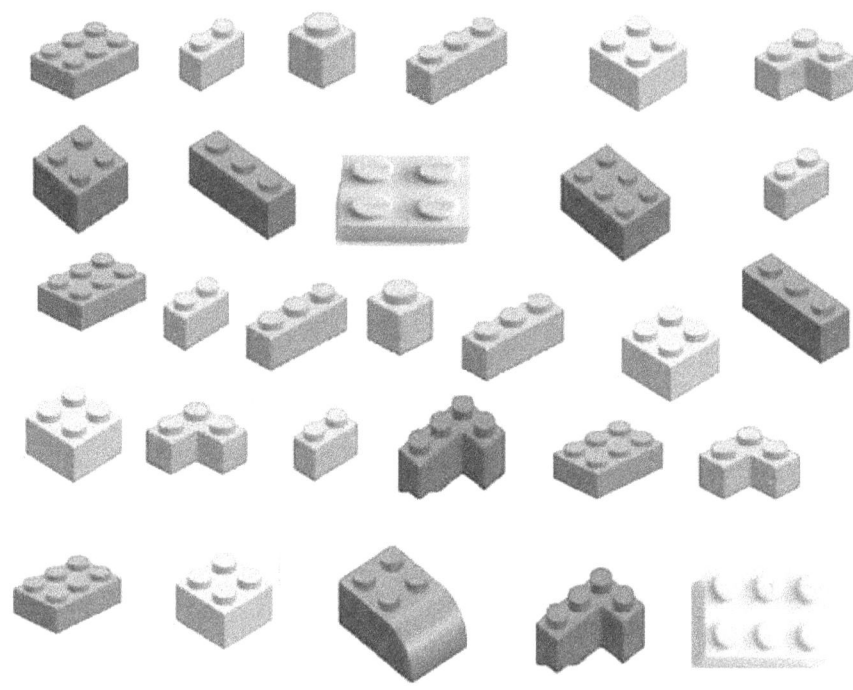

Subitizing

Circle the number of dots on each of the Lego® surface faces when its value is 4.   7

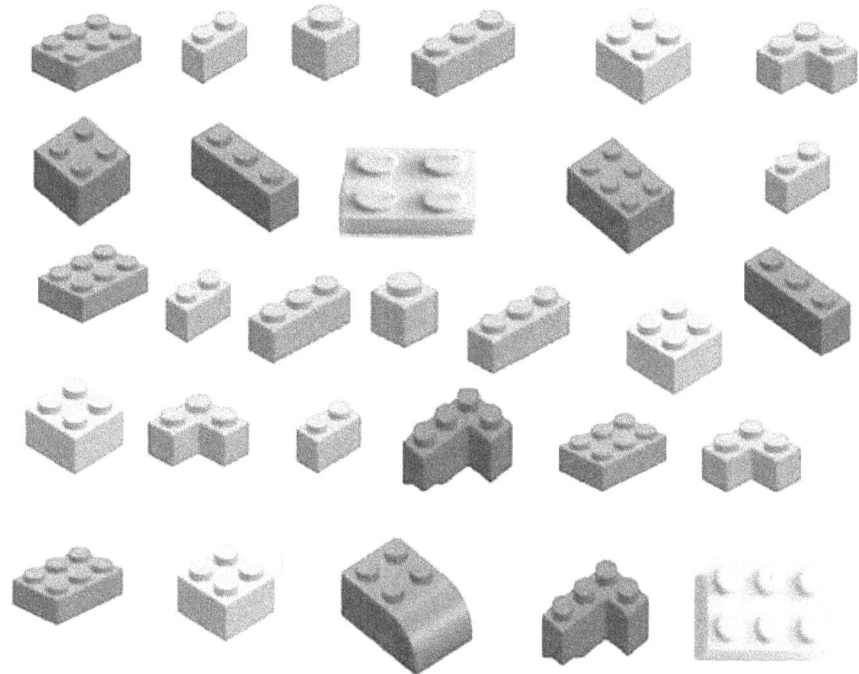

Subitizing

Circle the number of dots on each of the Lego® surface faces when its value is 1. 2

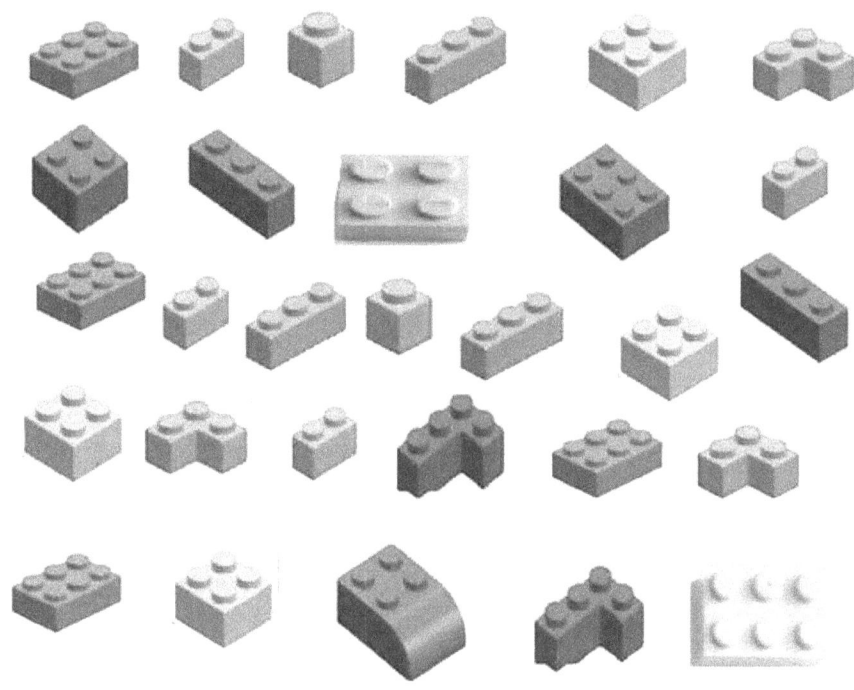

Counting and sequencing
计数及数列

In a pack of cards, there are four suits: Spade (♠), Club (♣), Heart (♥), and Diamond (♦).
Count the number of each suit for pink (or shaded) playing cards only and then write the number of each suit for pink playing cards as follows:
Spade _____ Heart _____ Club _____ Diamond _____
1, 4, 12, 3

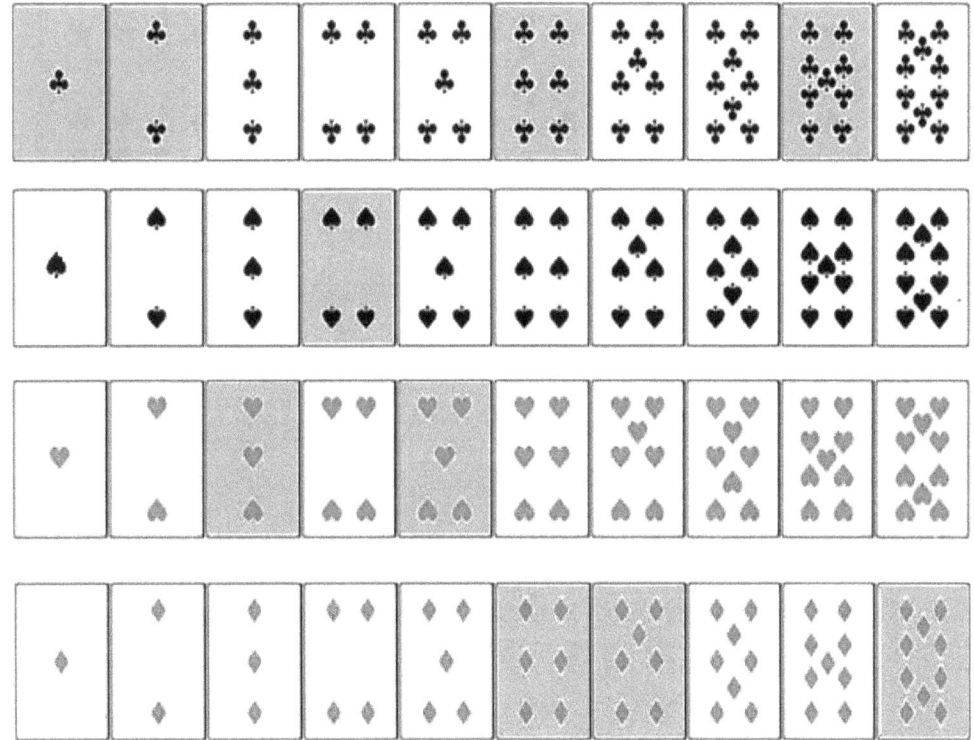

Ho Math Chess — Pre-K and Kindergarten Math

Counting and sequencing

In a pack of cards, there are four suits: Spade (♠), Club (♣), Heart (♥), and Diamond (♦).
Count the number of each suit for pink playing cards only and then write the number of each suit for pink playing cards as follows:

Spade _____ Heart _____ Club _____ Diamond _____
5 5 5 6

Ho Math Chess — Pre-K and Kindergarten Math

Counting beans

Without counting, can you tell which group has more beans? _____ group 1

Group 1 > Group 2 or Group 2 > Group 1? _____ 1 > 2

Group 1

Group 2

Without counting, can you tell which group has more beans? _____ group 1

Group 1 > Group 2 or Group 2 > Group 1? _____ 1 > 2

Group 1
34

Group 2
28

Copying numbers

Copy number according to the arrow direction and the copied number must match the previous number's exact location.

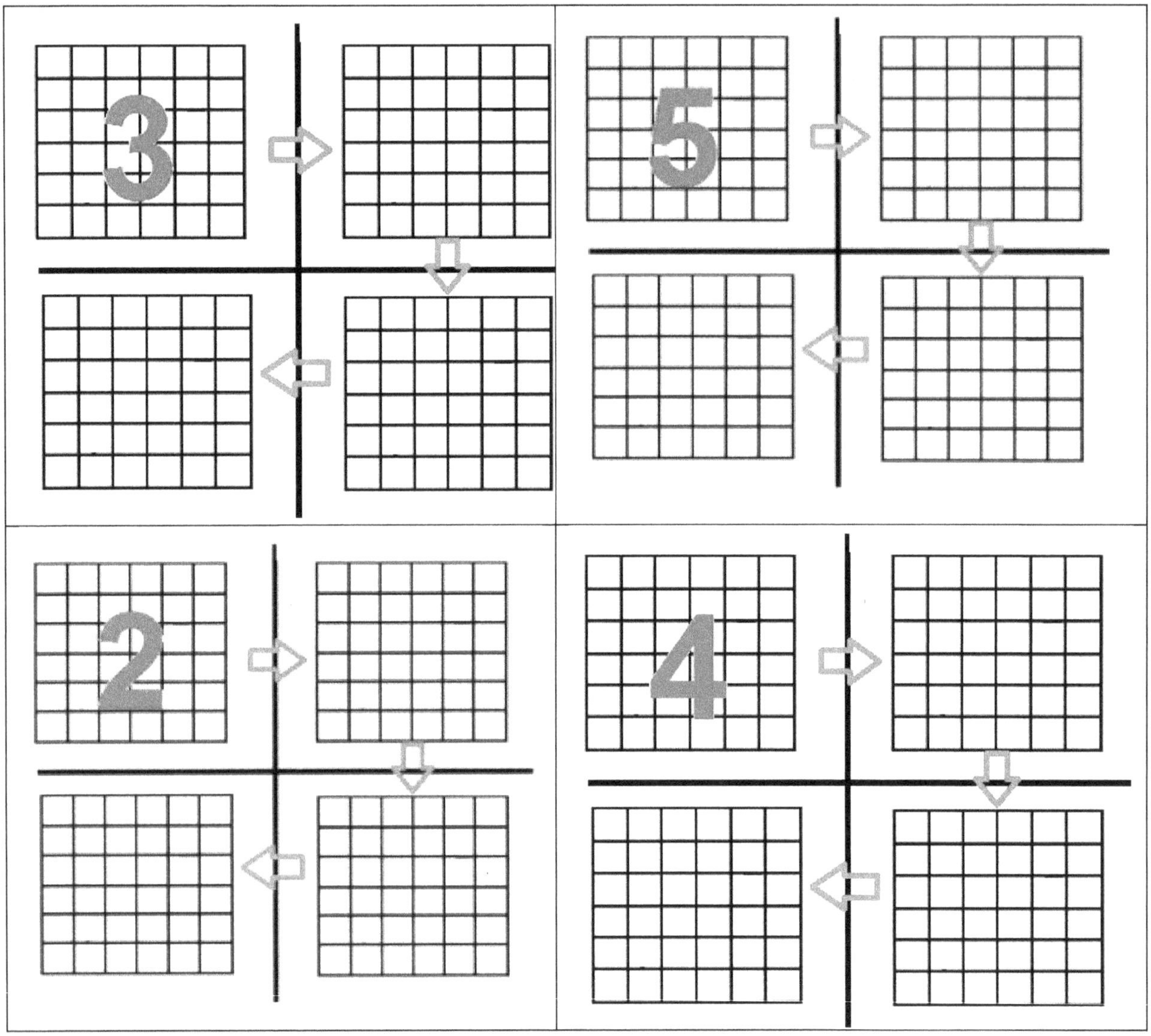

Tracing 1 2 3

1 2 3	1 2 3	1 2 3	1 2 3
3 1 2	3 1 2	3 1 2	3 1 2
2 1 3	2 1 3	2 1 3	2 1 3
3 1 2	3 1 2	3 1 2	3 1 2
2 1 3	2 1 3	2 1 3	2 1 3
1 3 2	1 3 2	1 3 2	1 3 2
2 3 1	2 3 1	2 3 1	2 3 1
3 2 1	3 2 1	3 2 1	3 2 1

Filling in the missing number using 1, 2, and 3

3 ☐ 1	1 2 ☐	☐ 2 1
☐ 2 1	3 ☐ 1	1 2 ☐
☐ 2 1	1 2 ☐	3 ☐ 1
1 2 ☐	3 ☐ 1	☐ 2 1

Fill in the missing number using 1, 2, and 3

1 2 ☐	☐ 2 1	3 ☐ 1
☐ 2 1	3 ☐ 1	1 2 ☐
3 ☐ 1	☐ 2 1	1 2 ☐
☐ 2 1	1 2 ☐	3 ☐ 1

Ho Math Chess   Pre-K and Kindergarten Math

Counting the number of smiling faces

Count the number of smiling faces and then write the counted number for each cell.

Count the number of smiling faces and then write the counted number for each cell.

Ho Math Chess   Pre-K and Kindergarten Math
何数棋谜   棋谜式幼儿健脑思维趣味数学
© 2012 – 2021 Frank Ho, Amanda Ho, Canada copyright 1095661, Trademark 771400

Count the number of smiling faces and then write the counted number for each cell.

Pattern of 123

Use the above figure pattern rule to fill in a number in each square.

312 231
231 213
213 231
312 321

Ho Math Chess — Pre-K and Kindergarten Math

何数棋谜 棋谜式幼儿健脑思维趣味数学

© 2012 – 2021 Frank Ho, Amanda Ho, Canada copyright 1095661, Trademark 771400

Figures and shapes

图及形狀

Square (one face) Can you trace the following square(s) without lifting your pencil?	Cube (6 faces) Can you trace the following cube without lifting your pencil?
□	cube
square with X	cube
square with inner square	cube

Yes, no
No, no
Yes, no

Ho Math Chess — Pre-K and Kindergarten Math

Basic Shapes
基本形狀

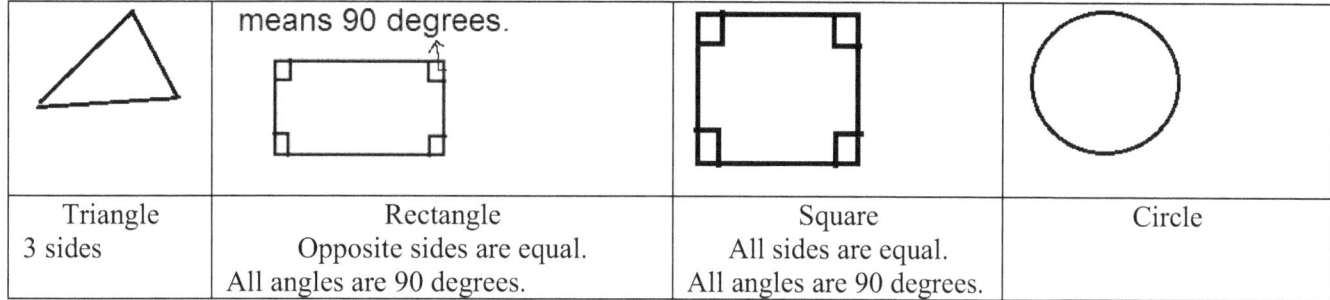

Triangle	Rectangle	Square	Circle
3 sides	Opposite sides are equal. All angles are 90 degrees.	All sides are equal. All angles are 90 degrees.	

Find the odd shape in each row and circle it.

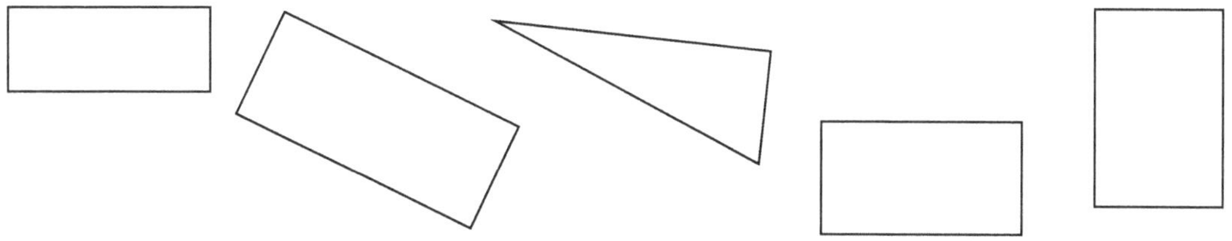

triangle

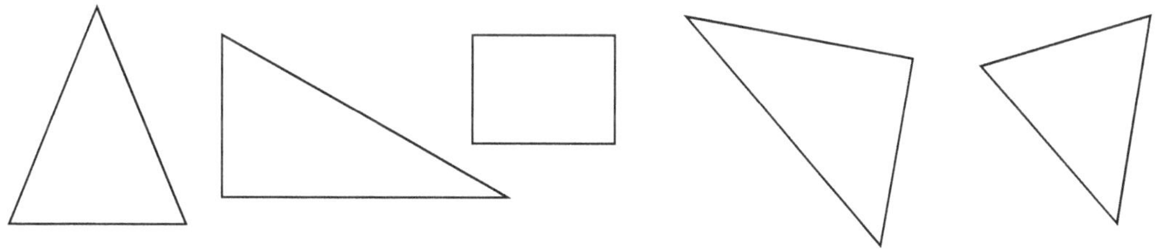

Rectangle

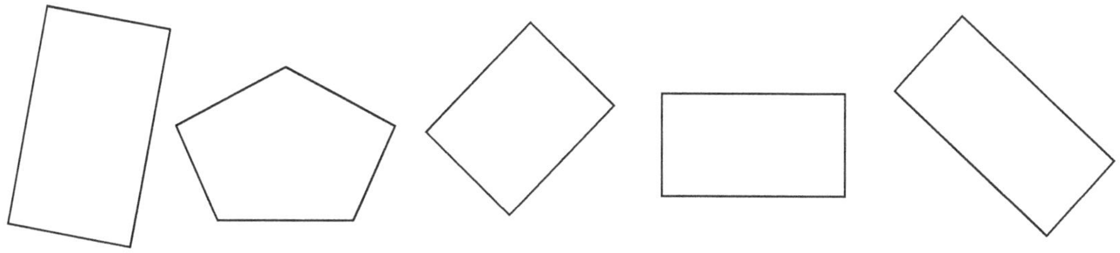

Pentagon

Ho Math Chess — Pre-K and Kindergarten Math

Find the odd shape in each row and circle it.

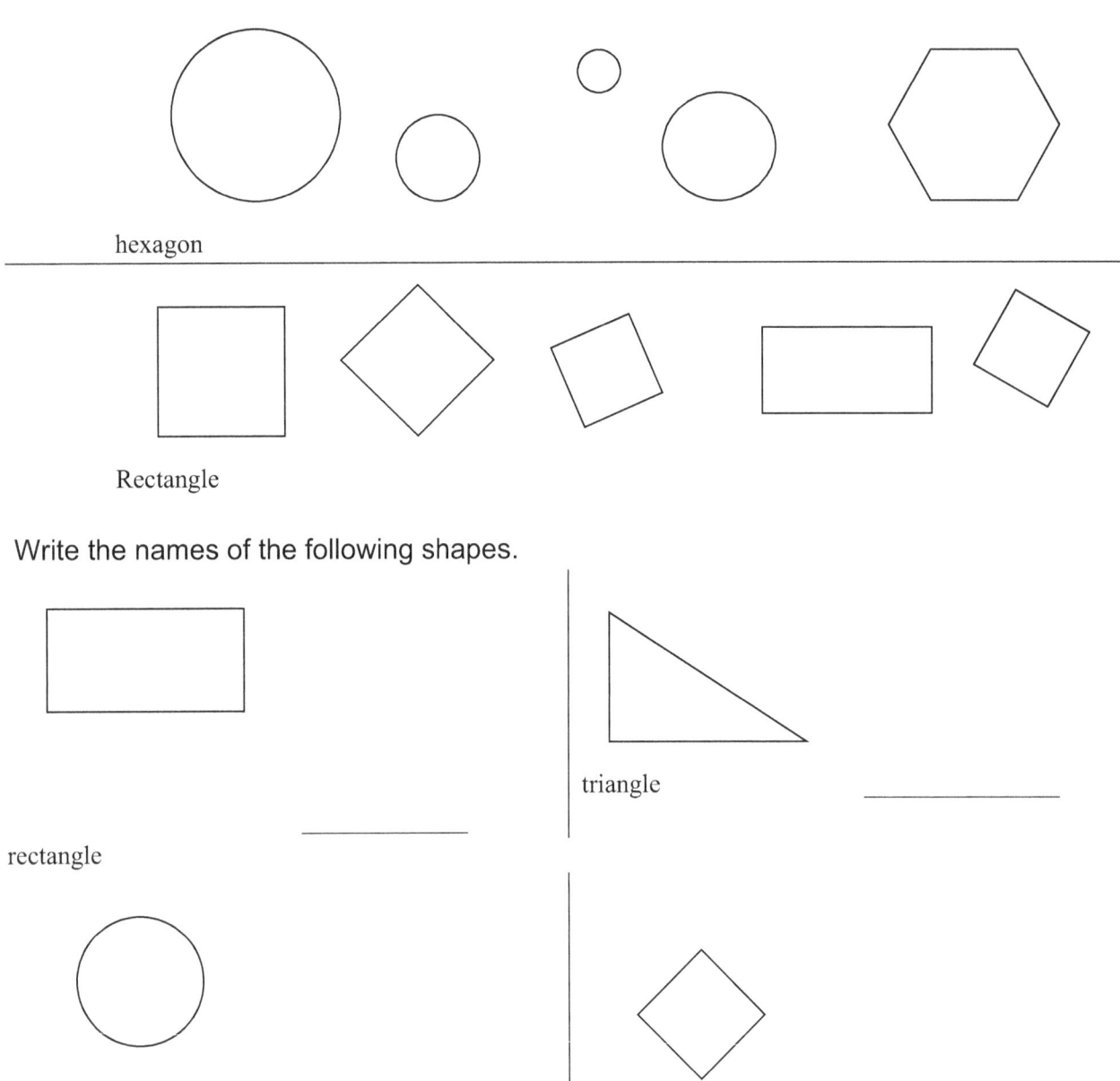

hexagon

Rectangle

Write the names of the following shapes.

rectangle

triangle

Circle

square

Write the names of the following shapes.

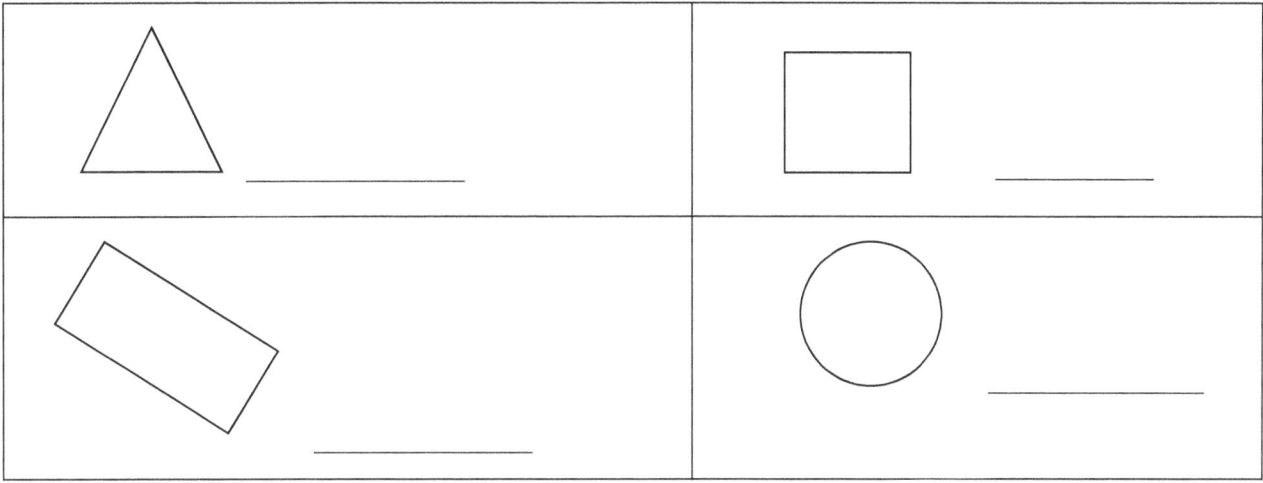

How many sides and corners of the following shapes (polygon)?

   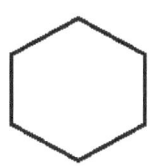

_____sides   _____sides   _____sides   _____sides   6

_____corners _____corners _____corners _____corners 6

Ho Math Chess — Pre-K and Kindergarten Math

Circle the shape on the right side, which is the same as the one on the left-most column?

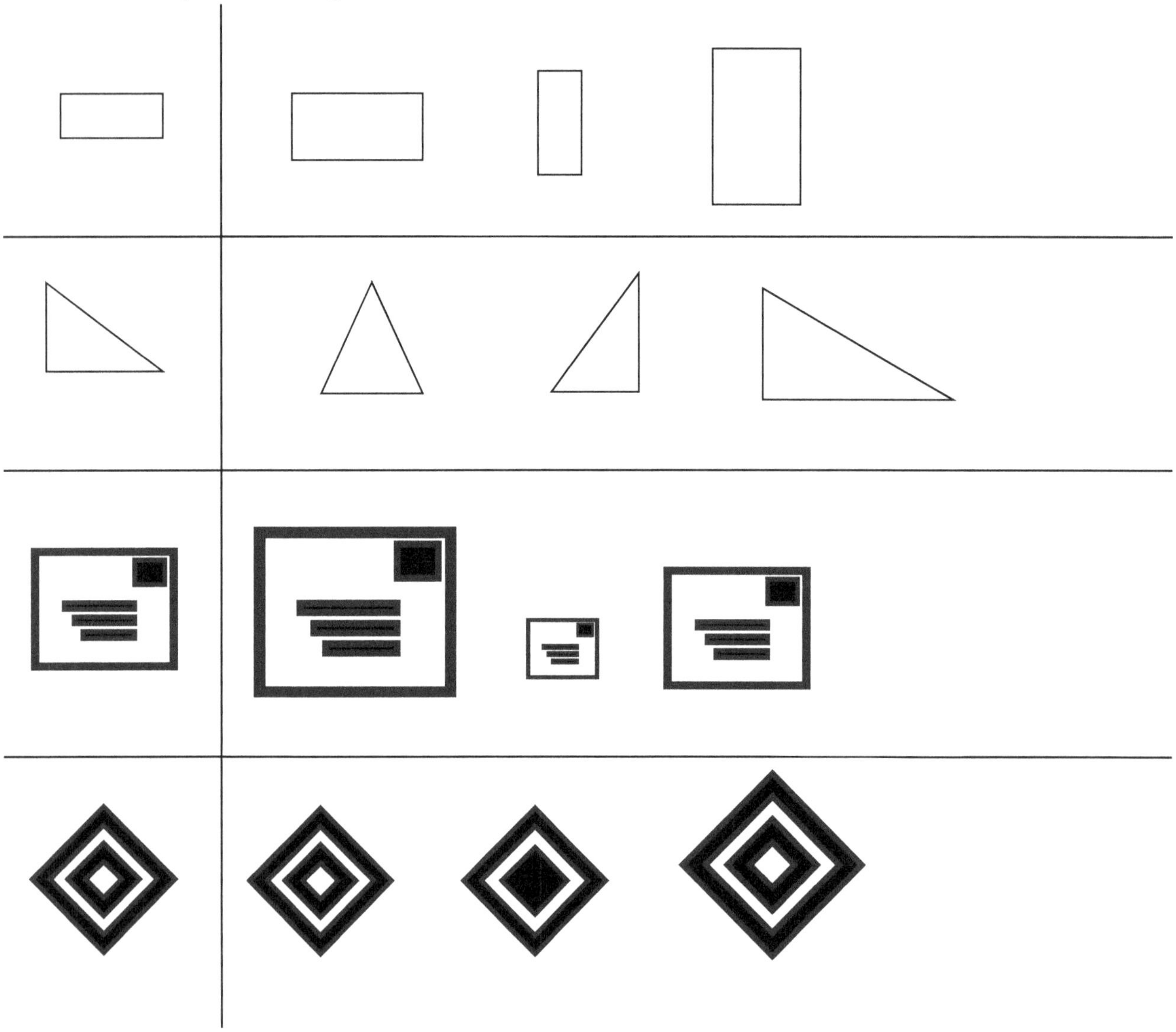

Weight
重量

Circle the heavier side of each scale.

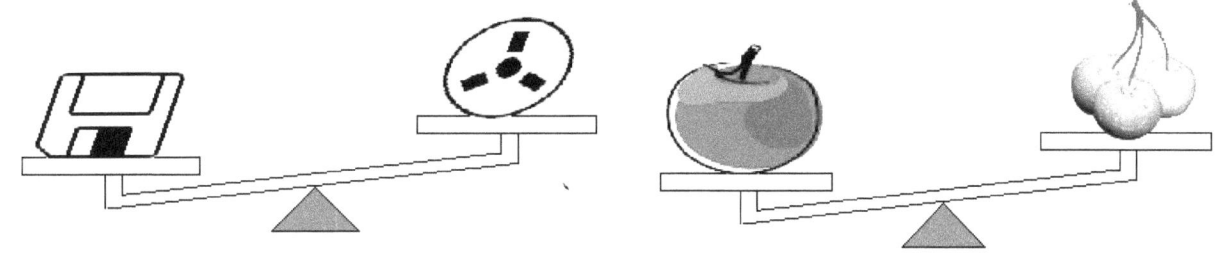

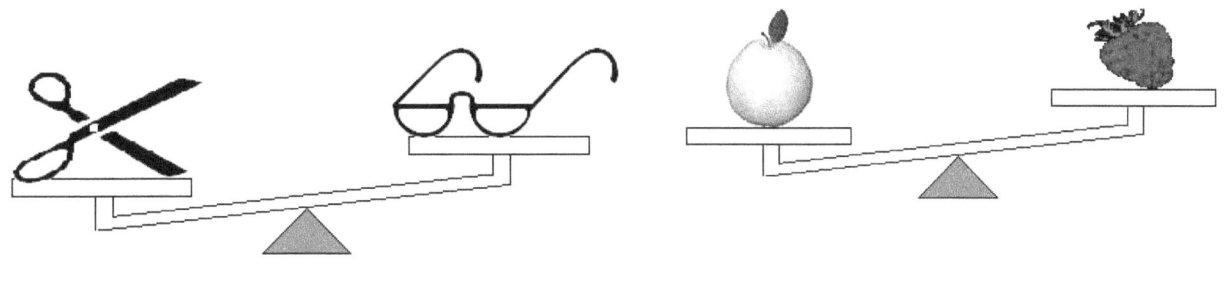

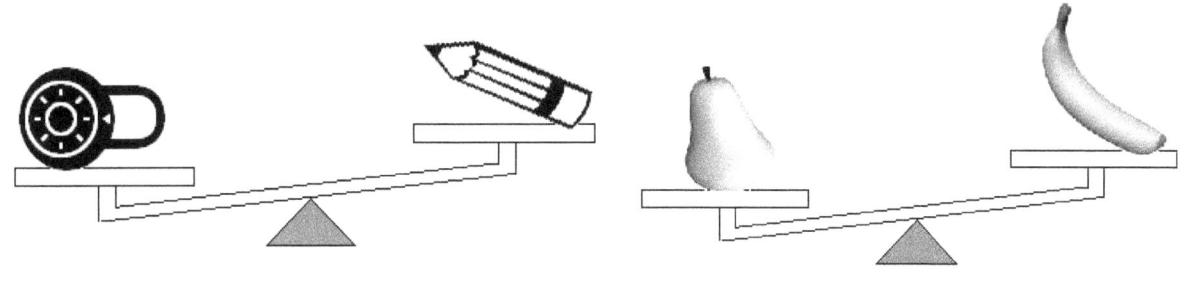

Length
長度

Use a ruler to measure the length of the following items.

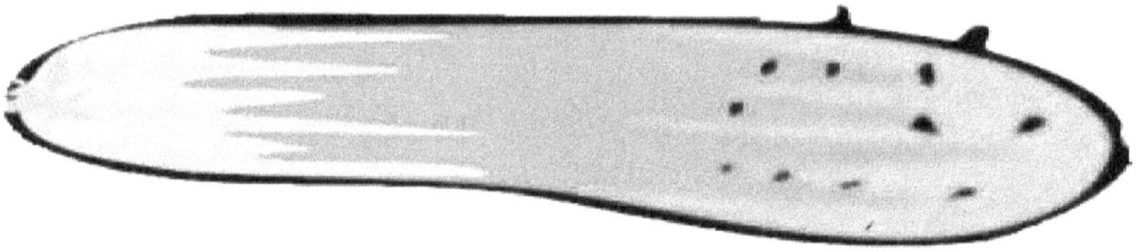

_____ cm 12.5

_____ cm 7.5

_____ cm 5.5

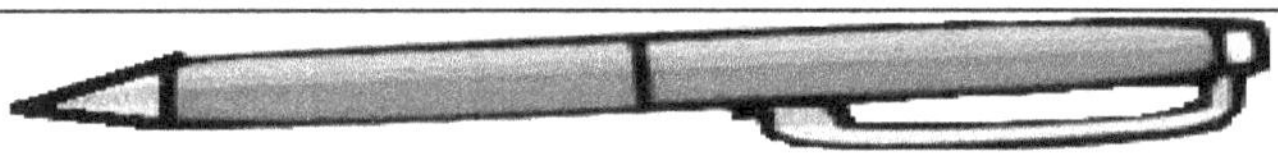

_____ cm 8.5

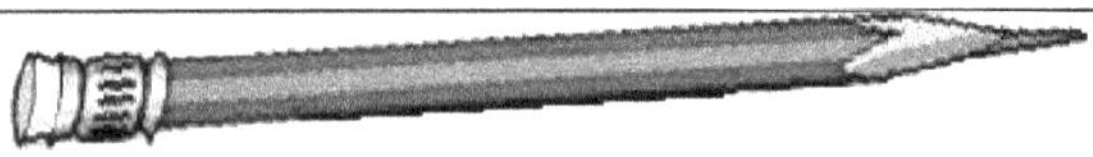

_____ cm 7.5

Time

时刻或时间

5 o'clock   or   5:00

9 or 9:00
_____ o'clock or _____

8 or 8:00
_____ o'clock or _____

3 or 3:00
_____ o'clock or _____

4 or 4:00
_____ o'clock or _____

6 or 6:00
_____ o'clock or _____

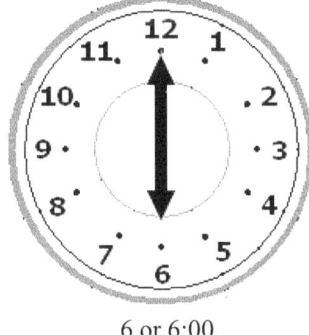
6 or 6:00
_____ o'clock or _____

10 or 10:00
_____ o'clock or _____

2 or 2:00
_____ o'clock or _____

123 writing

1, 2, or 3 are written on the 3 front faces of each cube. Write a number beside the question mark.

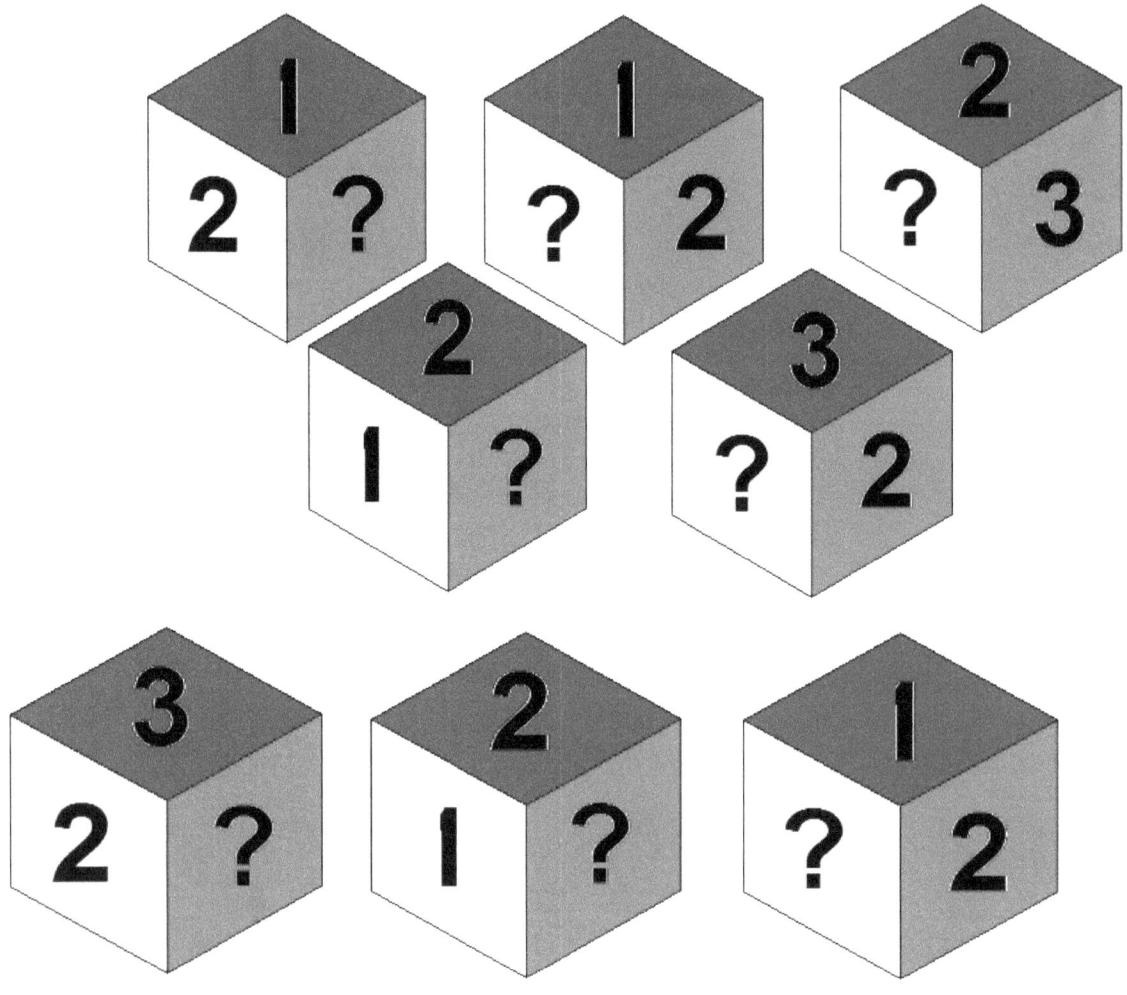

Ho Math Chess — Pre-K and Kindergarten Math

何数棋谜　棋谜式幼儿健脑思维趣味数学

© 2012 – 2021 Frank Ho, Amanda Ho, Canada copyright 1095661, Trademark 771400

Figure	Number
☝	
🖐(3)	12
✌	12
🖐(3)	12
✌	12
☝	

Ho Math Chess   Pre-K and Kindergarten Math

何数棋谜　棋谜式幼儿健脑思维趣味数学

© 2012 – 2021 Frank Ho, Amanda Ho, Canada copyright 1095661, Trademark 771400

Adding more circles to match each number above.

Writing a number to match its height.

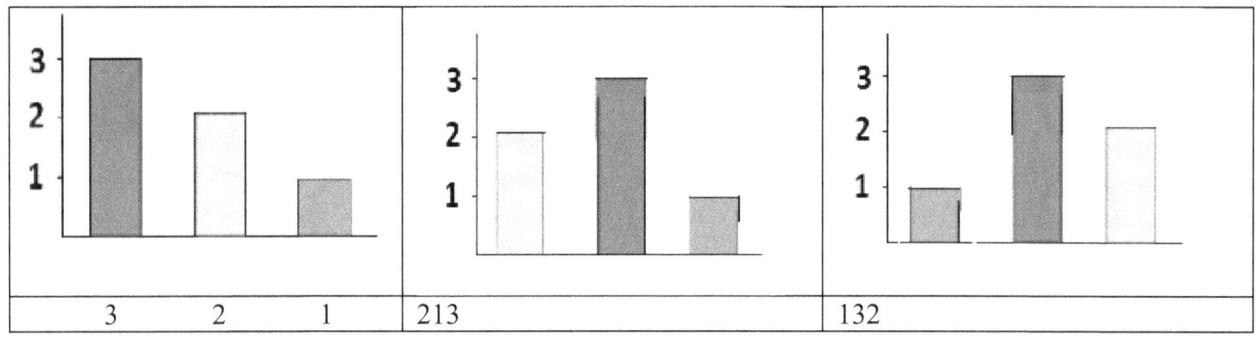

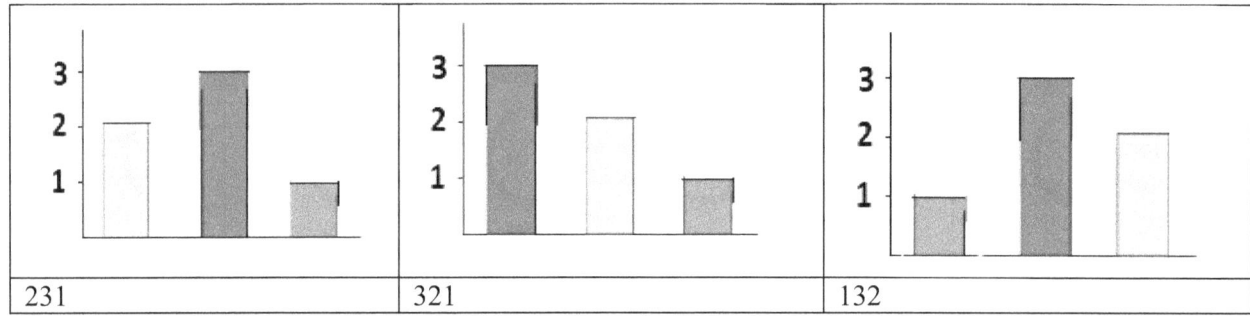

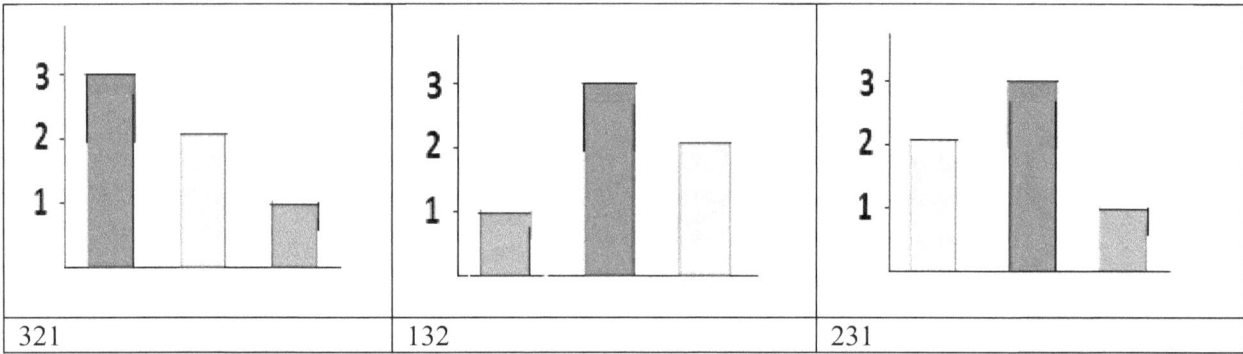

Ho Math Chess   Pre-K and Kindergarten Math
何数棋谜   棋谜式幼儿健脑思维趣味数学
© 2012 – 2021 Frank Ho, Amanda Ho, Canada copyright 1095661, Trademark 771400

Filling in 1, 2, or 3 from inner circle to outer circle

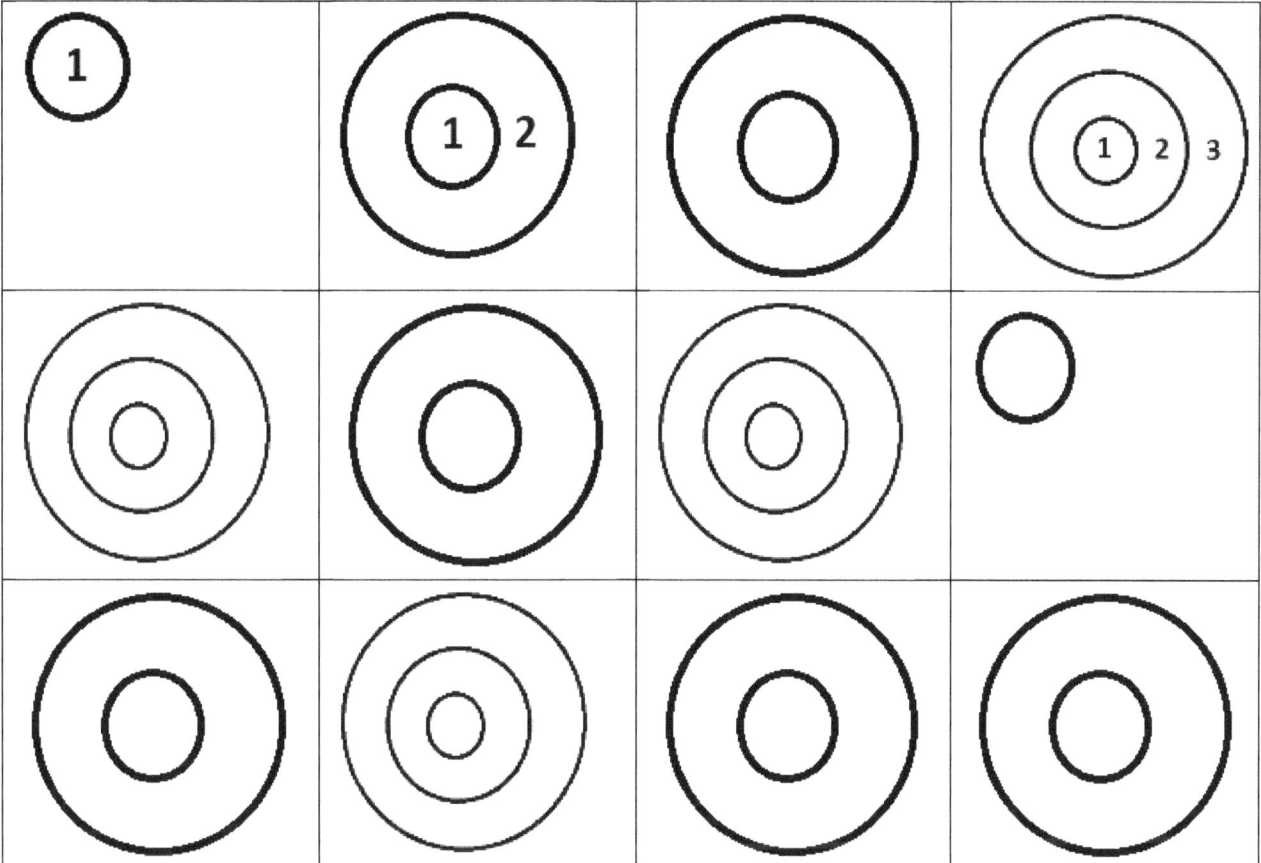

1, 12, 12, 123
123, 12, 123, 1
12, 123, 12, 12

Replacing each "?" by a number

1 2, 21

1 3, 3?   2

2?, 32   3

22, 2?   2

1?, 11   1

33, ?3   3

?2, 21   1

?3, 31   1

Correcting any of the following incorrect counting numbers

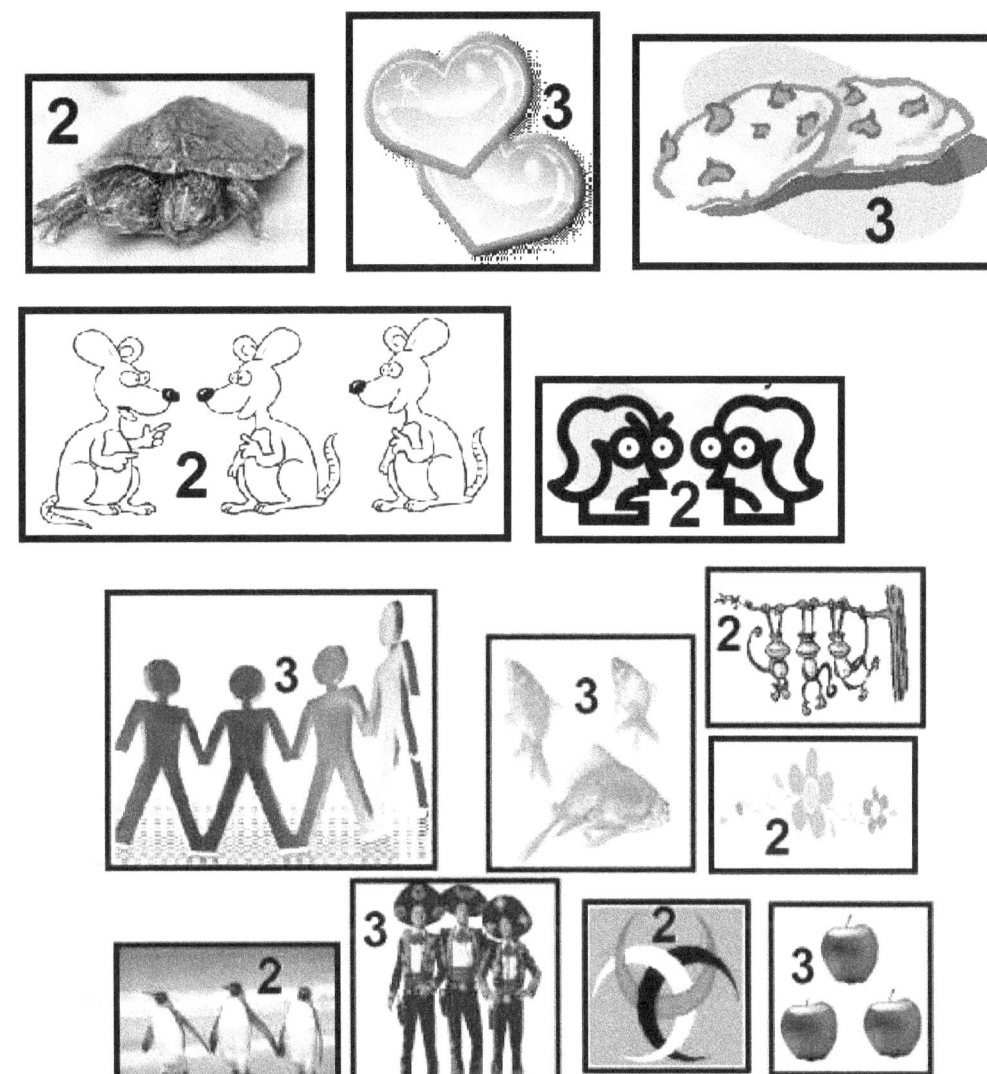

Ho Math Chess — Pre-K and Kindergarten Math

Filling in 1, 2, or 3 by pattern

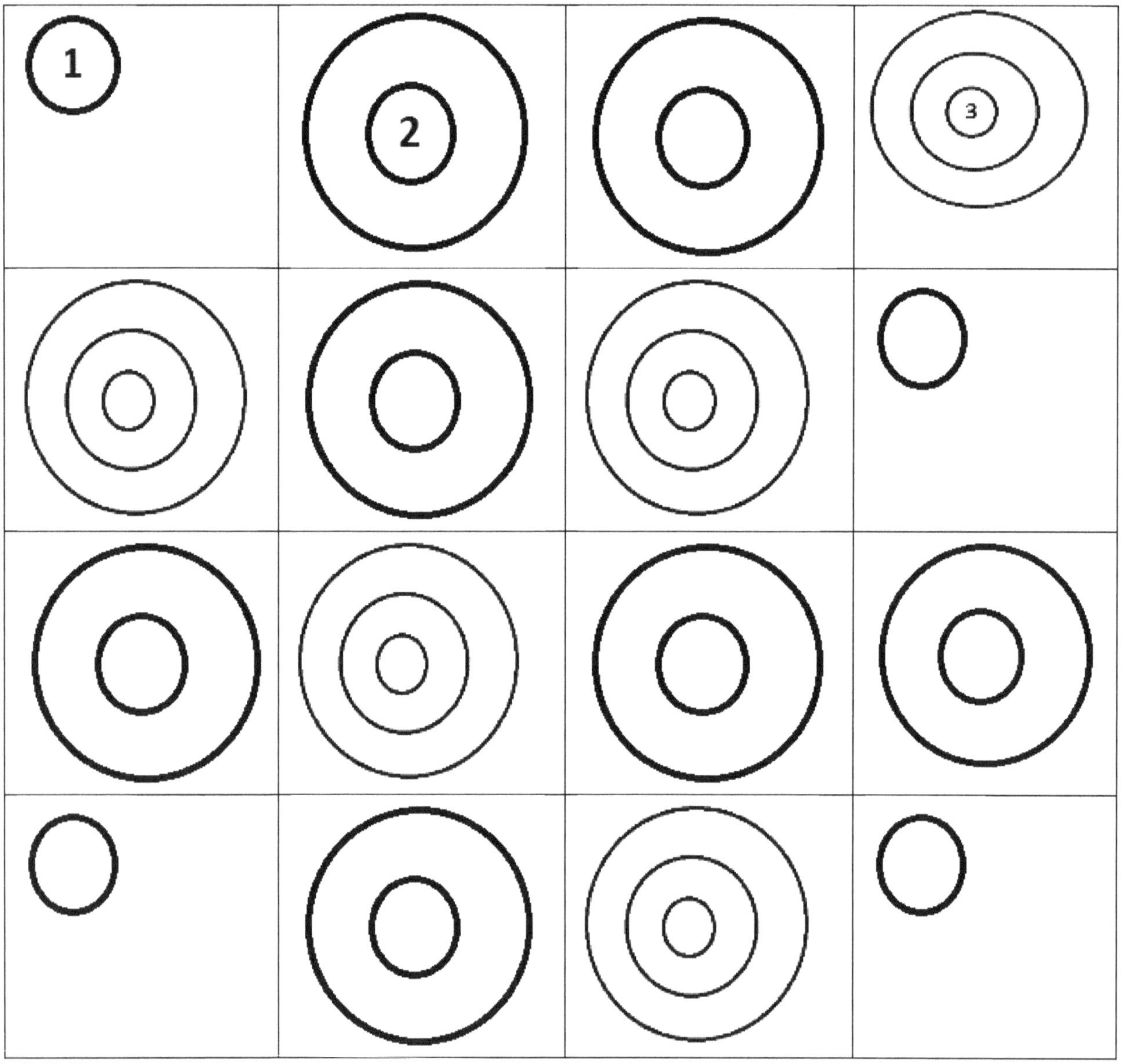

1, 2, 2, 3
3, 2, 3, 1
2, 3, 2, 2
1, 2, 3, 1

Filling in 1, 2, or 3

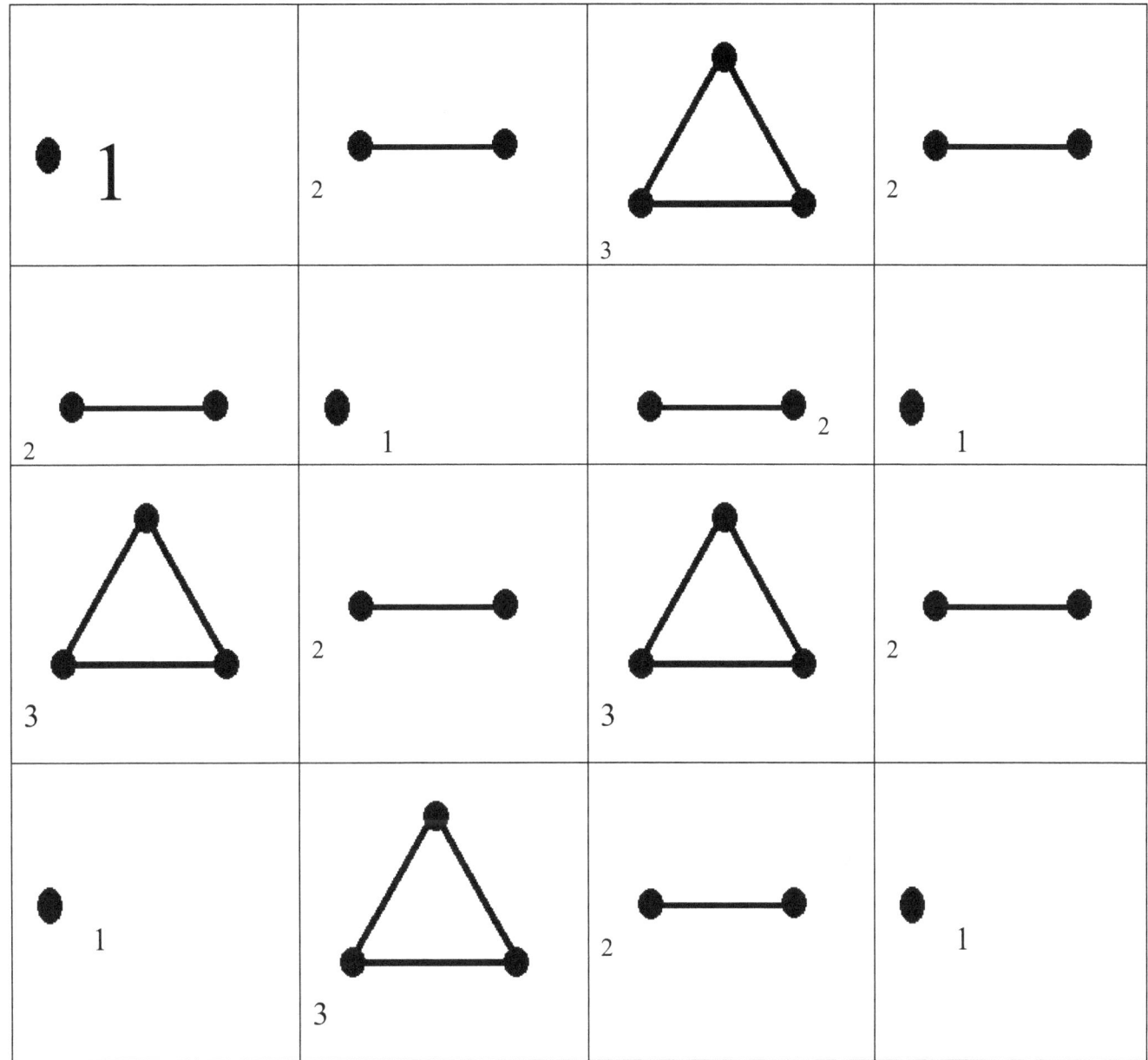

Ho Math Chess   Pre-K and Kindergarten Math

Counting number and writing number on each number line

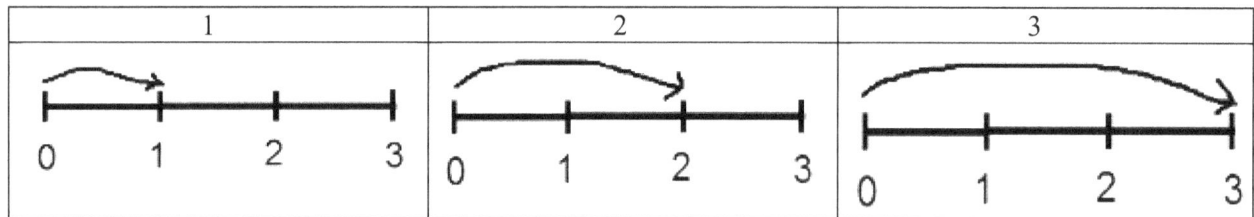

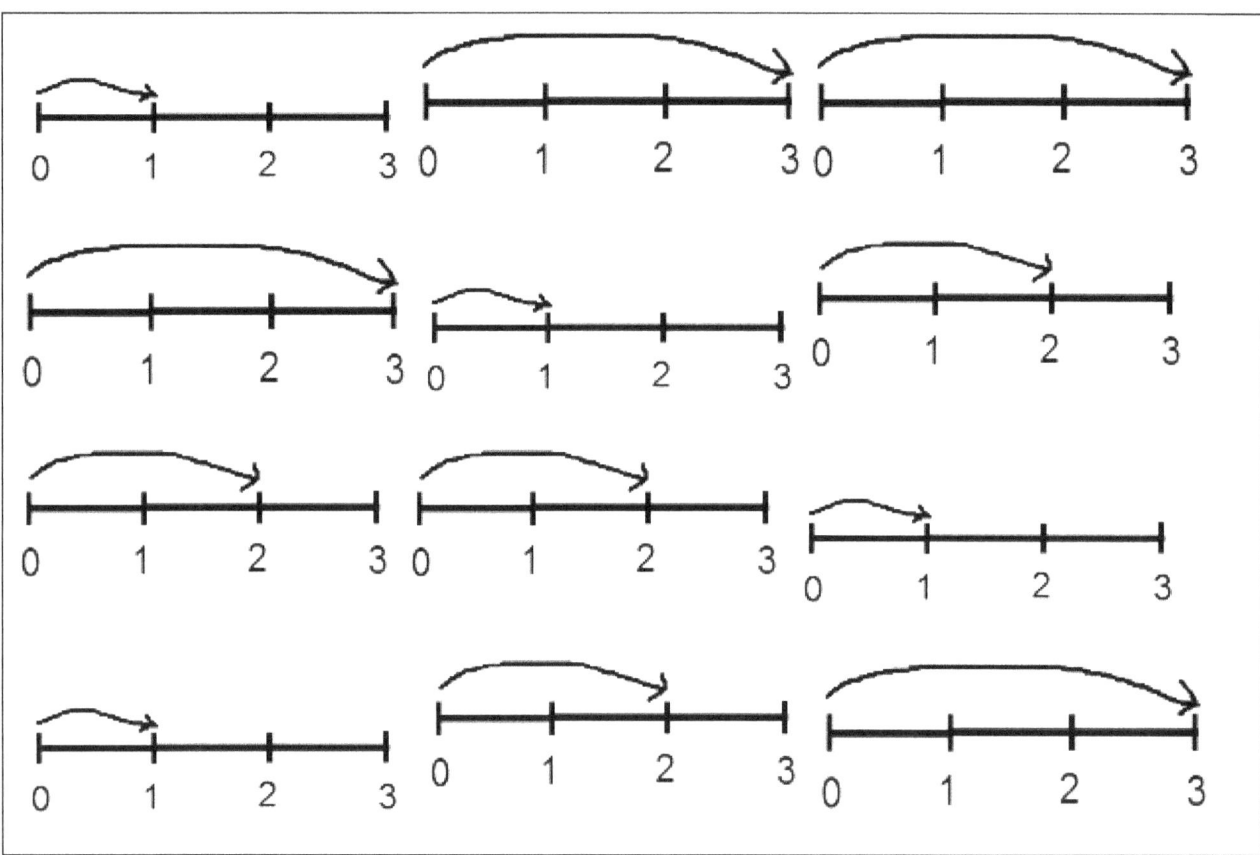

1, 2, 3
1, 3, 3
3, 2, 2
2, 2, 1
1, 2, 3

Counting number and writing number on each number line

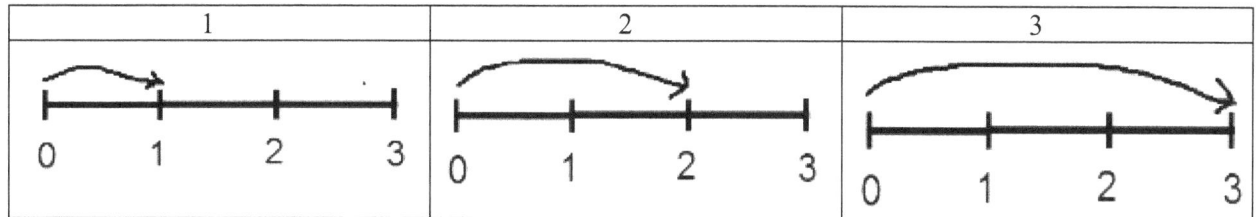

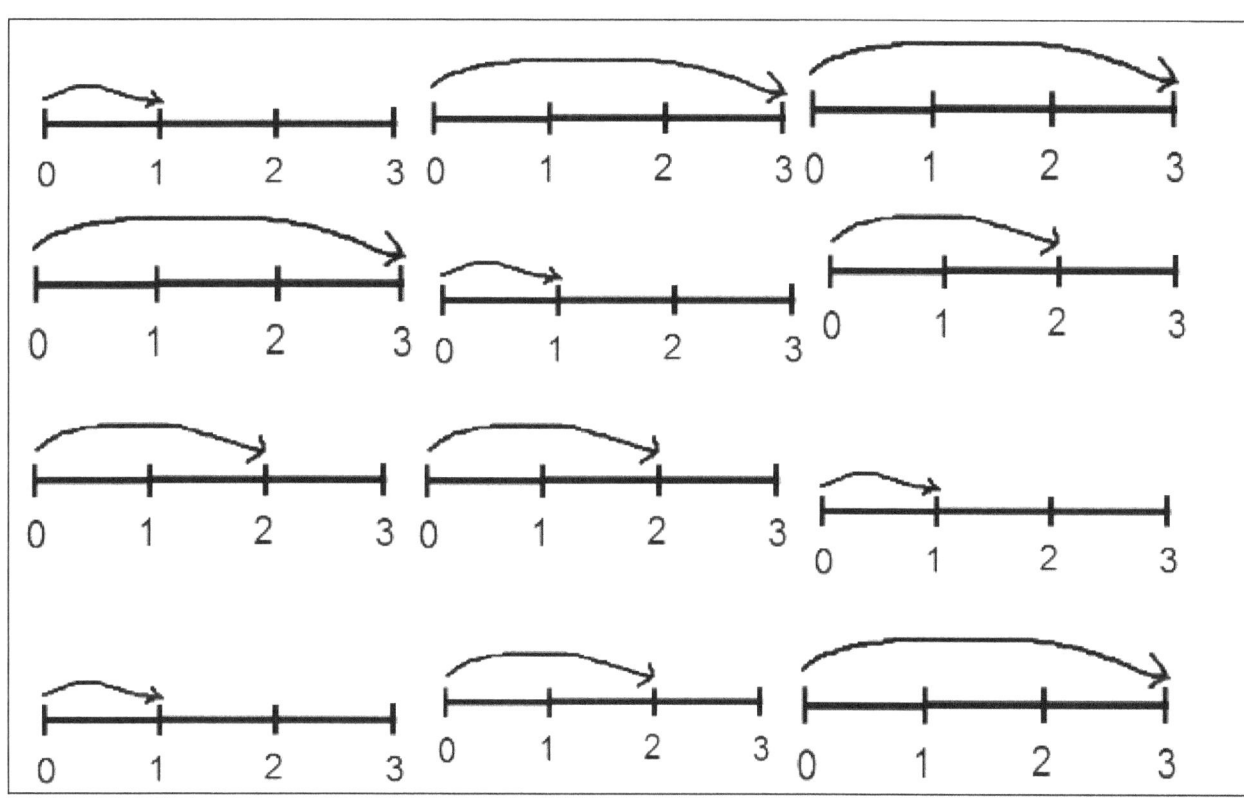

Counting number and writing number on each number line

1	2	3
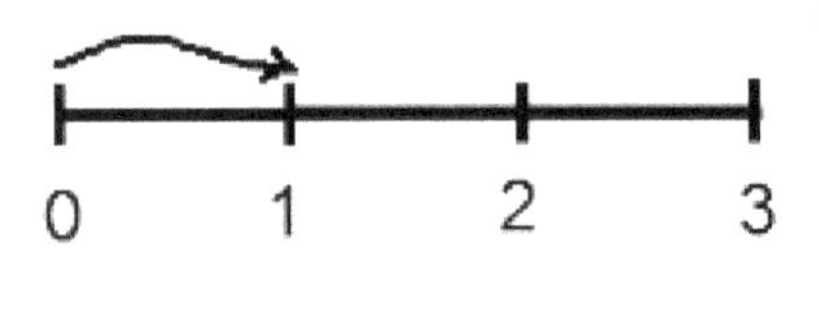 0 1 2 3	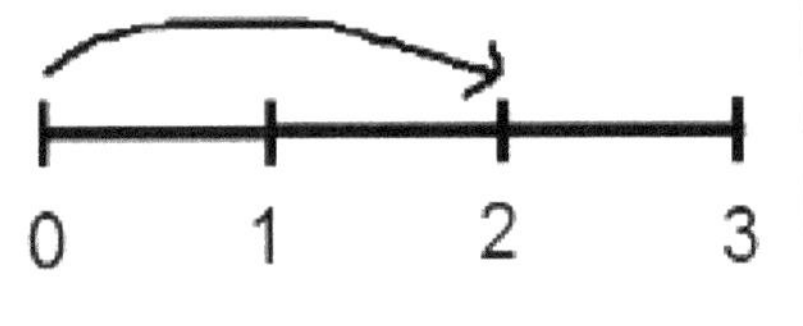 0 1 2 3	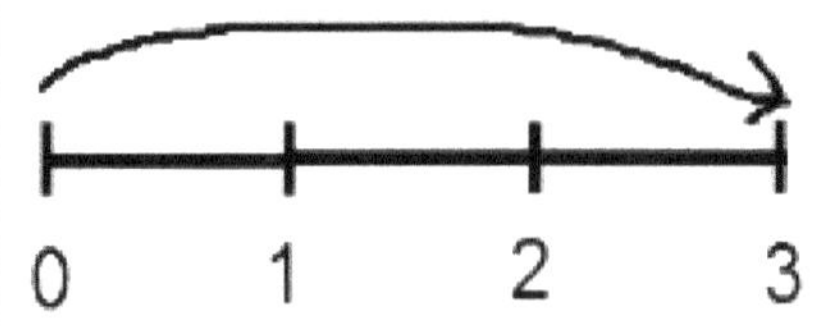 0 1 2 3

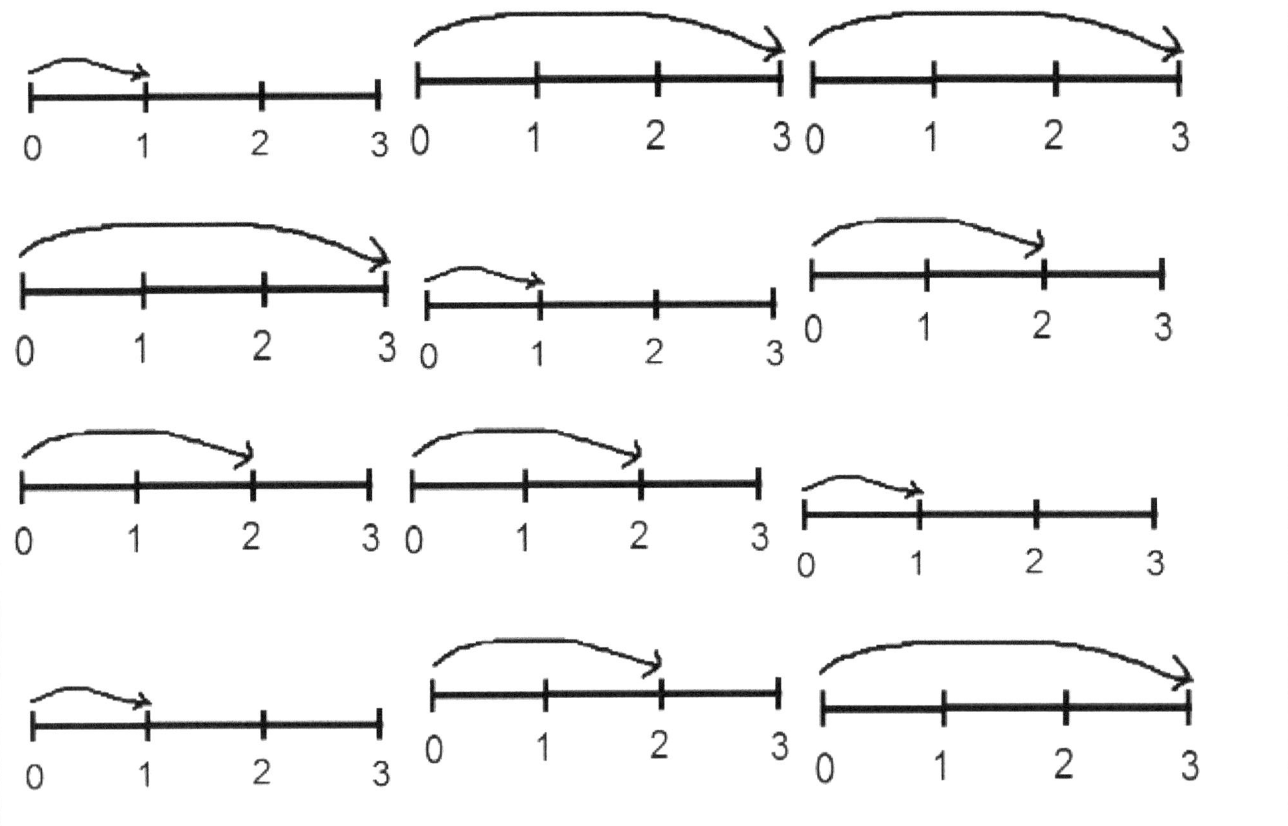
0 1 2 3
0 1 2 3
0 1 2 3
0 1 2 3
0 1 2 3
0 1 2 3
0 1 2 3
0 1 2 3
0 1 2 3
0 1 2 3
0 1 2 3
0 1 2 3

Colouring 1, 2 and 3

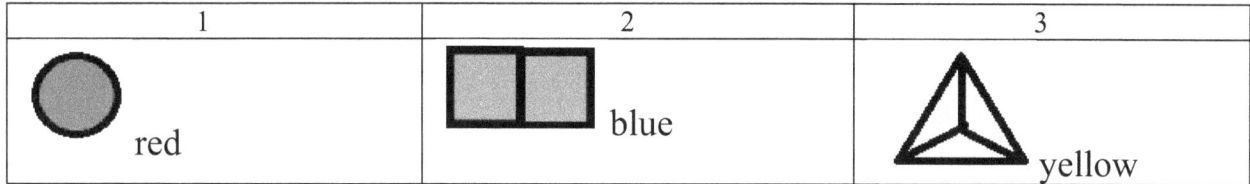

Use the above colour table to colour the following shapes.

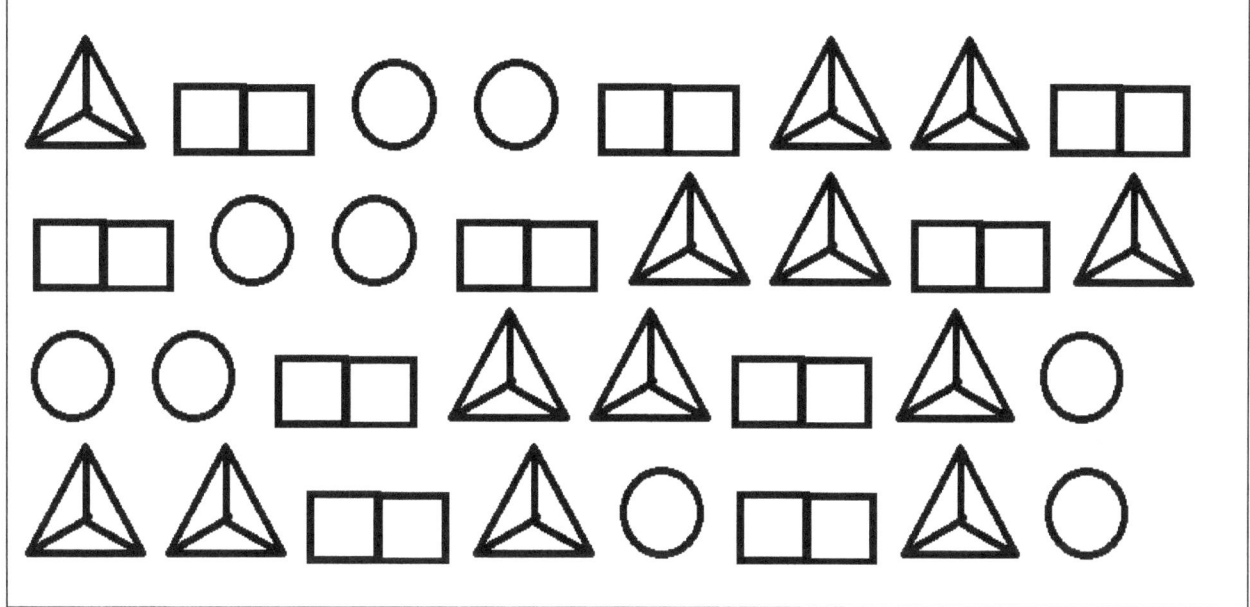

1	2	3
◯ blue	▢ red	△ yellow

Use the above colour table to colour the following shapes.

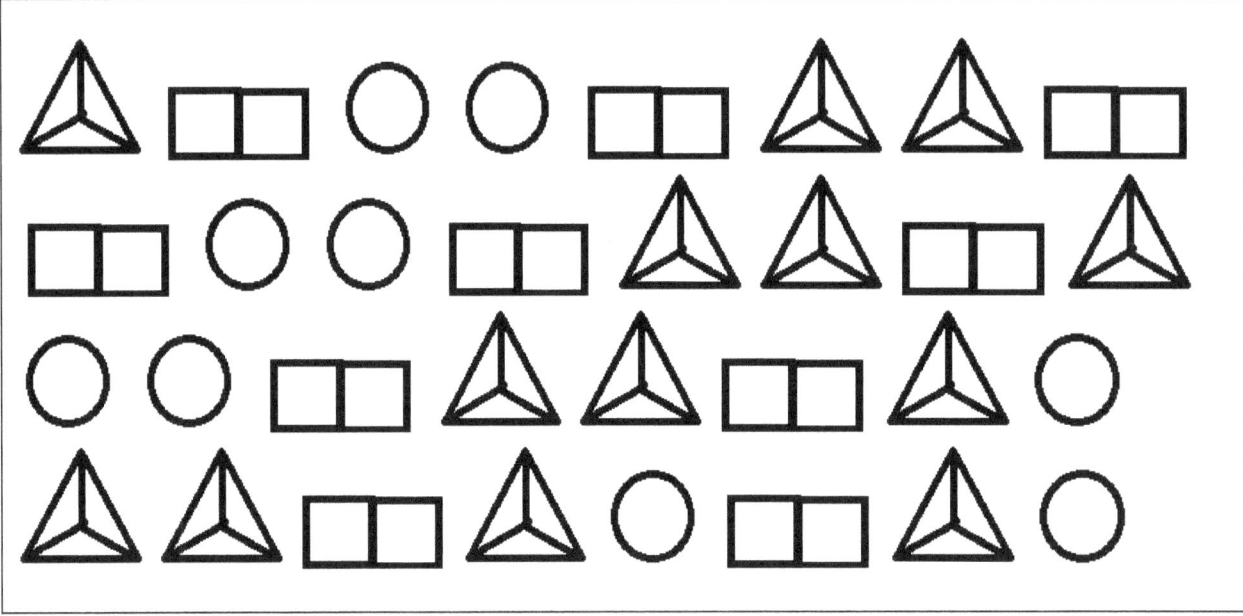

Ho Math Chess — Pre-K and Kindergarten Math

何数棋谜　棋谜式幼儿健脑思维趣味数学

© 2012 – 2021 Frank Ho, Amanda Ho, Canada copyright 1095661, Trademark 771400

1	2	3
○ yellow	▭ red	△ blue

Use the above colour table to colour the following shapes.

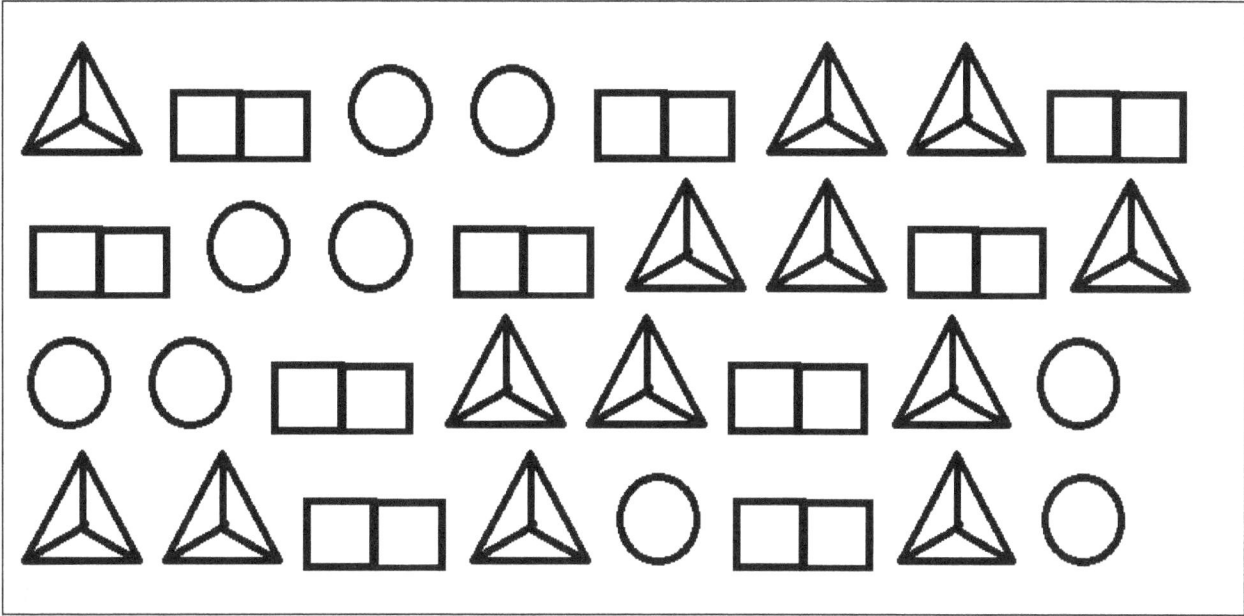

Ho Math Chess — Pre-K and Kindergarten Math

Finding the missing number

Only 1, 2, and 3 can be used in each cell.

1 2 __ 3	3 1 __ 2	2 __ 3 1	1 __ 3 2
1 __ 2	1 __ 3	__ 3 2	__ 2 1
3 2 __	2 3 __	1 __ 3	__ 1 2
1 __ 2	__ 3 1	1 2 __	2 __ 3
3 __ 1	2 3 __	1 3 __	3 __ 2
__ 2 1	2 __ 3	2 1 __	__ 2 3
2 3 __	3 __ 2	__ 2 1	1 3 __

Weight

Scale	Write weight in gram.
(pencil on scale, dial points to 4)	4
(book on scale, dial points to 3)	3

Weight

Scale	Write weight in gram.
	6
	4

Ho Math Chess — Pre-K and Kindergarten Math

何数棋谜　棋谜式幼儿健脑思维趣味数学

© 2012 – 2021 Frank Ho, Amanda Ho, Canada copyright 1095661, Trademark 771400

What are row and column? Matching by drawing
列及行

Figure		Name
☐ (square)		Circle
○ (circle)		Triangle
△ (triangle)		Table with 3 rows and 2 columns
(3x3 grid)		Square
(2x3 grid)		Cube with 3 rows and 3 columns on each face.
(Rubik's cube)		Table with 3 rows and 3 columns

Circling the missing part

Complete figure	Incomplete figure Circle the following part, which is missing when compared to the left-hand side figure.	Complete figure	Incomplete figure Circle the following part, which is missing when compared to the left-hand side figure.

answer

Ho Math Chess — Pre-K and Kindergarten Math

Sizes

How many of the following shapes are in the diagram?

▢ (large square) Answer ____ 3	▫ (small square) Answer ____ 2	◯ (large circle) Answer ____ 2
○ (small circle) Answer ____ 2	△ (large triangle) Answer ____ 2	△ (small triangle) Answer ____ 1

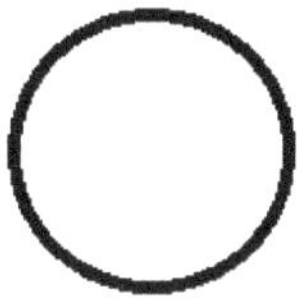

Sizes

How many of the following shapes are in the diagram?

☐ (large square outline)	Answer ____ 2
☐ (small square outline)	Answer ____ 4
○ (large circle)	Answer ____ 2
○ (small circle)	Answer ____ 2
▲ (large gray triangle)	Answer ____ 2
■ (small gray square)	Answer ____ 1

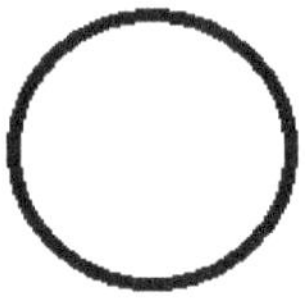

Ho Math Chess — Pre-K and Kindergarten Math

Connecting numbers

Connect numbers from the smallest to the largest in straight or curved lines.	Connect numbers from the smallest to the largest in straight or curved lines.	Connect numbers from the smallest to the largest in straight or curved lines.
1 2 3 4	3 2 1 4	1 2 4 3
Connect numbers from the smallest to the largest in straight or curved lines.	Connect numbers from the smallest to the largest in straight or curved lines.	Connect numbers from the smallest to the largest in straight or curved lines.
1 2 3 4	2 1 3 4	4 2 1 3

www.homathchess.com

Connecting numbers

Connect numbers from the smallest to the largest in straight or curved lines. 1 2 3 4	Connect numbers from the smallest to the largest in straight or curved lines. 3 2 1 4	Connect numbers from the smallest to the largest in straight or curved lines. 1 2 4 3
Connect numbers from the smallest to the largest in straight or curved lines. 1 2 3 4	Connect numbers from the smallest to the largest in straight or curved lines. 2 1 3 4	Connect numbers from the smallest to the largest in straight or curved lines. 4 2 1 3

Connecting shapes

Connect numbers from the smallest to the largest in straight or curved lines. (×6)

Connecting numbers

Connect numbers from the smallest to the largest in straight or curved lines.	Connect numbers from the smallest to the largest in straight or curved lines. 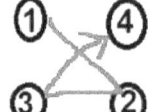	Connect numbers from the smallest to the largest in straight or curved lines. 
Connect numbers from the smallest to the largest in straight or curved lines.  	Connect numbers from the smallest to the largest in straight or curved lines.  	Connect numbers from the smallest to the largest in straight or curved lines.   

Column and row

The number of rows = _____ 5 The number of columns = _____ 5	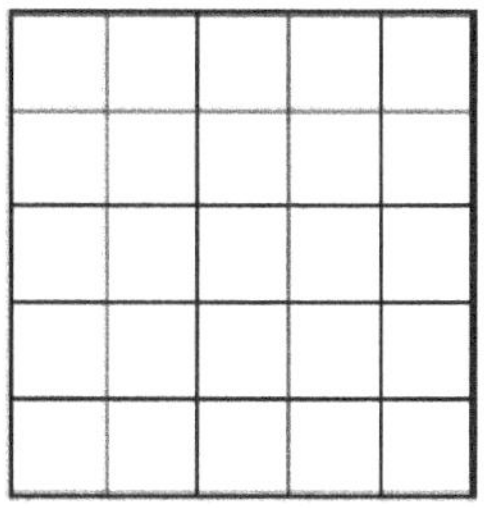 (5×5 grid)
The number of rows = _____ 3 The number of columns = _____ 3	(3×3 grid)
The number of rows = _____ 2 The number of columns = _____ 3	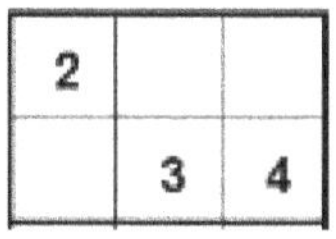 (2×3 grid with 2, 3, 4)
The number of rows = _____ 4 The number of columns = _____ 3	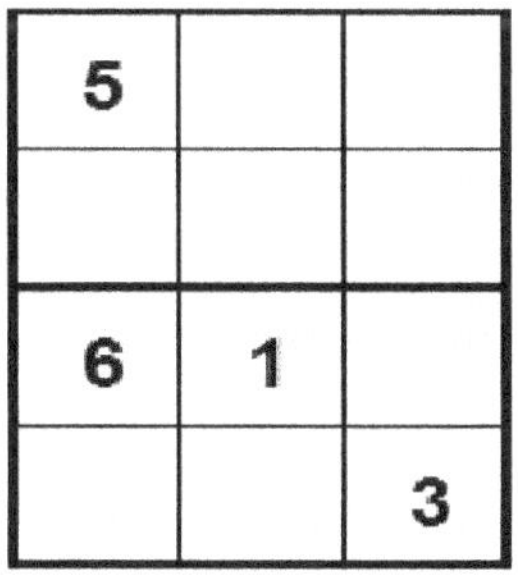 (4×3 grid with 5, 6, 1, 3)

Left, right, top, down.
左右上下

Count how many squares on the left column? _____	5
Count how many squares on the right column? _____	12
Count how many circles on the right column? _____	2
Count how many circles on the left column? _____	2
Count how many triangles on the right column? _____	3
Count how many triangles on the left column? _____	11

Ho Math Chess — Pre-K and Kindergarten Math

Pattern

Fill in the answers.

Sequence	Answer
1, 2, 2, 3, 1, 2, 2, 3, 1, 2, 2, 3, 1, ___, 2, 3	2
1, 3, 3, 1, 3, 3, 1, 3, 3, 1, ___, 3	1
1, 1, 1, 1, 2, 2, 2, 2, 3, 3, ___, 3	3
3, 3, 2, 2, 1, 1, 3, ___, 2, ___, 1, 1	3 2
1, 1, 2, 2, 3, 3, 1, ___, 2, ___, 3, 3	1 2
1, 2, 2, 3, 1, 2, 2, ___, 1, ___, 2, 3	3 2
3, 2, 1, 3, ___, 1, 3, ___, 1, ___, 2, 1, 3, 2, 1	2 2 3

Find the answer to replace the question mark(s).

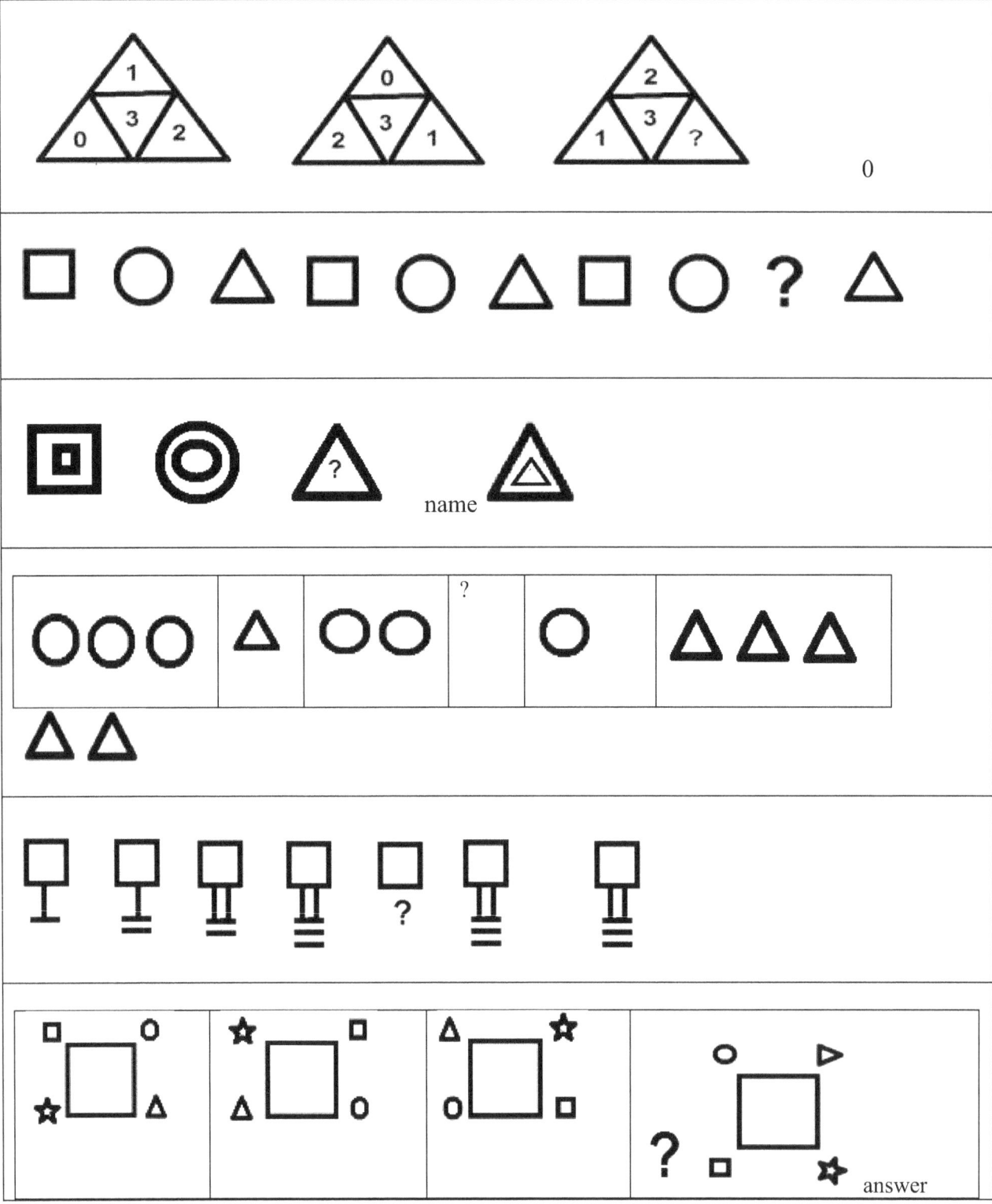

Maze
数迷

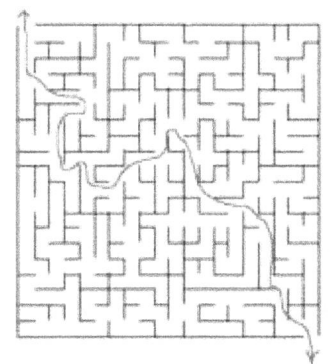
answer

Maze

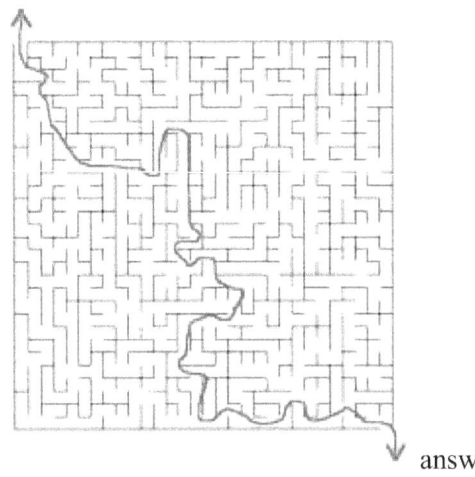
answer

Maze

Answer

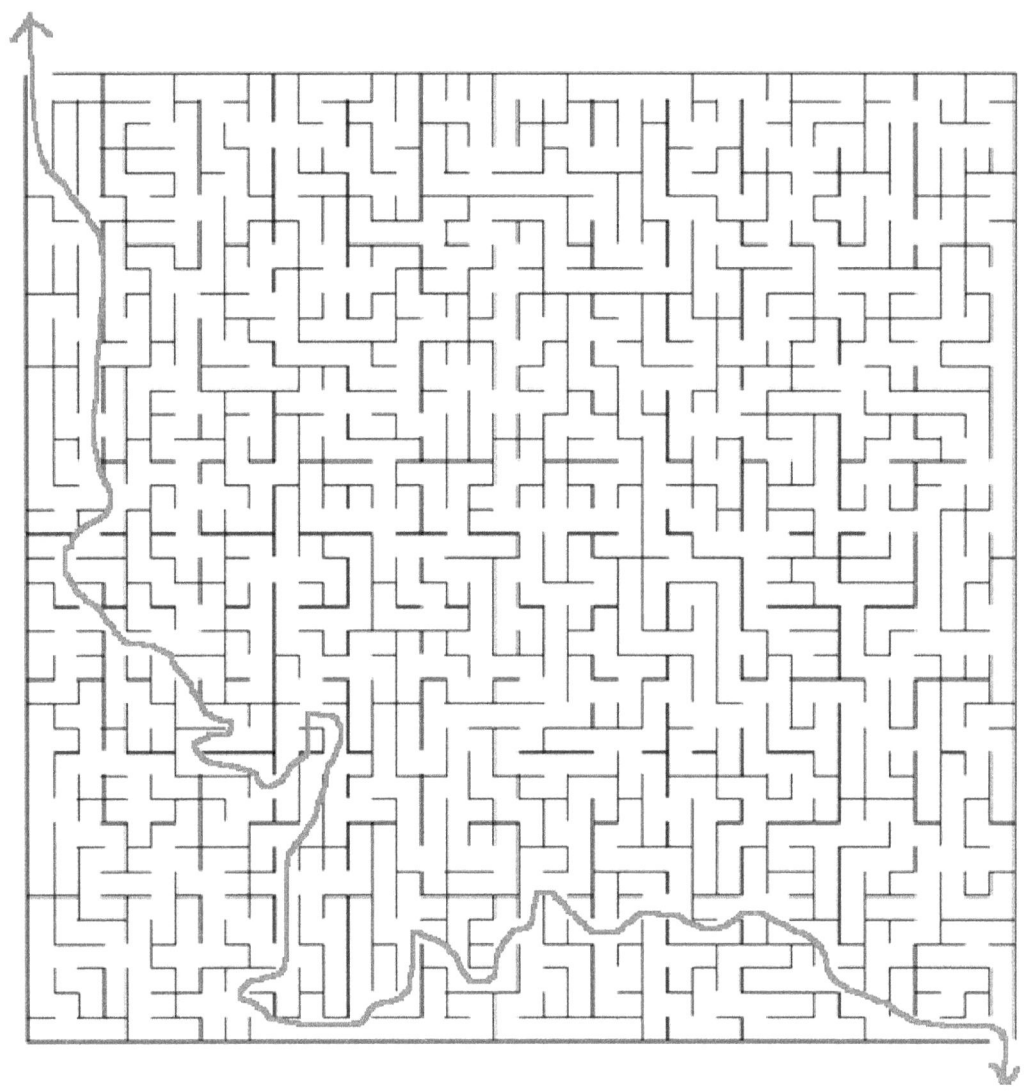

Answer

Pattern

Fill in each "?" by an answer.

Number	Figure	Symbol	Dots	Line segment
0				0 1 2 3 4 5 6 7
1	🍎		●	0 1 2 3 4 5 6 7
2	🍎🍎		●●	0 1 2 3 4 5 6 7
3	🍎🍎🍎		●●●	0 1 2 3 4 5 6 7
4	?		●●●●	0 1 2 3 4 5 6 7
5	🍎🍎🍎🍎🍎		?	0 1 2 3 4 5 6 7
6	?		●●●●●●	0 1 2 3 4 5 6 7
7	🍎🍎🍎🍎🍎🍎🍎	?	●●●●●●●	0 1 2 3 4 5 6 7

Similarities

类比求答案

Replace each question mark with a number.

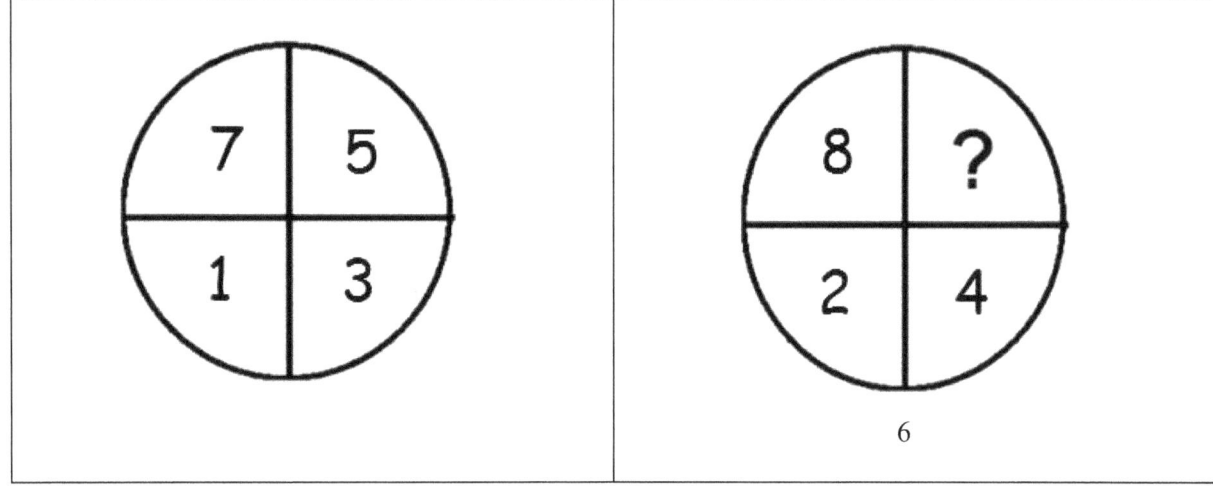

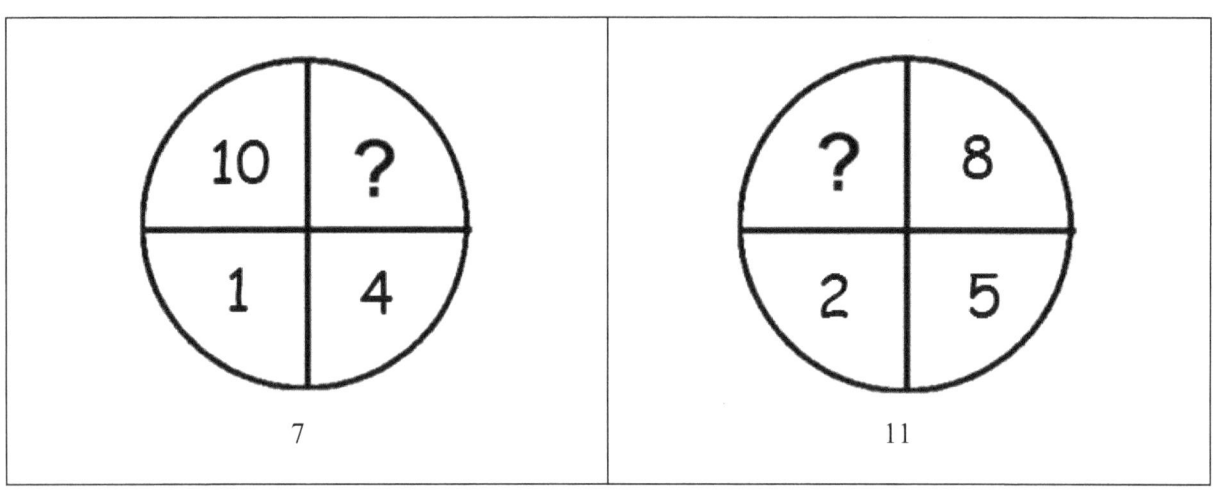

Tracing each number.

0123456789
0123456789
0123456789
0123456789
0123456789
0123456789
0123456789
0123456789
0123456789

0123456789
0123456789
0123456789
0123456789
0123456789

Counting the number of circles in each square

Write the result beside each square.

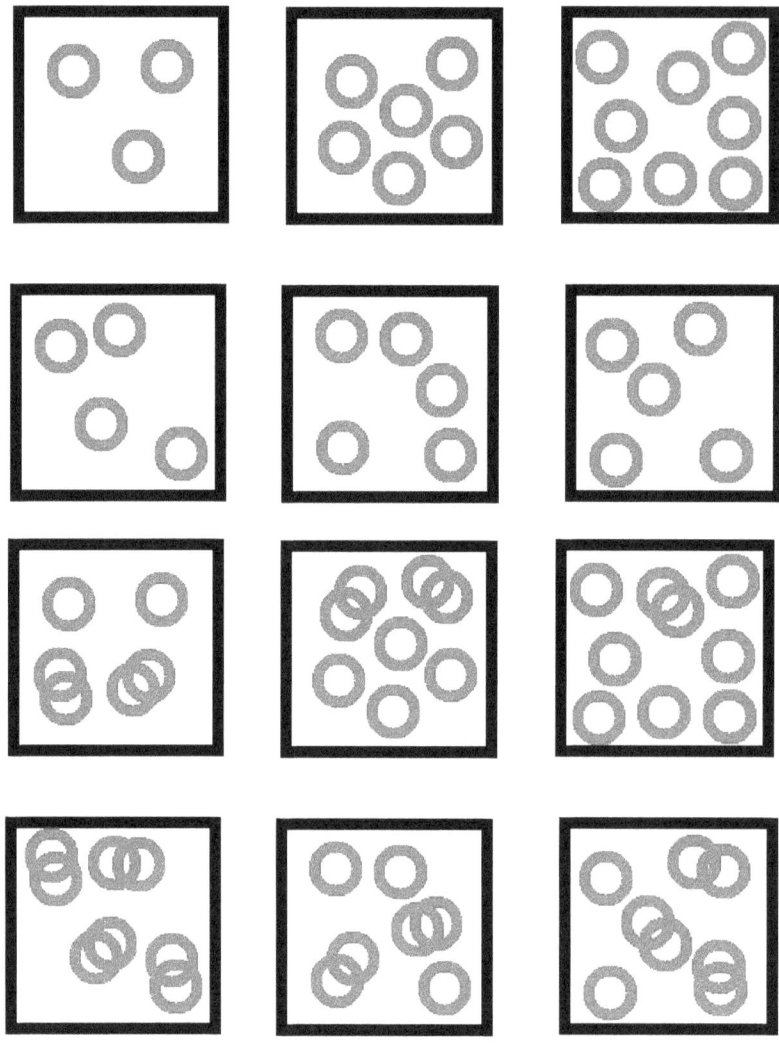

3 6 8
4 5 5
6 8 10
8 7 8

Write the number of red circles within each square beside each square.

0 3 3
3 4 4
5 5 6

Count the numbers of 1's in each square box and then write the answer of the number of 1's beside each square box.

 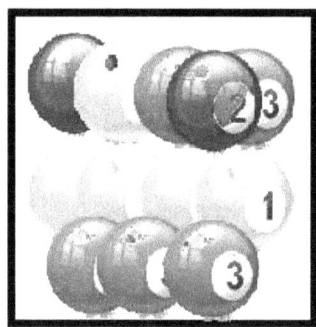 

2 5 3

Count the number of 2's in each square box and then write the number of 2's beside each square box.

2 2 6

Count the number of 3's in each square box and then write the number of 3's beside each square box.

 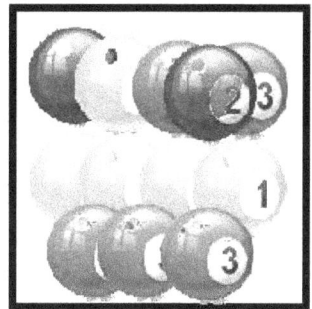 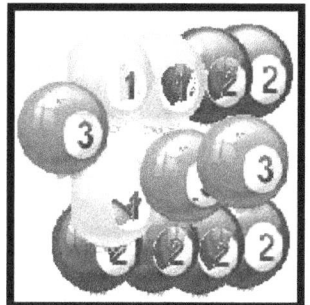

353

Count the number of 3's in the following rectangle box, then write the answer beside the rectangle box.

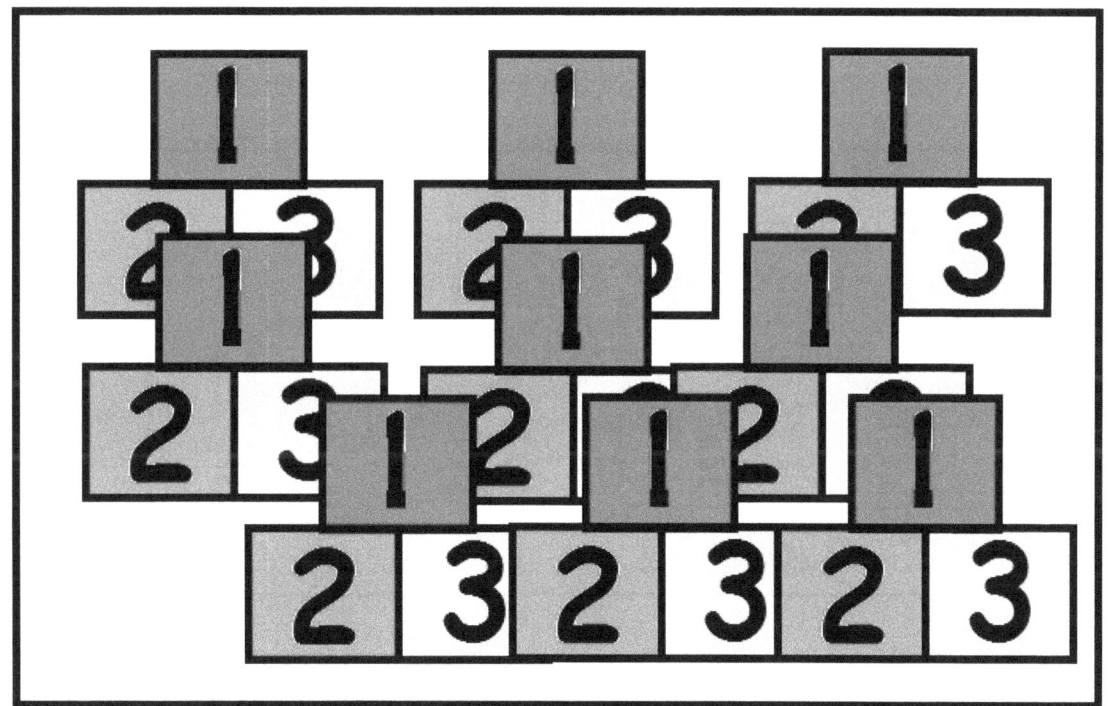

9

Count the number of 1's in the following rectangle box, then write the answer beside the rectangle box.

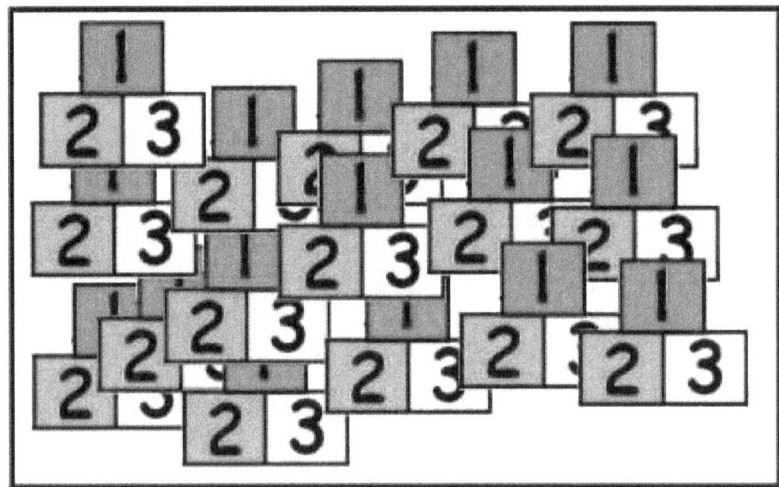

16

Count the number of 2's in the following rectangle box, then write the answer beside the rectangle box.

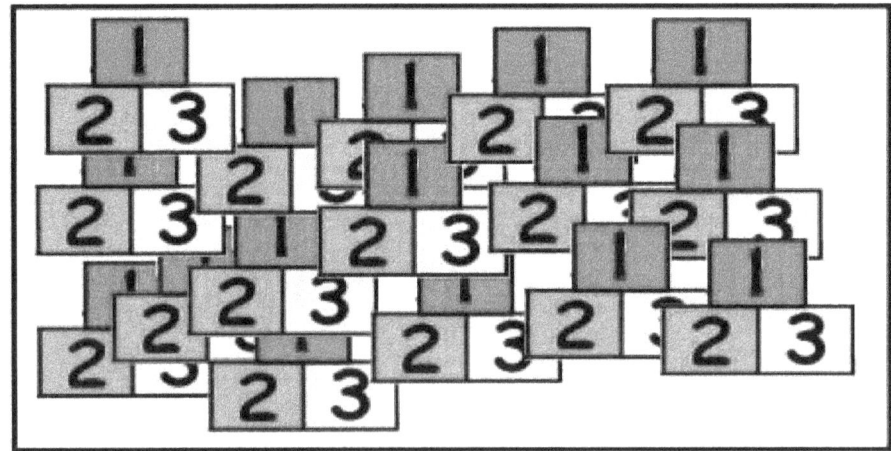

16

Counting numbers

	0	1	2	3	4	5	6	7	8	9
Numbers in total										

5755531332

Count numbers and information analysis.

	0	1	2	3	4	5	6	7	8	9
Numbers in total										

4 5 10 9 6 2 4 2 3

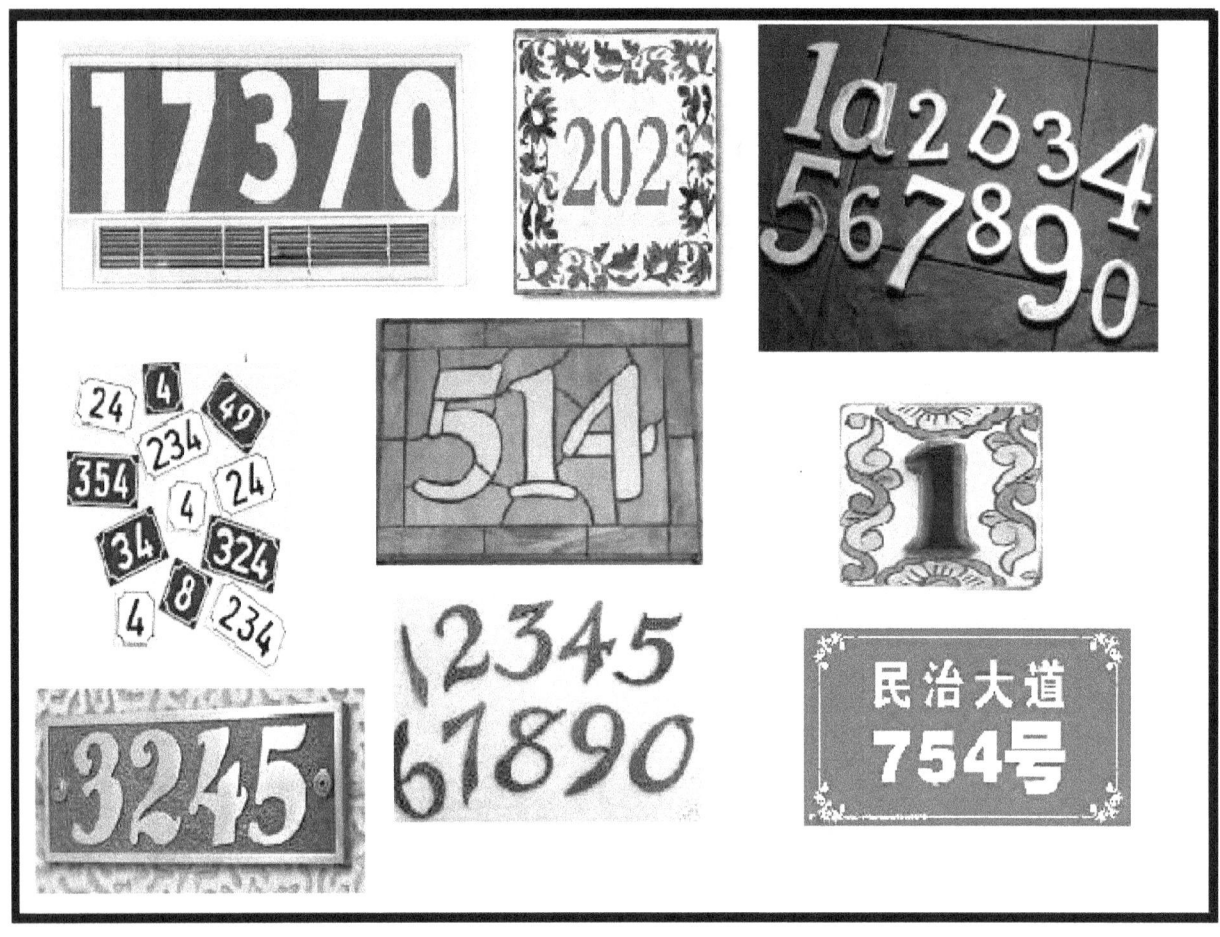

Ho Math Chess — Pre-K and Kindergarten Math

何数棋谜　棋谜式幼儿健脑思维趣味数学

© 2012 – 2021 Frank Ho, Amanda Ho, Canada copyright 1095661, Trademark 771400

Maze
数迷

Ho Math Chess   Pre-K and Kindergarten Math
何数棋谜   棋谜式幼儿健脑思维趣味数学
© 2012 – 2021 Frank Ho, Amanda Ho, Canada copyright 1095661, Trademark 771400

Answer

answer

Matrix reasoning

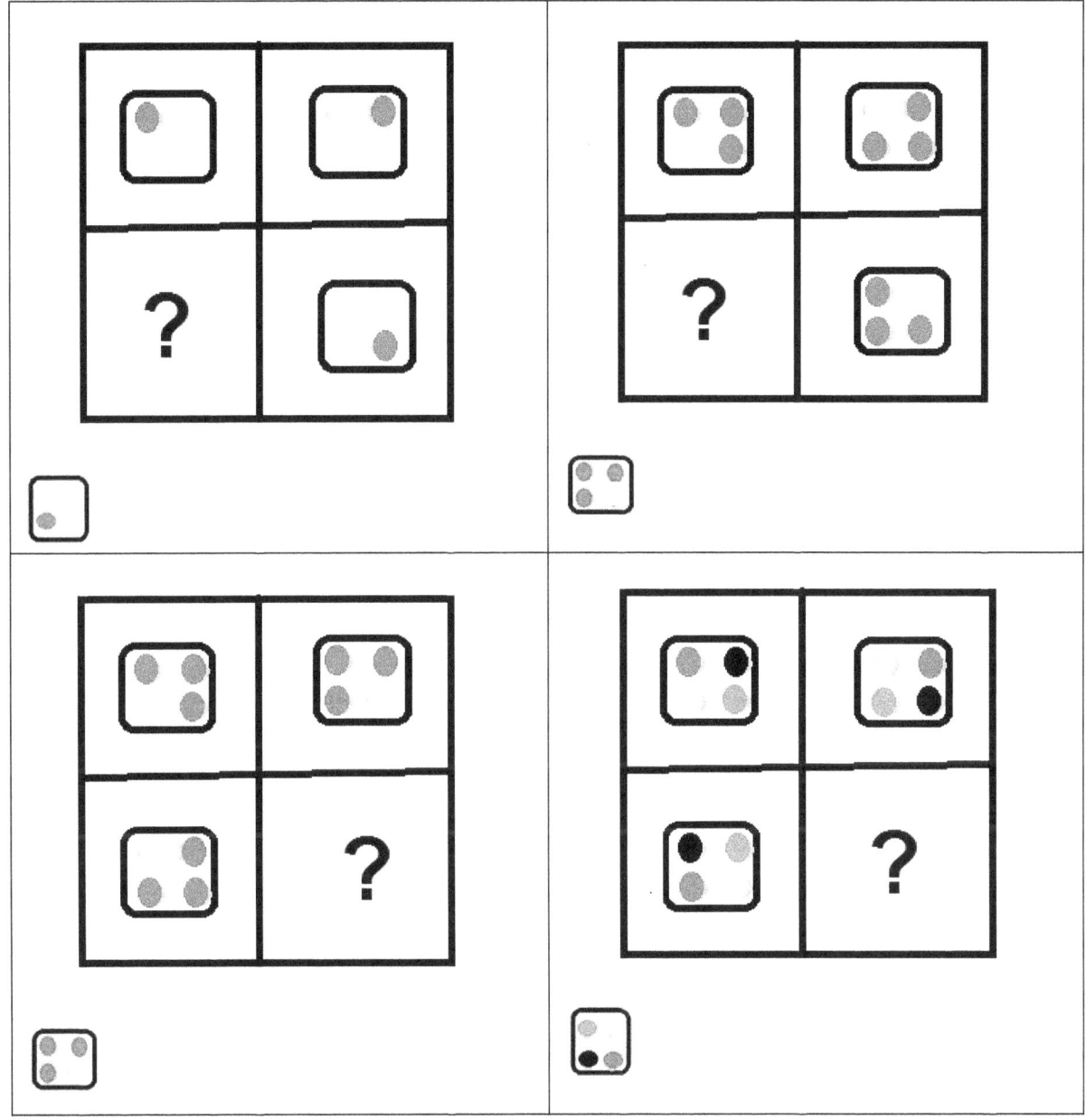

Ho Math Chess — Pre-K and Kindergarten Math

何数棋谜　棋谜式幼儿健脑思维趣味数学

© 2012 – 2021 Frank Ho, Amanda Ho, Canada copyright 1095661, Trademark 771400

Pattern

Find an answer for each question mark.

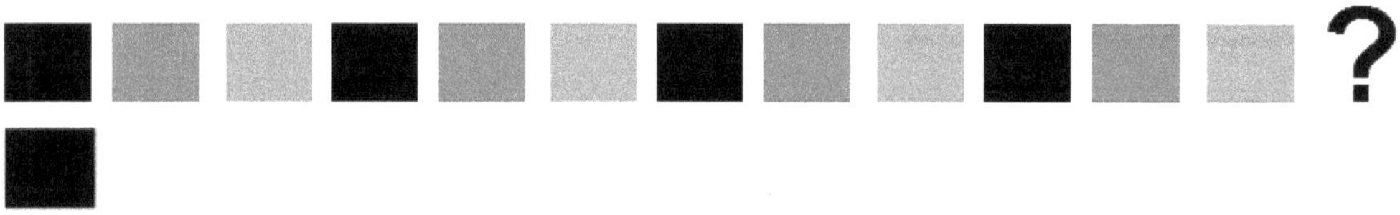

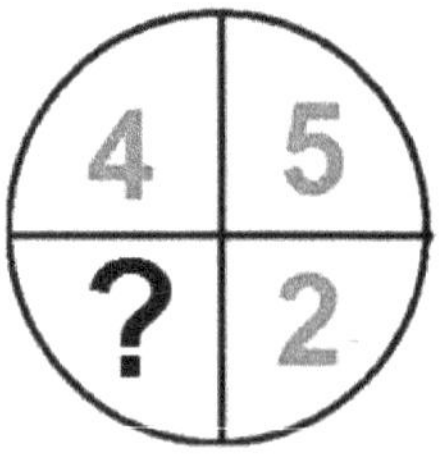

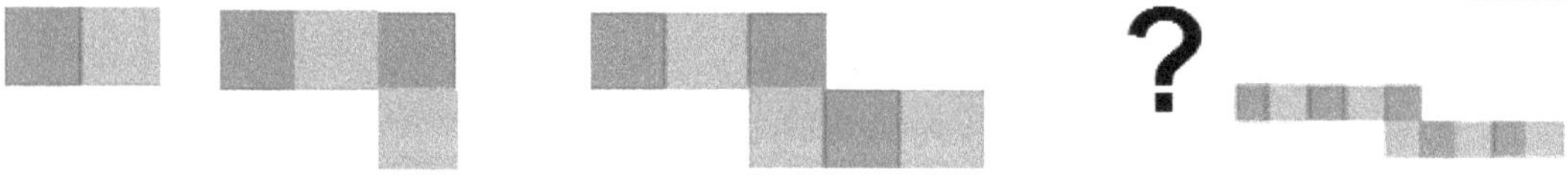

Ho Math Chess — Pre-K and Kindergarten Math

何数棋谜　棋谜式幼儿健脑思维趣味数学

© 2012 – 2021 Frank Ho, Amanda Ho, Canada copyright 1095661, Trademark 771400

Sudoku 数独

Every number from 1 to 4 shall appear only once in every row and column of the following 4 by 4 large square and also within each 2 by 2 small square.

Puzzle 1:
1243
4312
3124
2431

Puzzle 2:
2134
4321
1243
3412

Puzzle 3:
1234
3421
4312
2143

Puzzle 4:
3124
4213
2341
1432

www.homathchess.com

Ho Math Chess — Pre-K and Kindergarten Math

何数棋谜　棋谜式幼儿健脑思维趣味数学

© 2012 – 2021 Frank Ho, Amanda Ho, Canada copyright 1095661, Trademark 771400

Multi-task, multi-step, multi-concept

Note that the black and white version may not show the numbers correctly. If this is true, then skip this problem.

You are a chess piece located at c3.

All numbers are single digits.

1	2	3	1	2	3	1	2	3
?	?	?	?	?	?	?	?	?

401222022

1	2	3	1	2	3	1	2	3
?	?	?	?	?	?	?	?	?

102251002

1	2	3	1	2	3	1	2	3
?	?	?	?	?	?	?	?	?

111212022

1	2	3	1	2	3	1	2	3
?	?	?	?	?	?	?	?	?

2041122021

www.homathchess.com

Ho Math Chess — Pre-K and Kindergarten Math

© 2012 – 2021 Frank Ho, Amanda Ho, Canada copyright 1095661, Trademark 771400

Multi-task, multi-step, multi-concept

Note that the black and white version may not show the numbers correctly. If this is true, then skip this problem.

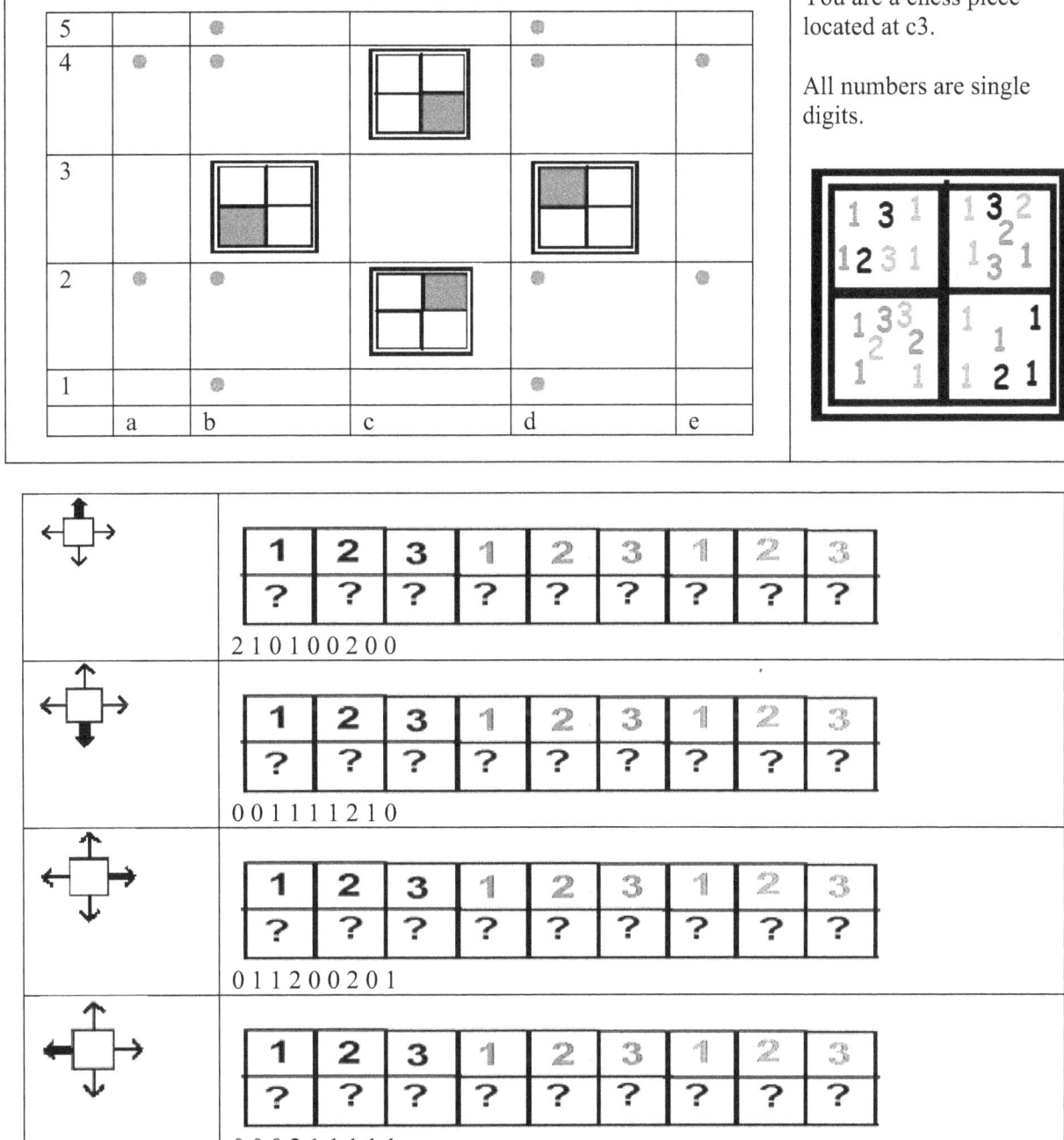

Sudoku 数独

Every number from 1 to 3 shall appear only once in every row and column of the following 3 by 3 large squares.

		1
	2	

231 312 123

		1
3		

231 123 312

		3
1		

213 321 132

3		
	2	

312 231 123

1		
	3	

231 123 312

2		
	1	

213 321 132

	2	
1		

321 213 132

	1	
		3

312 231 123

		1
	2	

312 231 123

Memory and computation training
记忆及计算训练

To teach these "Memory and Computation Training" problems, the instructor will show each problem in only 3 seconds to the students. After 3 seconds, the instructor will immediately cover the problems and ask the students to write the answers.

The instructor will show the problem for 3 seconds again and again to the students if the answer is not correct so that the student can have a chance to get the correct answer.

Memory and computation training

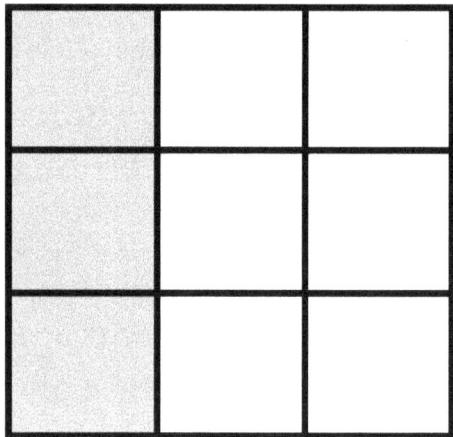 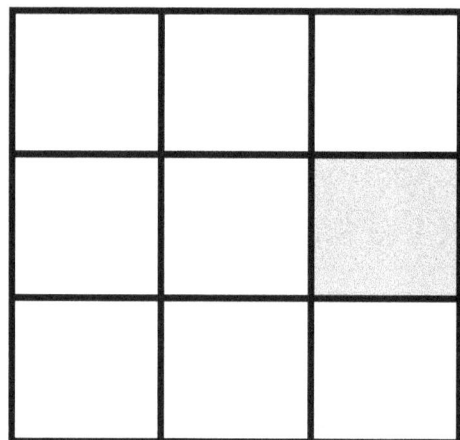

_____ + _____ = _____

3+1=4

Memory and computation training

_____ + _____ = _____

3+2=5

Memory and computation training

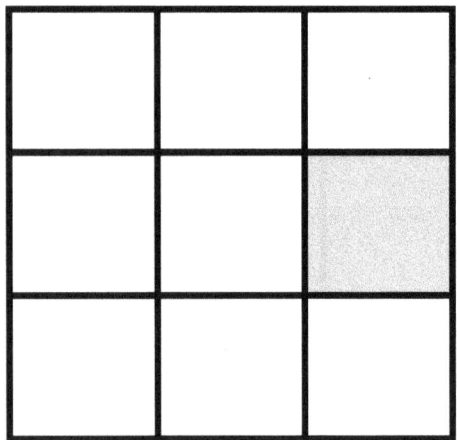 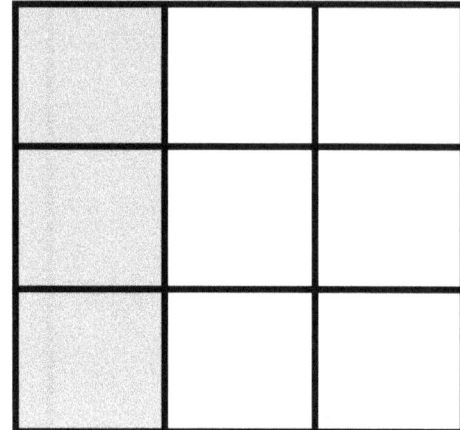

1+3=4

_____ + _____ = _____

Memory and computation training

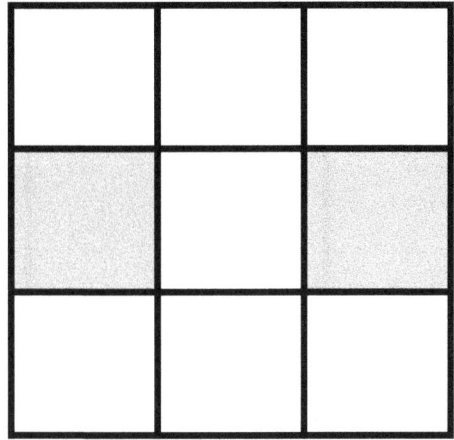 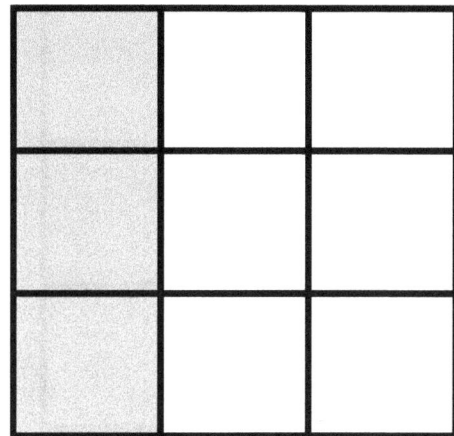

2+3=5

Memory and computation training

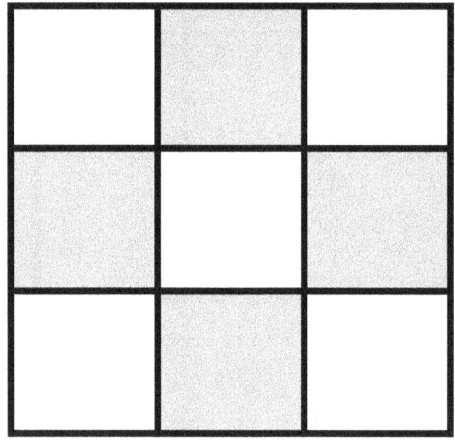 

4+1=5

_____ + _____ = _____

Memory and computation training

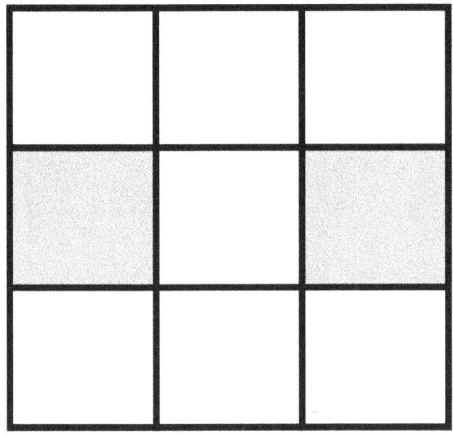 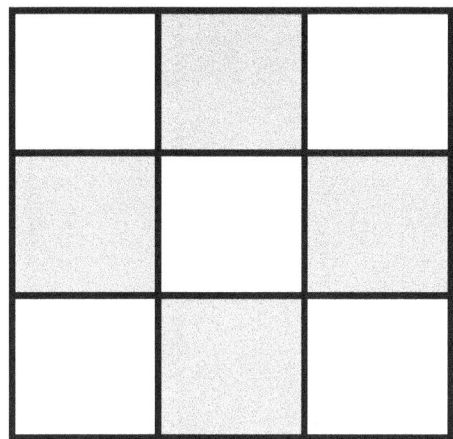

2+4=6

_____ + _____ = _____

Memory and computation training

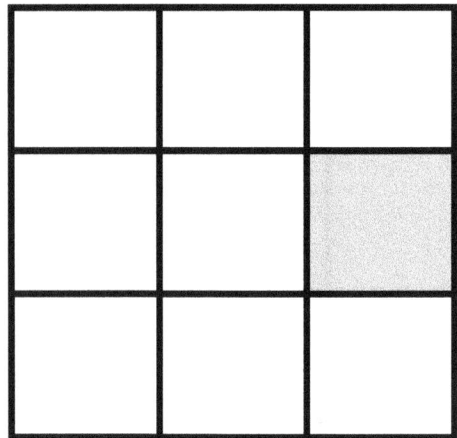 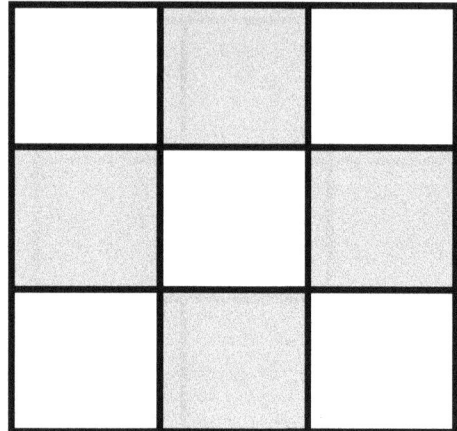

1+4=5

_____ + _____ = _____

Memory and computation training

5	6	7
8	✸	9
2	3	4

1	2	2
1	2	1
2	1	2

6+2=8

_____ + _____ = _____

Ho Math Chess — Pre-K and Kindergarten Math

Memory and computation training

5	6	7
8	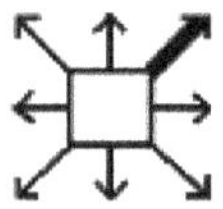 ✧	9
2	3	4

1	2	2
1	2	1
2	1	2

7+2=9

_____ + _____ = _____

Memory and computation training

5	6	7
8	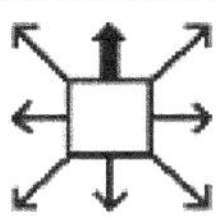 ✸	9
2	3	4

1	2	2
1	2	1
2	1	2

6+1=7

_____ + _____ = _____

Ho Math Chess Pre-K and Kindergarten Math

Memory and computation training

5	6	7
8	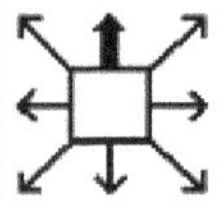 ✦	9
2	3	4

1	2	2
1	2	1
2	1	2

6+2=8

_____ + _____ = _____

Memory and computation training

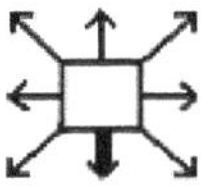

3+1=4

_____ + _____ = _____

Ho Math Chess — Pre-K and Kindergarten Math

何数棋谜　棋谜式幼儿健脑思维趣味数学

© 2012 – 2021 Frank Ho, Amanda Ho, Canada copyright 1095661, Trademark 771400

Memory and computation training

5	6	7
8		9
2	3	4

1	2	2
1	2	1
2	1	2

7+2=9

_____ + _____ = _____

Memory and computation training

5	6	7
8	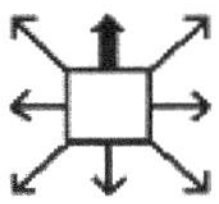 ✴	9
2	3	4

3	2	1
2	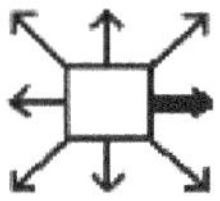 ✴	1
8	7	6

6+1=7

_____ + _____ = _____

Memory and computation training

5	6	7
8	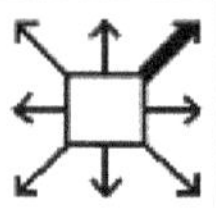	9
2	3	4

3	2	1
2	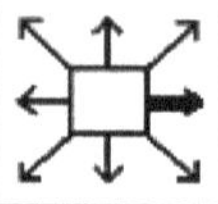	1
8	7	6

7+1=8

_____ + _____ = _____

Memory and computation training

6+2=8

_____ + _____ = _____

Memory and computation training

5	6	7
8	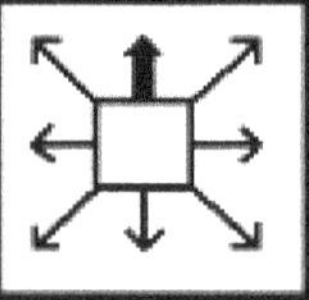 ✦	9
2	3	4

3	2	1
2	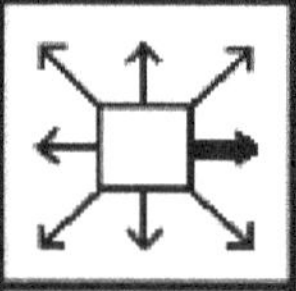 ✦	1
8	7	6

6+1=7

_____ + _____ = _____

Memory and computation training

5	6	7
8	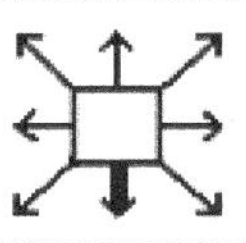	9
2	3	4

3	2	1
2	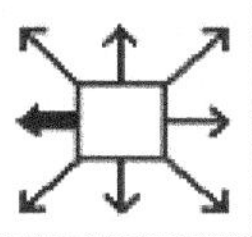	1
8	7	6

3+2=5

_____ + _____ = _____

Ho Math Chess   Pre-K and Kindergarten Math

何数棋谜   棋谜式幼儿健脑思维趣味数学

© 2012 – 2021 Frank Ho, Amanda Ho, Canada copyright 1095661, Trademark 771400

Memory and computation training

5	6	7
8	✦	9
2	3	4

3	2	1
8	✦	1
2	7	6

7+2=9

_____ + _____ = _____

Memory and computation training

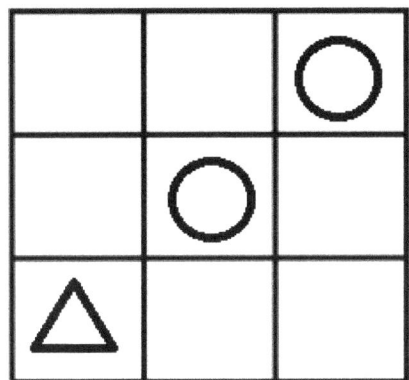 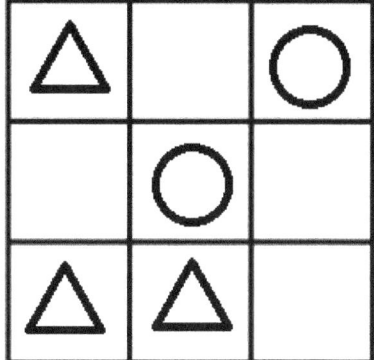

_____ △'s + _____ △'s = _____

Memory and computation training

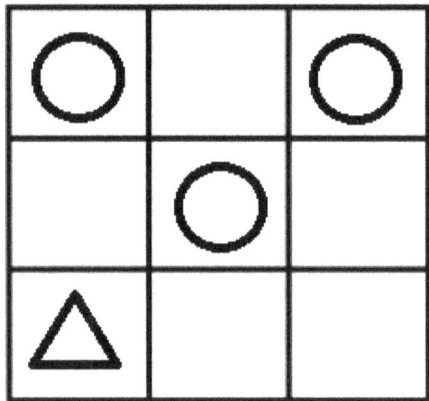 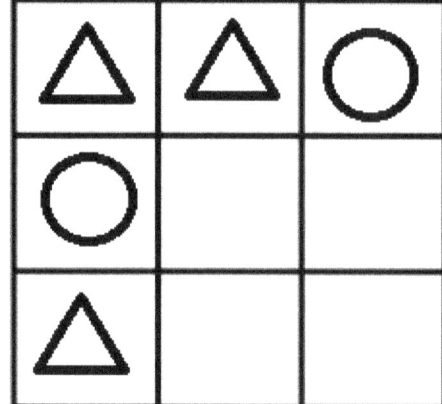

1+3=4

_____ △'s + _____ △'s = _____

Ho Math Chess — Pre-K and Kindergarten Math

何数棋谜　棋谜式幼儿健脑思维趣味数学

© 2012 – 2021 Frank Ho, Amanda Ho, Canada copyright 1095661, Trademark 771400

Memory and computation training

Find out how many pairs of symmetric numbers at the bishop. Consider the line with ? only.

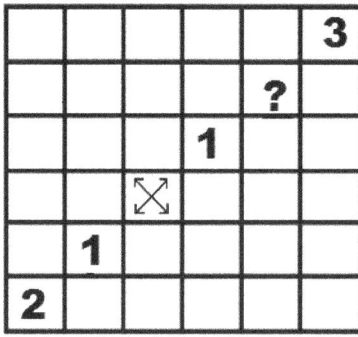

2+2=4

Ho Math Chess — Pre-K and Kindergarten Math

何数棋谜　棋谜式幼儿健脑思维趣味数学

© 2012 – 2021 Frank Ho, Amanda Ho, Canada copyright 1095661, Trademark 771400

Memory and computation training

Find out how many pairs of symmetric numbers at the bishop. Consider the line with ? only.

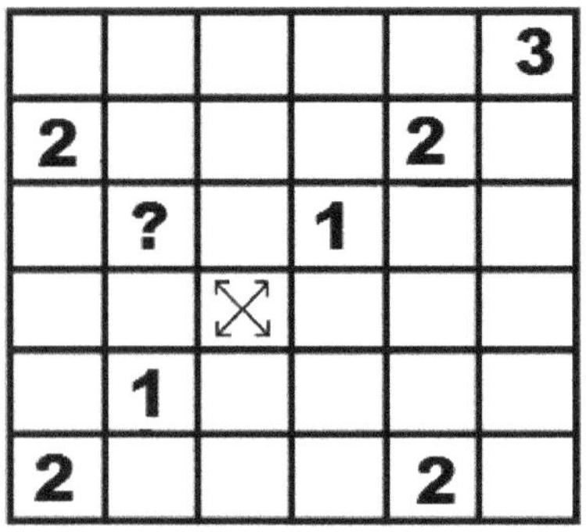

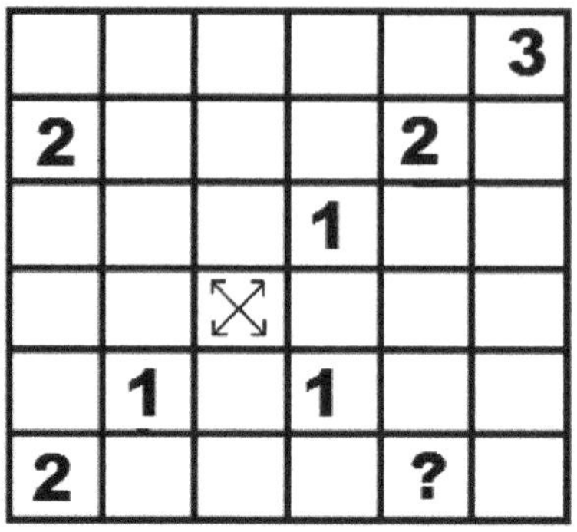

1+2=3

_____ + _____ = _____

Memory and computation training

3+3=6

_____ dogs + _____ dogs = _____ dogs

Memory and computation training

4+4=8

_____ dogs + _____ dogs = _____ dogs

Memory and computation training

3+4=7

_____ dogs + _____ dogs = _____ dogs

Memory and computation training

3+2=5

_____cats + _____cats = _____cats

Memory and computation training

4+3=7

_____dogs + _____cats = _____animals

Knight moves

Mark an "X" on a square to show where each 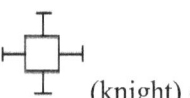 (knight) can move.

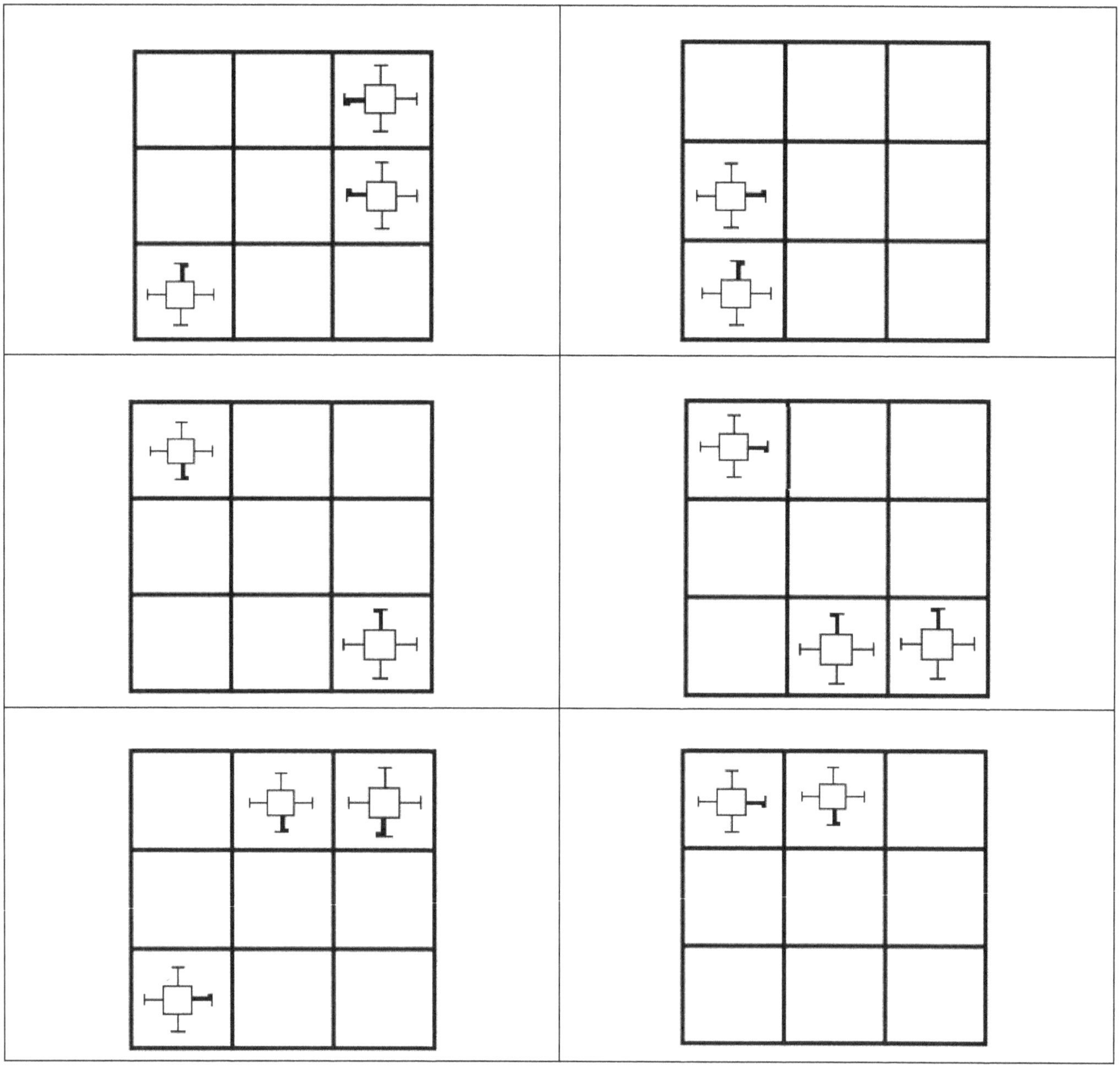

Ho Math Chess   Pre-K and Kindergarten Math

何数棋谜   棋谜式幼儿健脑思维趣味数学

© 2012 – 2021 Frank Ho, Amanda Ho, Canada copyright 1095661, Trademark 771400

Knight moves.
马的走法

Mark an "X" on a square to show where each  (knight) can move to.

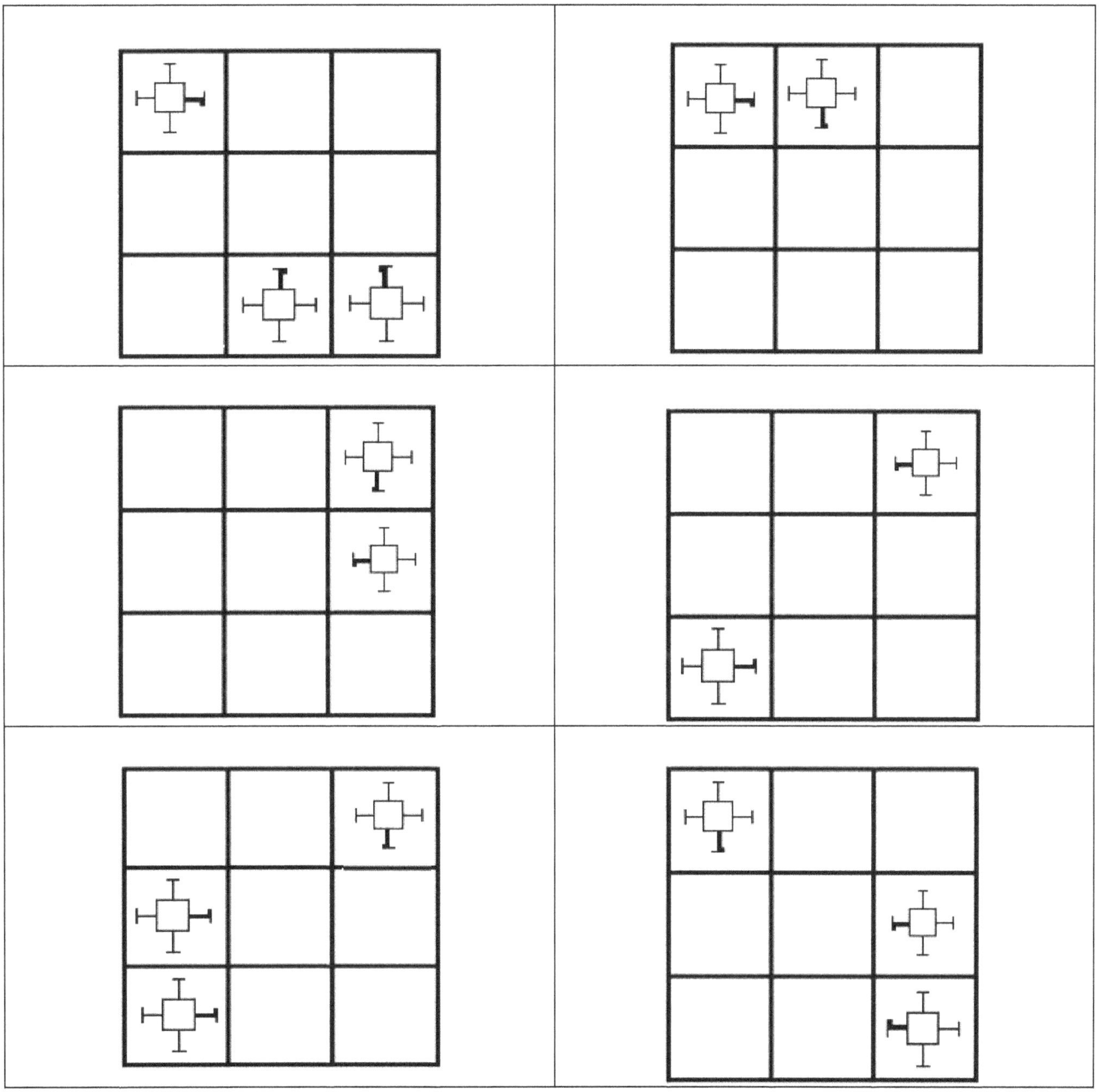

Ho Math Chess — Pre-K and Kindergarten Math

何数棋谜　棋谜式幼儿健脑思维趣味数学

© 2012 – 2021 Frank Ho, Amanda Ho, Canada copyright 1095661, Trademark 771400

Knight moves

Mark an "X" on a square to show where each ♞ (knight) can move to.

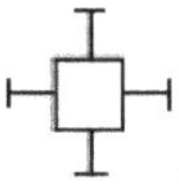

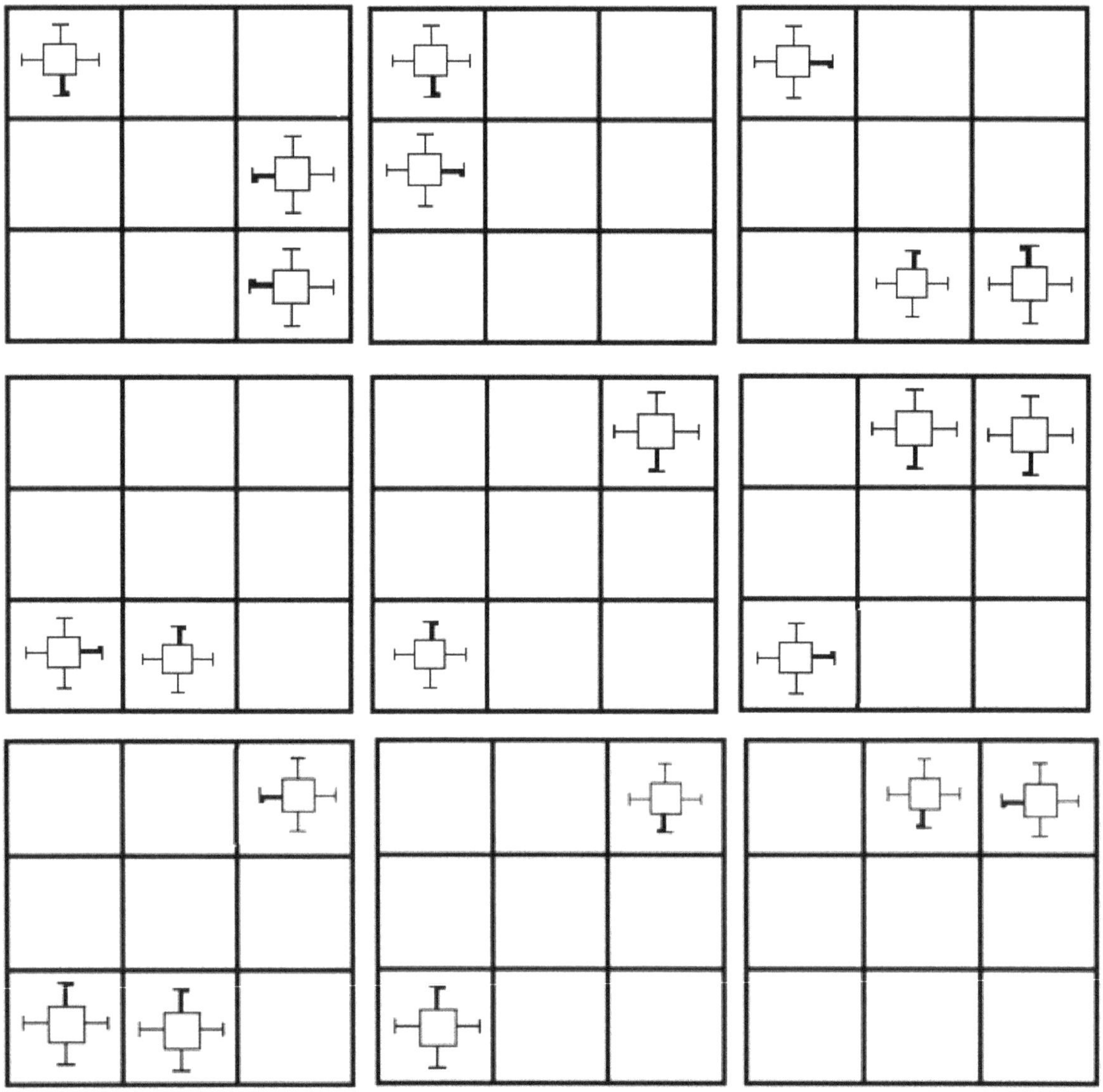

Knight moves

Mark an "X" on a square to show where each (knight) can move to.

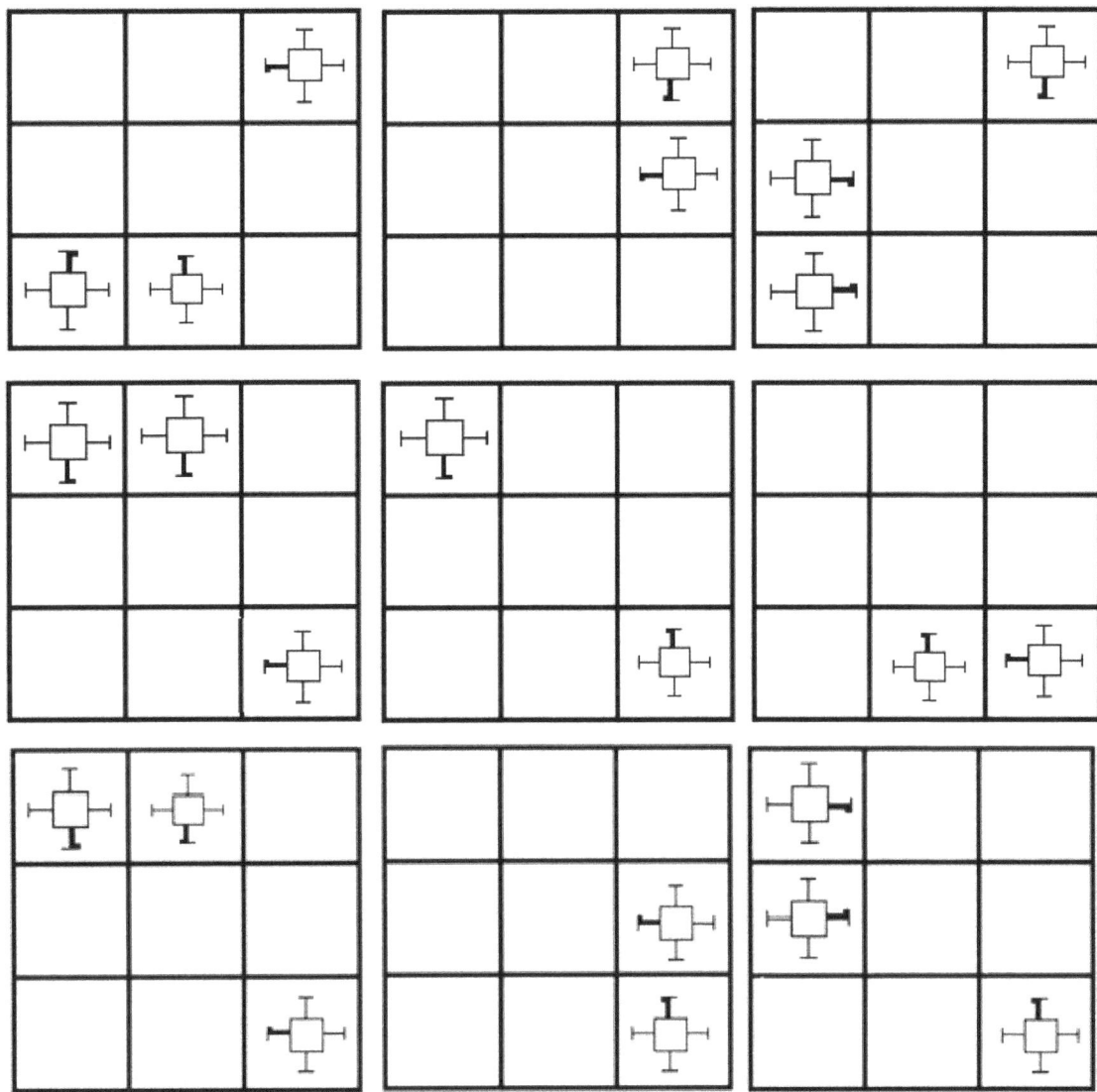

Adding 1, 2, 3

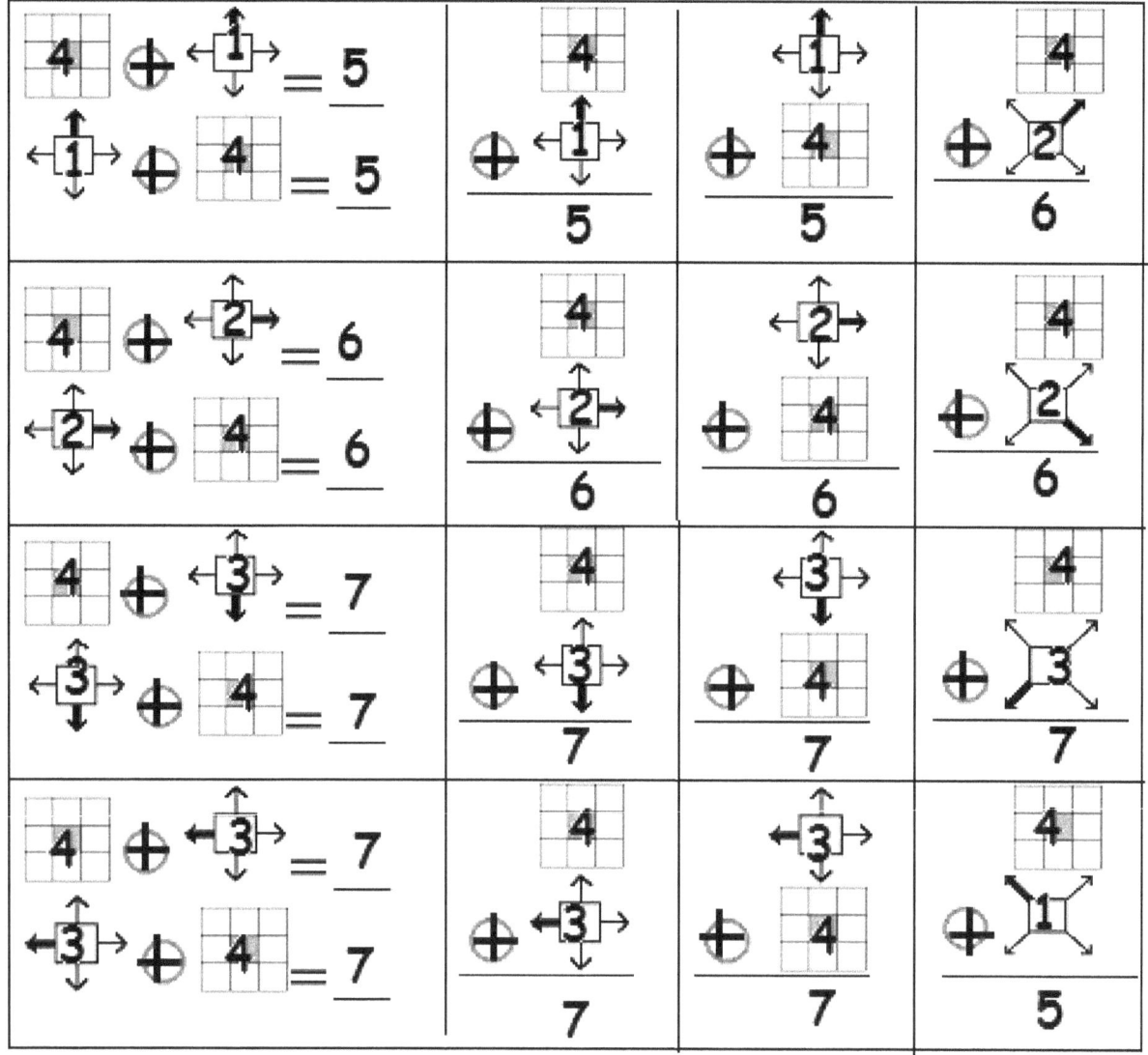

Ho Math Chess — Pre-K and Kindergarten Math

Adding up to 9 Example

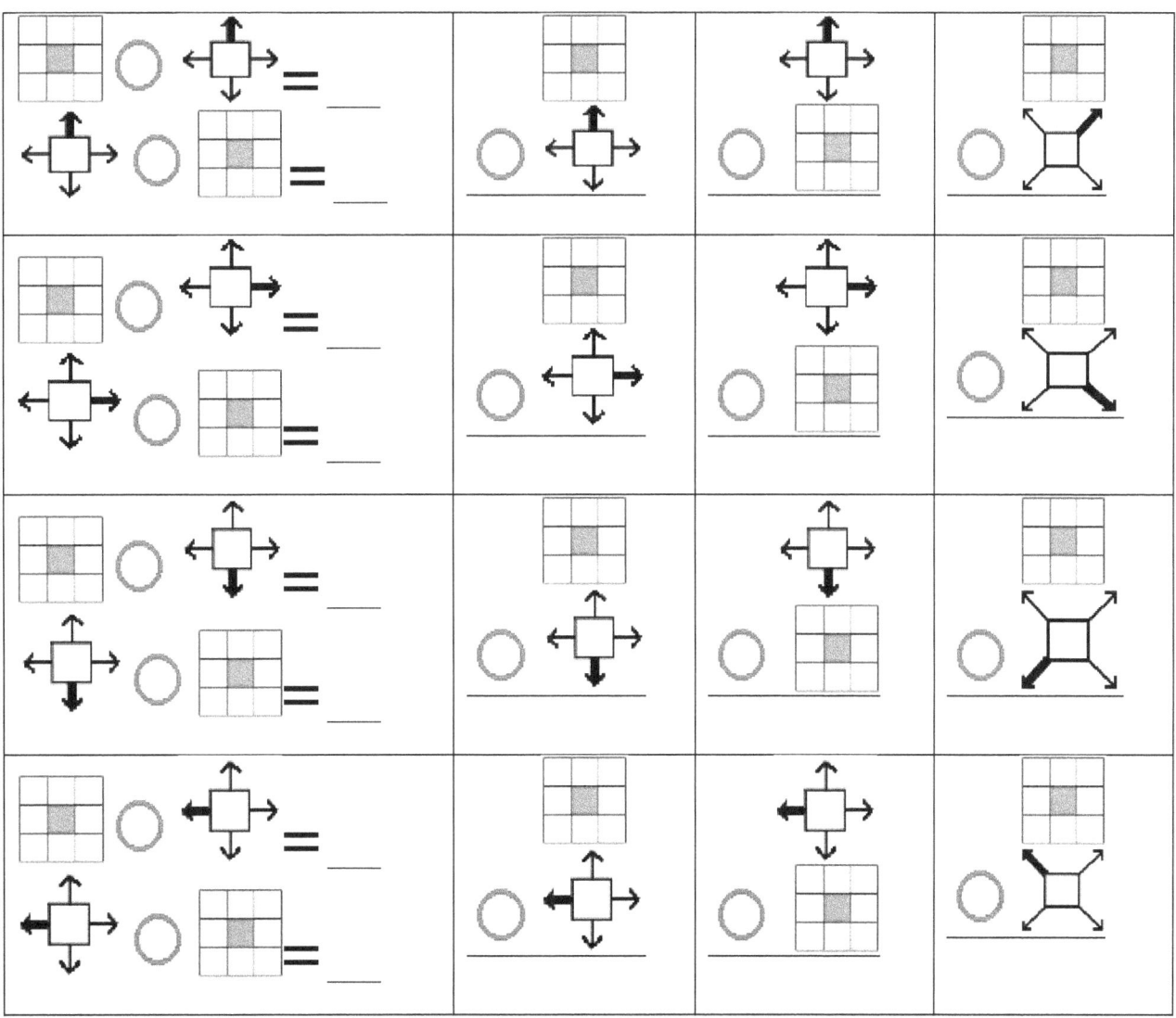

Ho Math Chess — Pre-K and Kindergarten Math

何数棋谜　棋谜式幼儿健脑思维趣味数学

© 2012 – 2021 Frank Ho, Amanda Ho, Canada copyright 1095661, Trademark 771400

Adding up to 9

You are a chess piece located at a square indicated by a shaded square.

4	4	4	5
5	5	5	5
6	6	6	6
6	6	6	4

answer

○ = +

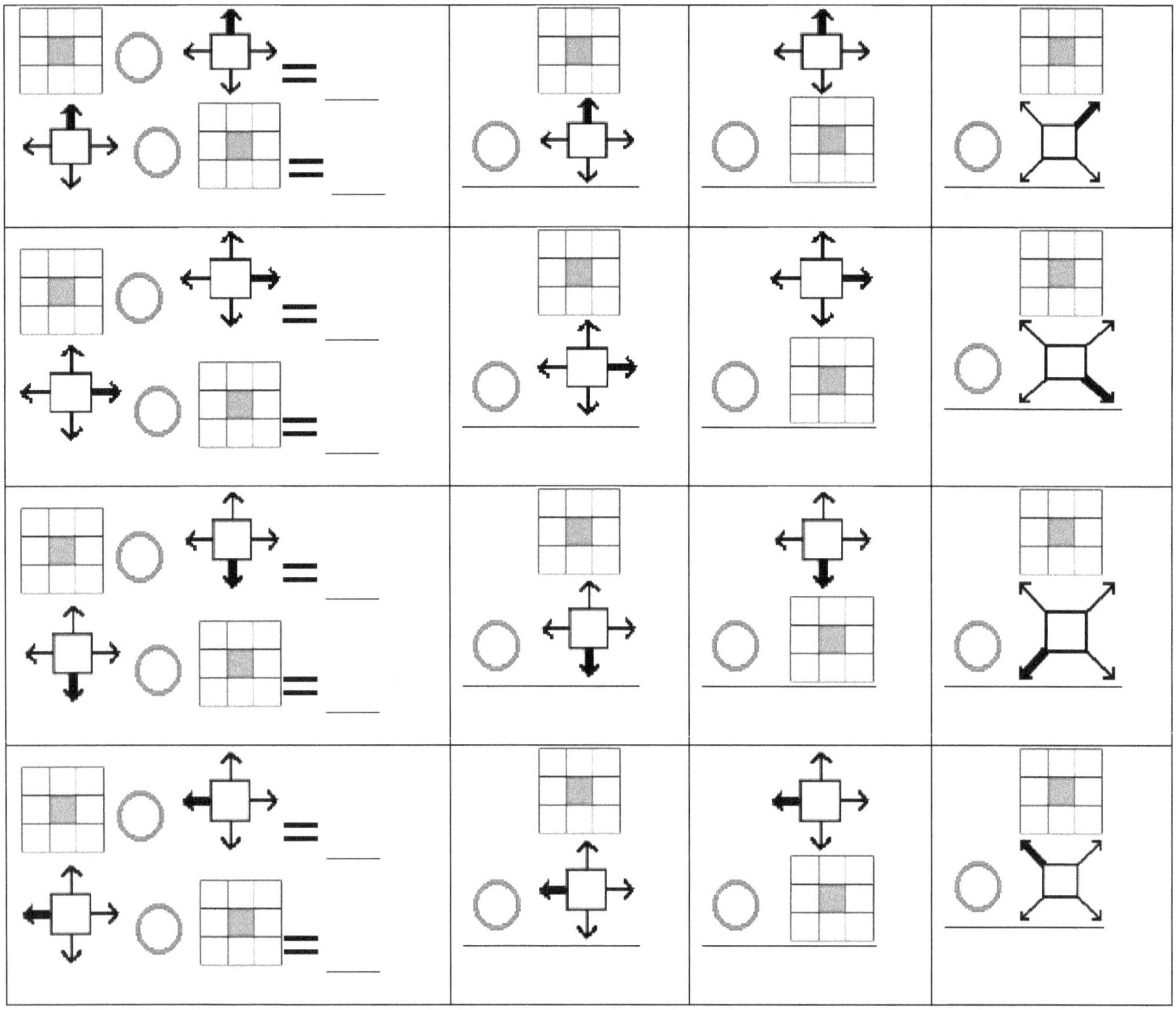

Adding up to 9

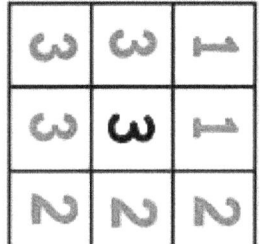

You are a chess piece located at a square indicated by a shaded square.

6	6	6	4
4	4	4	5
5	5	5	5
6	6	6	6

answer

◯ = +

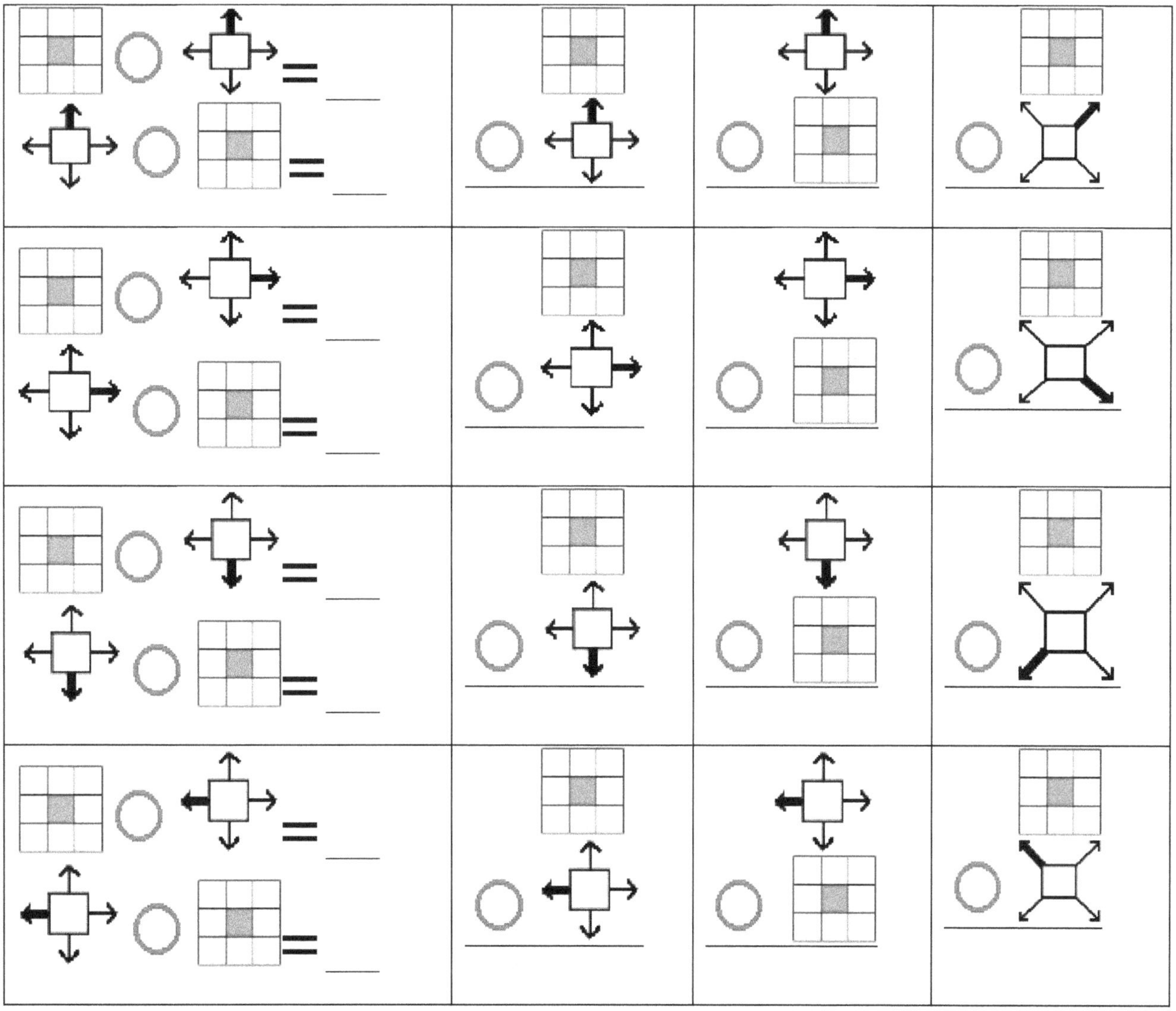

Ho Math Chess — Pre-K and Kindergarten Math

何数棋谜　棋谜式幼儿健脑思维趣味数学

© 2012 – 2021 Frank Ho, Amanda Ho, Canada copyright 1095661, Trademark 771400

Adding up to 9

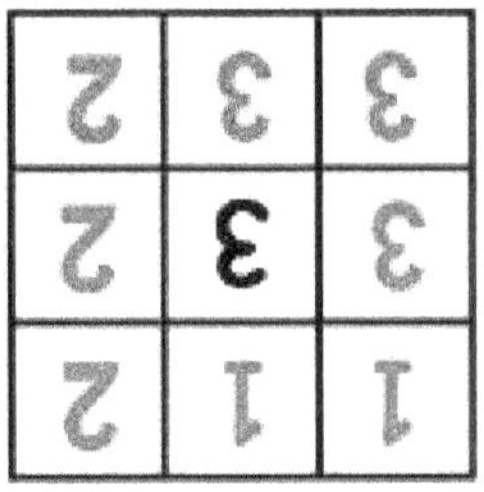

You are a chess piece located at a square indicated by a shaded square.

6	6	6	6
6	6	6	4
6	6	6	5
5	5	5	5

answer

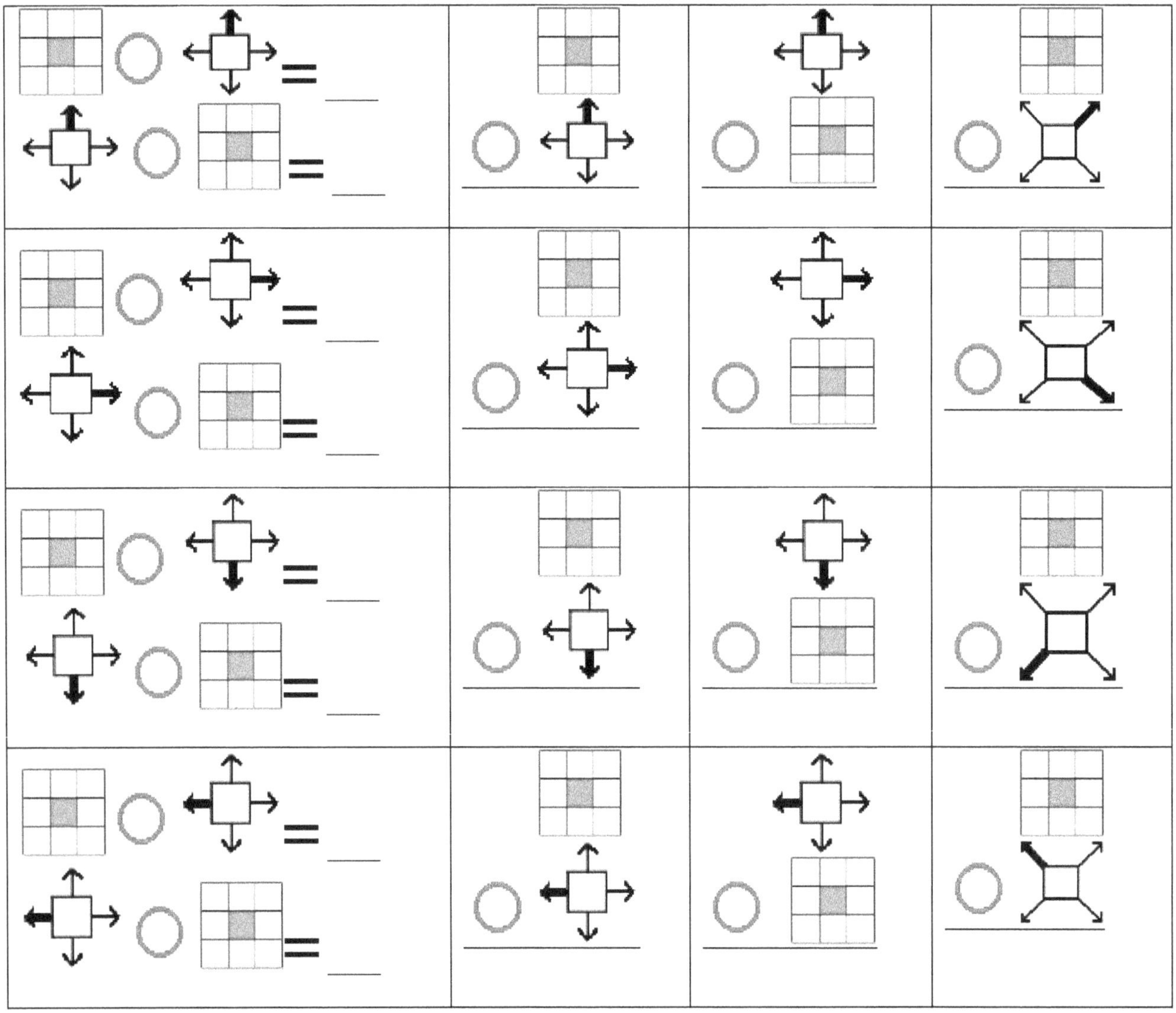

◯ = +

Ho Math Chess — Pre-K and Kindergarten Math

Adding up to 9

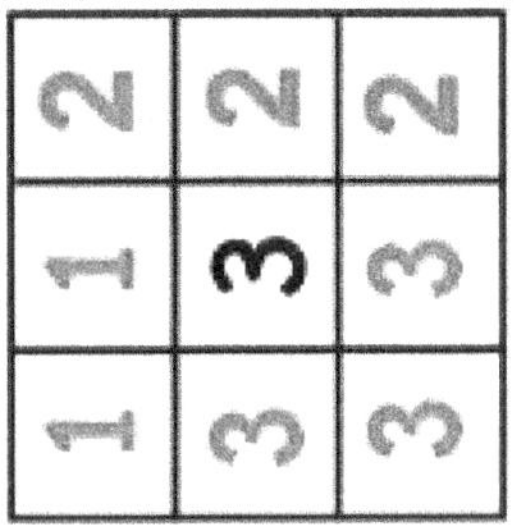

You are a chess piece located at a square indicated by a shaded square.

5	5	5	5
6	6	6	6
6	6	6	4
4	4	4	5

answer

○ = +

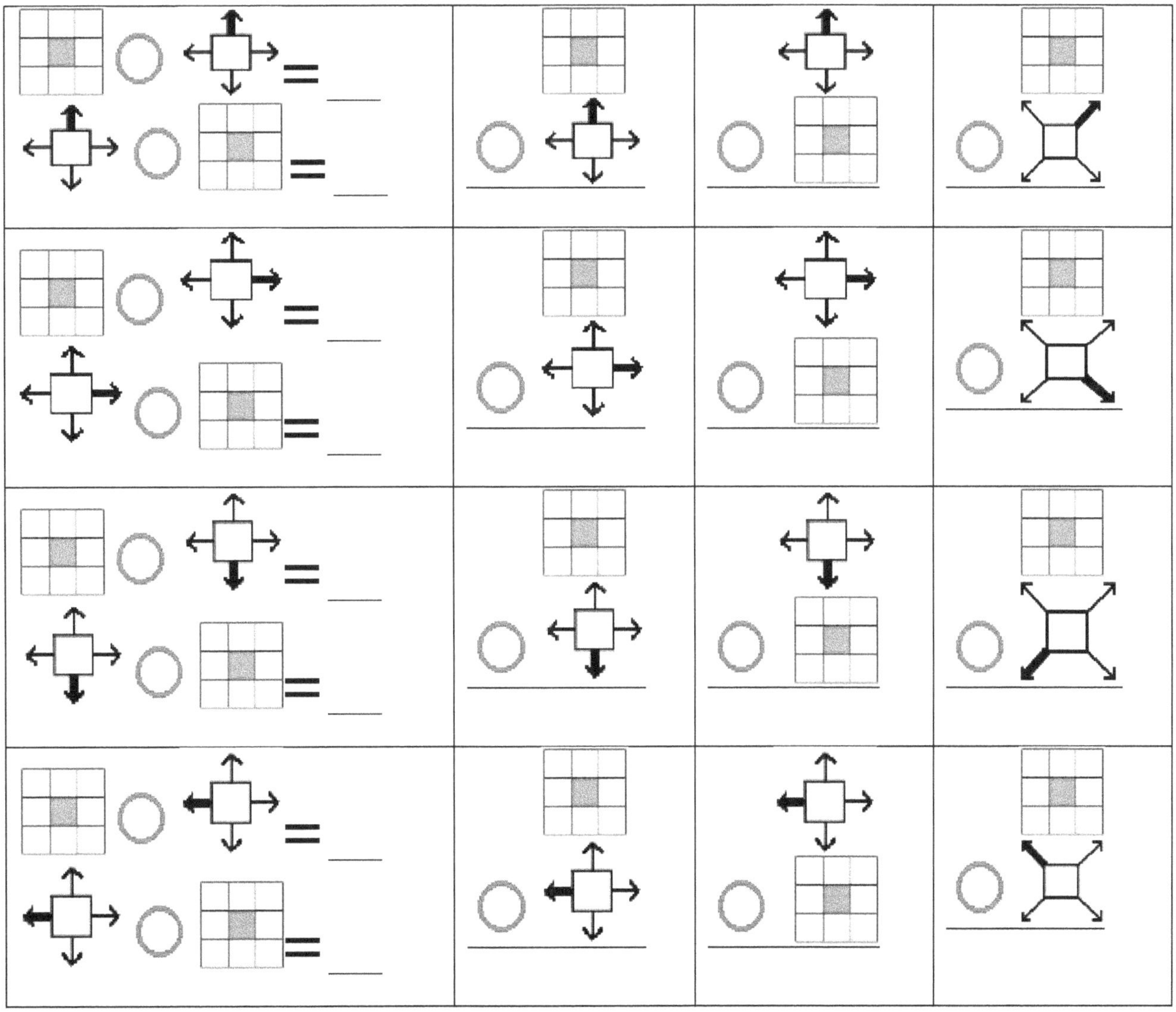

Ho Math Chess — Pre-K and Kindergarten Math

Adding up to 9

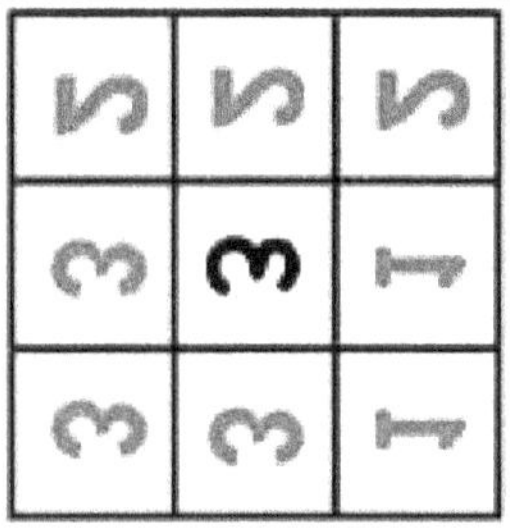

You are a chess piece located at a square indicated by a shaded square.

5	5	5	5
4	4	4	4
6	6	6	5
6	6	6	5

answer

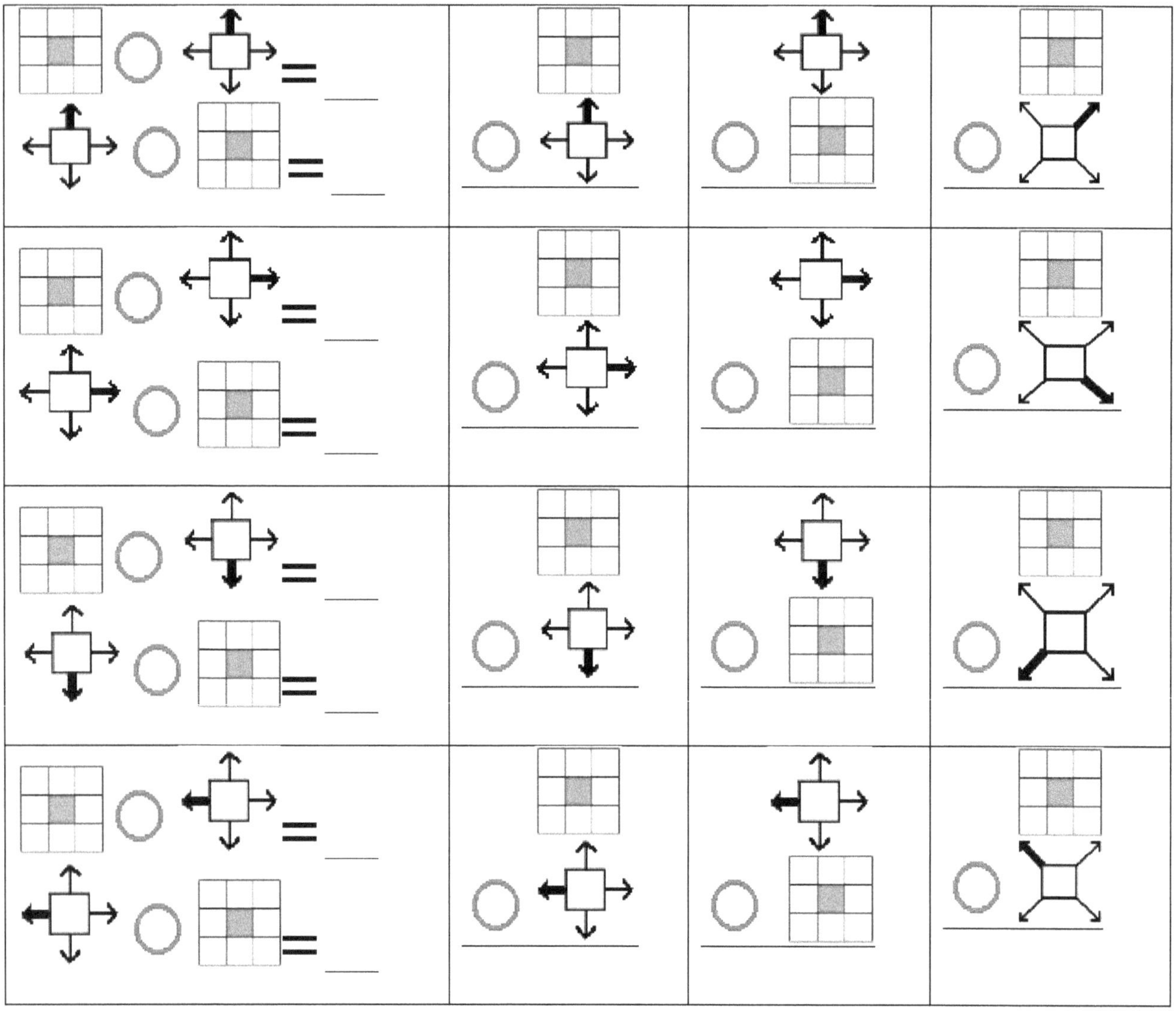

○ = +

Ho Math Chess — Pre-K and Kindergarten Math

Adding up to 9

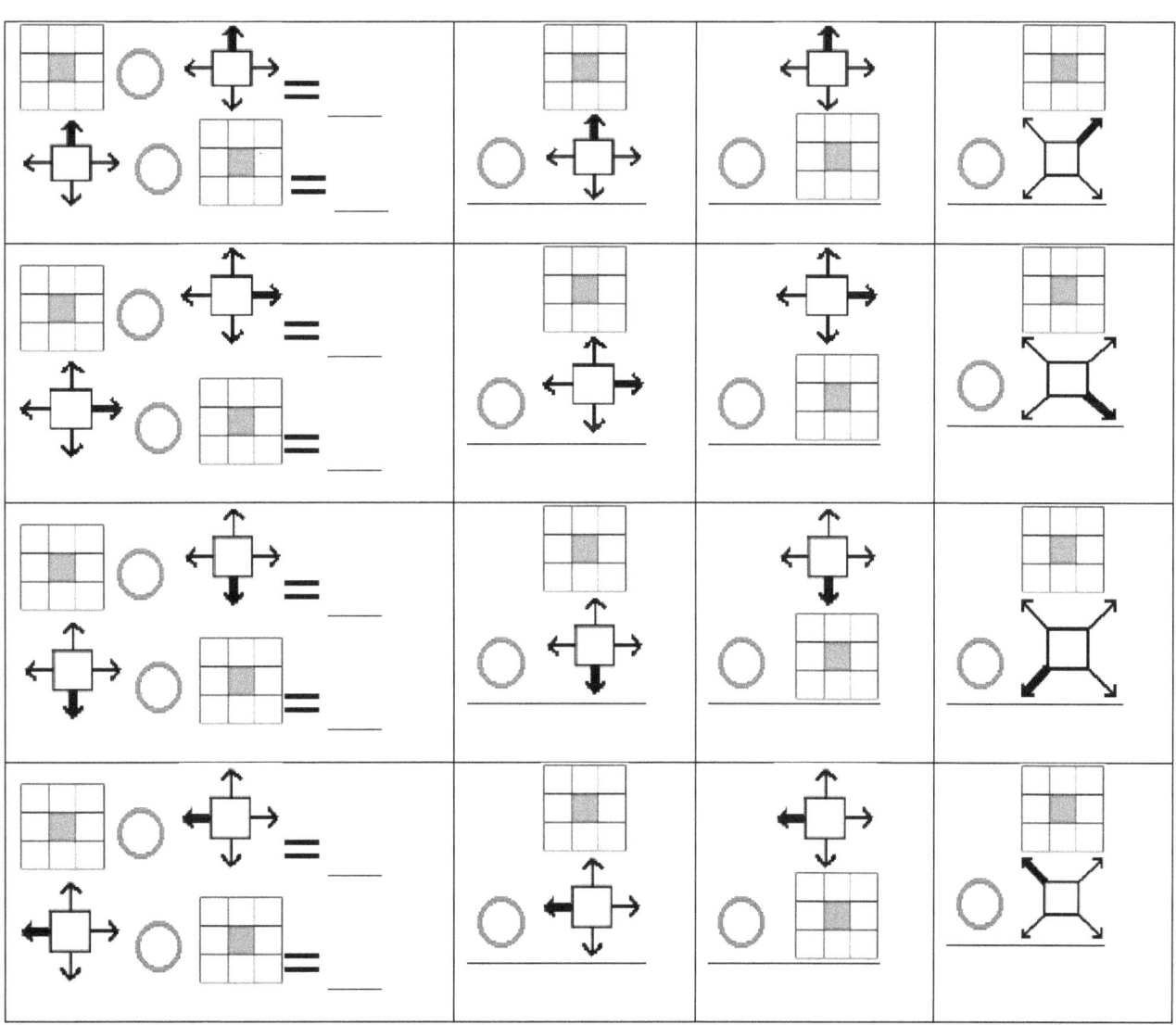

Ho Math Chess — Pre-K and Kindergarten Math

何数棋谜　棋谜式幼儿健脑思维趣味数学

© 2012 – 2021 Frank Ho, Amanda Ho, Canada copyright 1095661, Trademark 771400

Adding up to 9

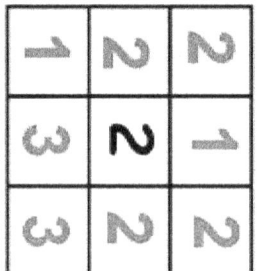

You are a chess piece located at a square indicated by a shaded square.

4	4	4	4
3	3	3	4
4	4	4	5
5	5	5	3

answer

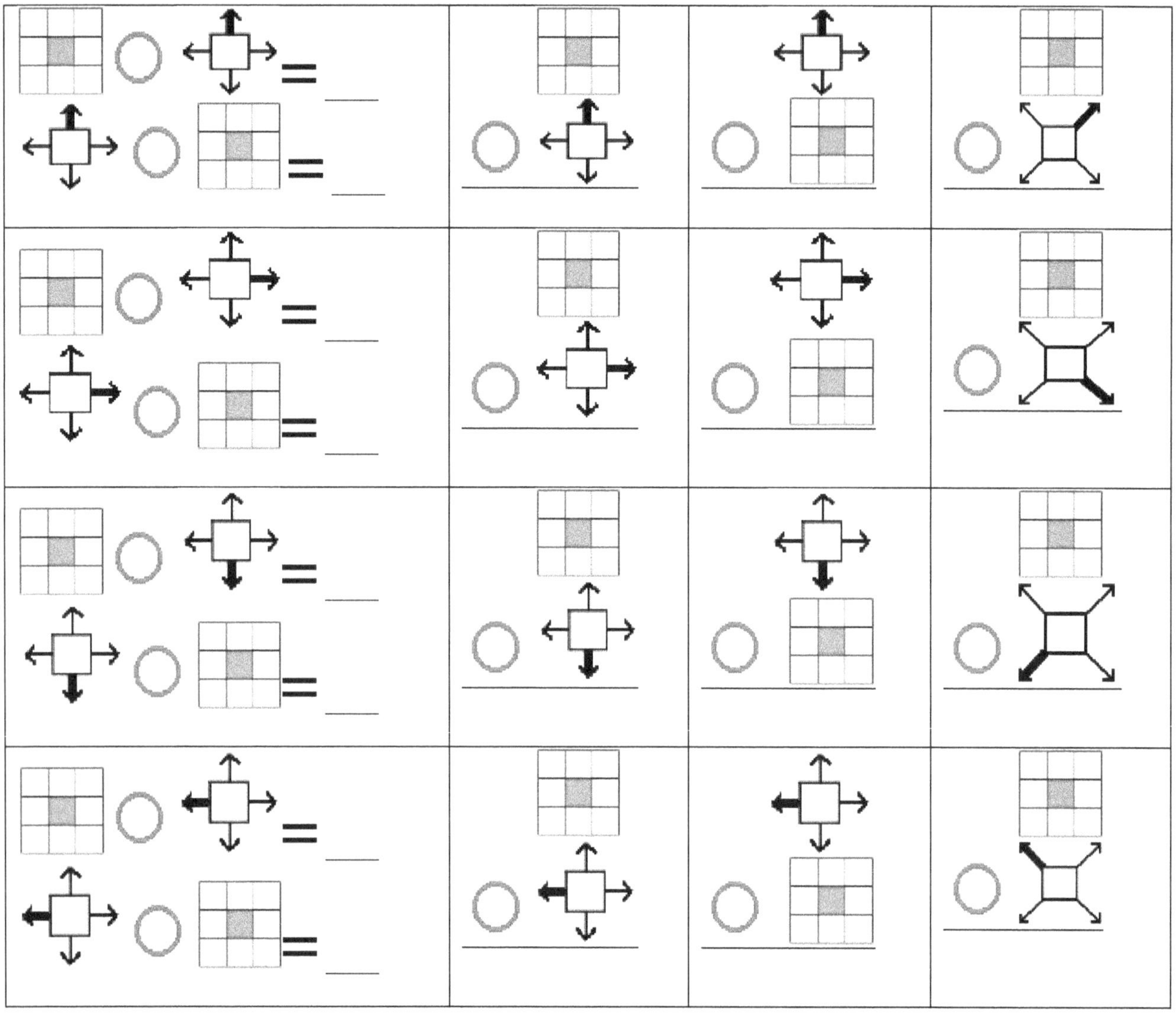

○ = +

Adding up to 9

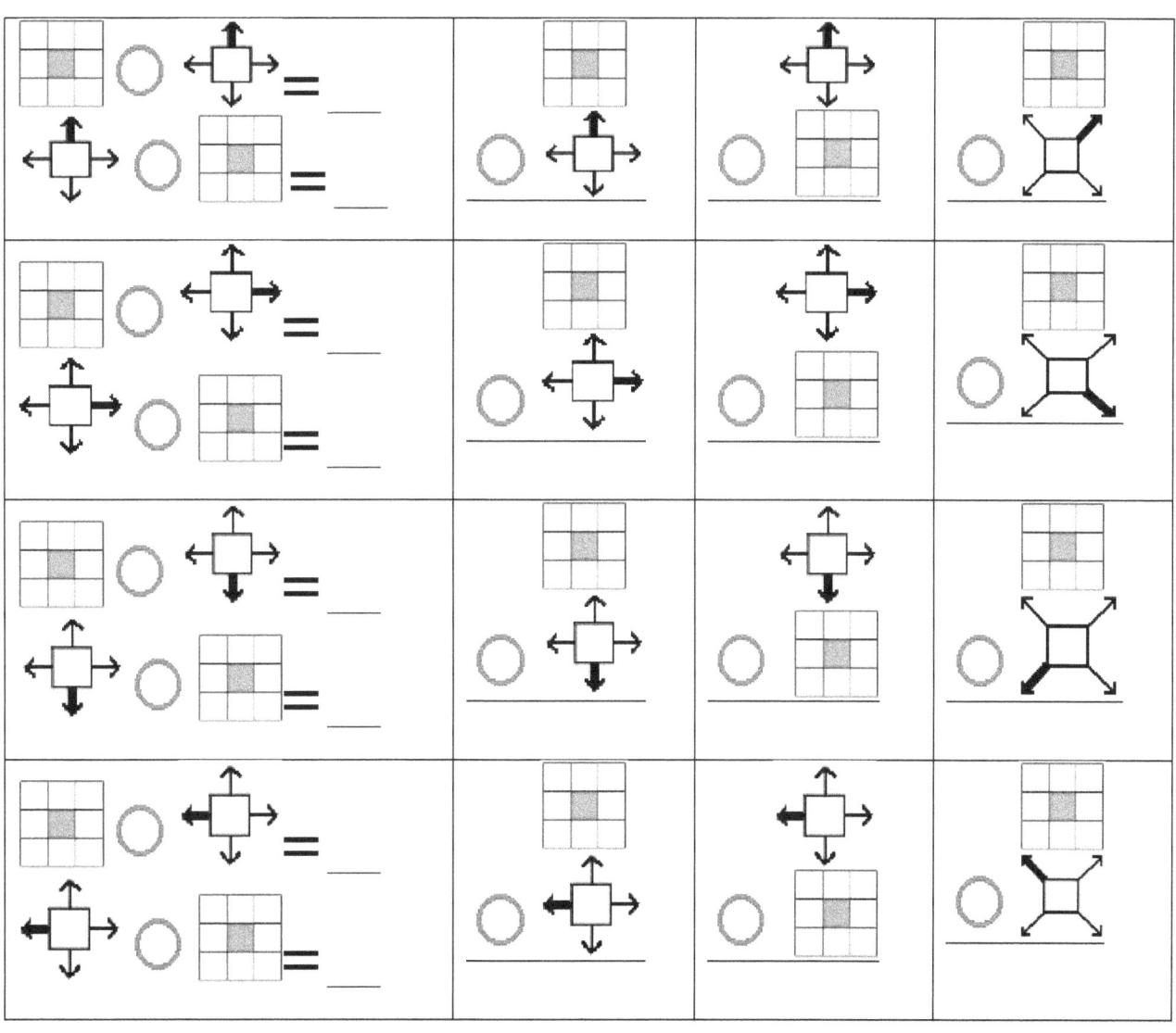

Ho Math Chess — Pre-K and Kindergarten Math

Adding up to 9

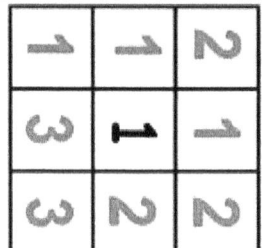

You are a chess piece located at a square indicated by a shaded square.

2	2	2	3
2	2	2	3
3	3	3	4
4	4	4	2

answer

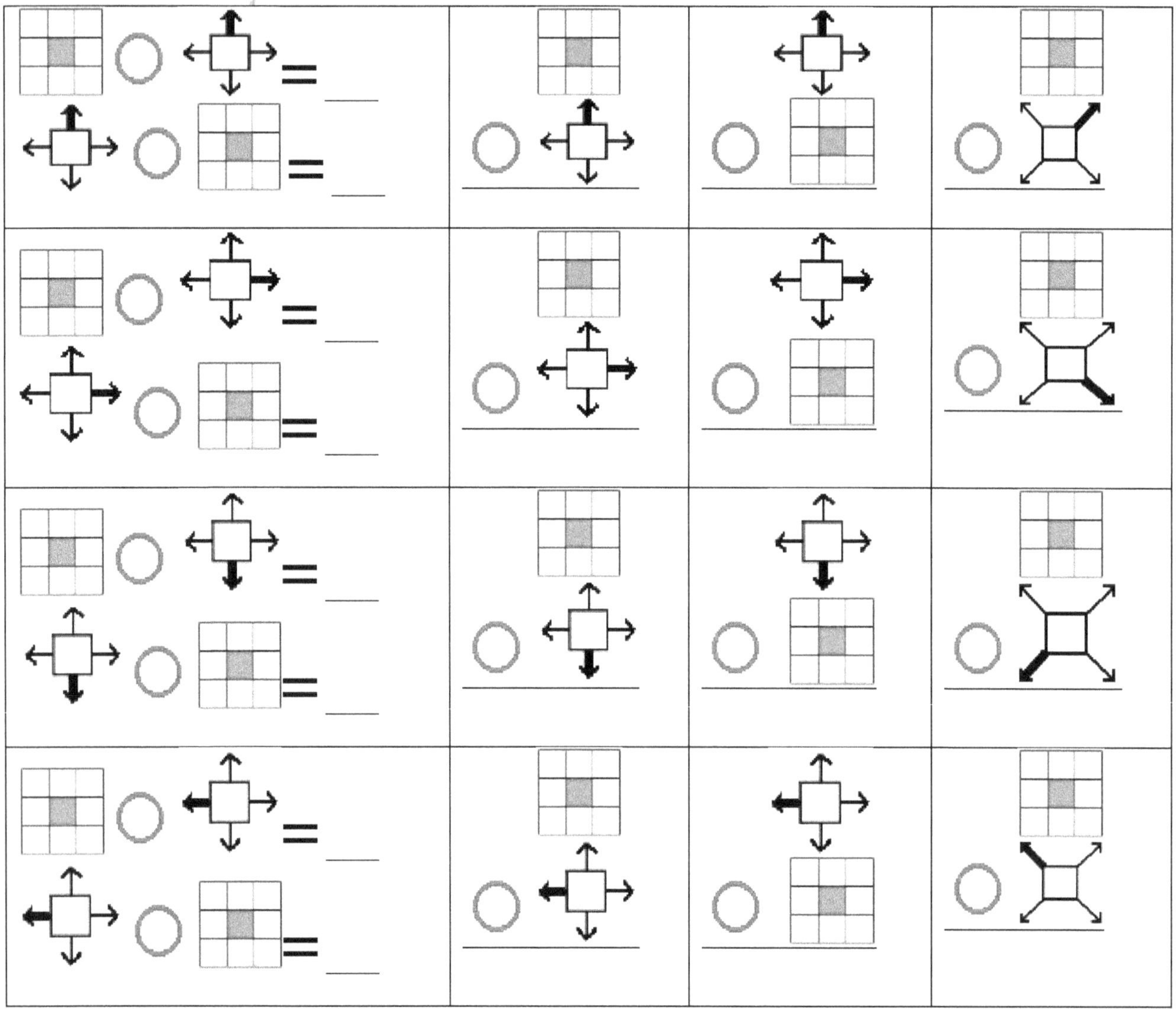

○ = +

Ho Math Chess — Pre-K and Kindergarten Math

Adding up to 9

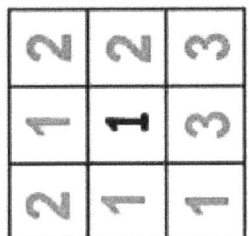

You are a chess piece located at a square indicated by a shaded square.

3	3	3	4
4	4	4	2
2	2	2	3
2	2	2	3

answer

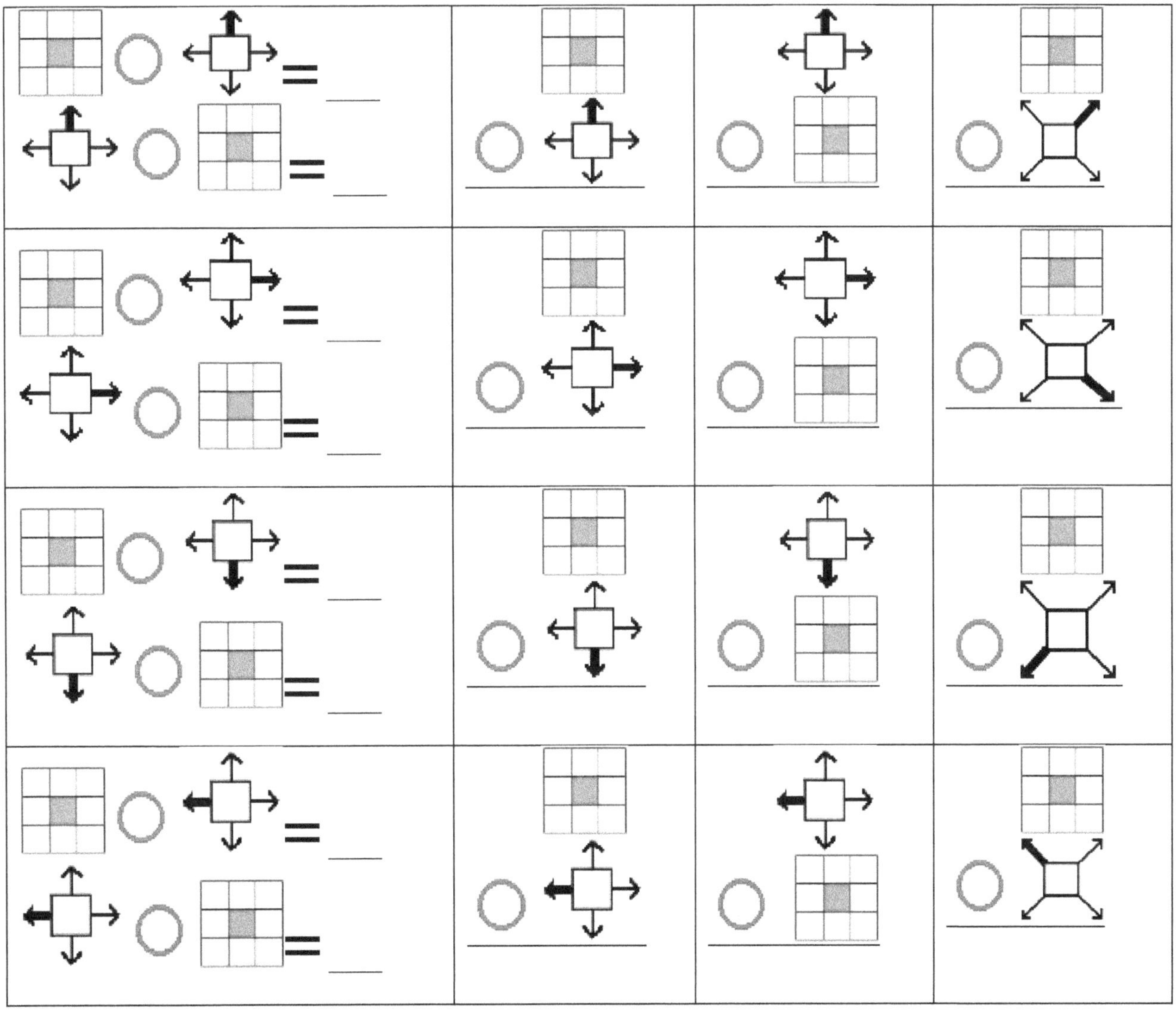

Adding up to 9

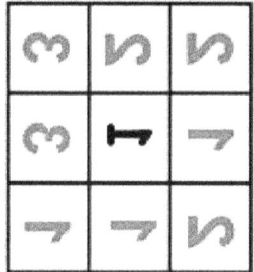

You are a chess piece located at a square indicated by a shaded square.

3	3	3	3
2	2	2	3
2	2	2	2
4	4	4	4

answer

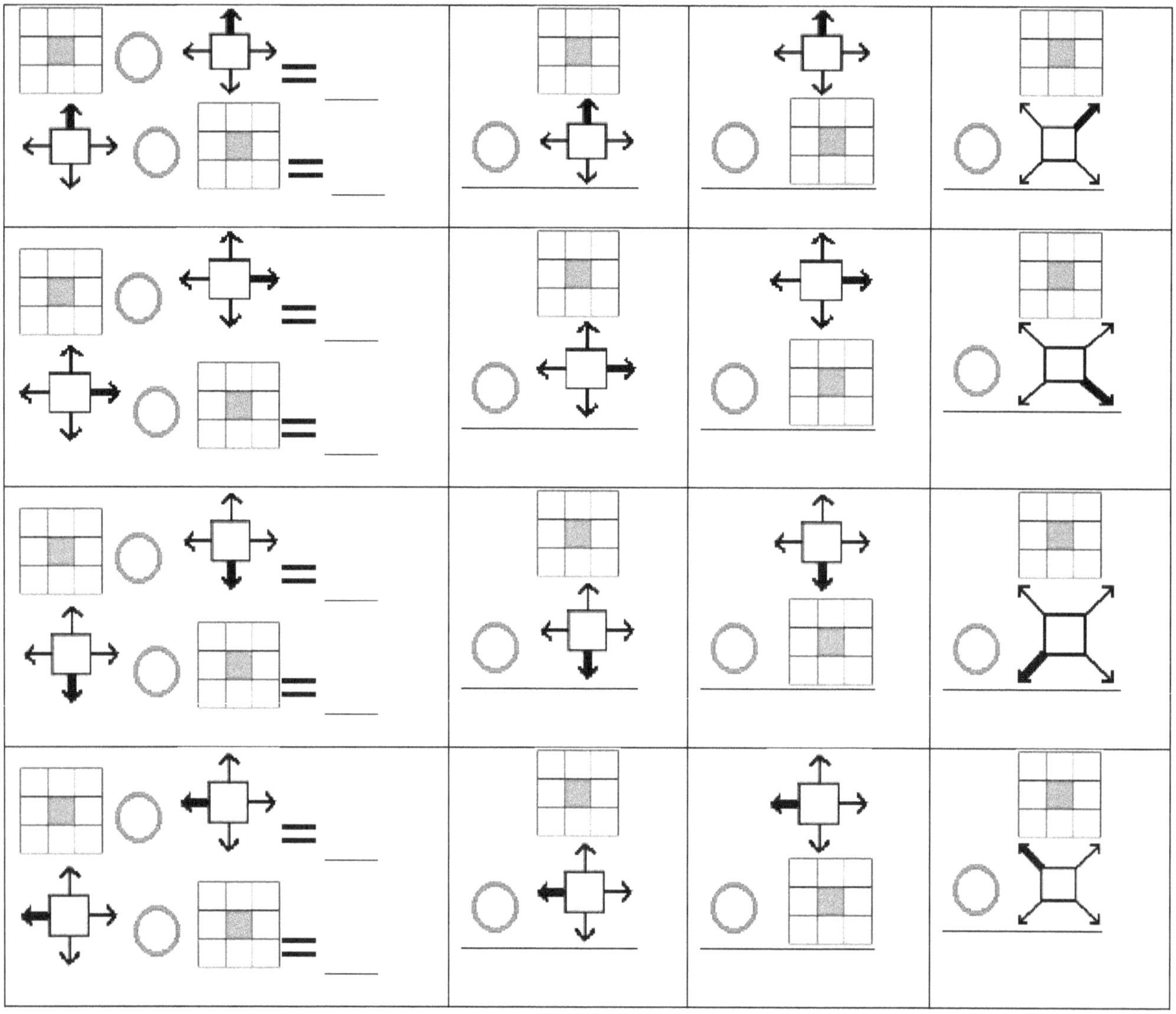

Adding up to 9

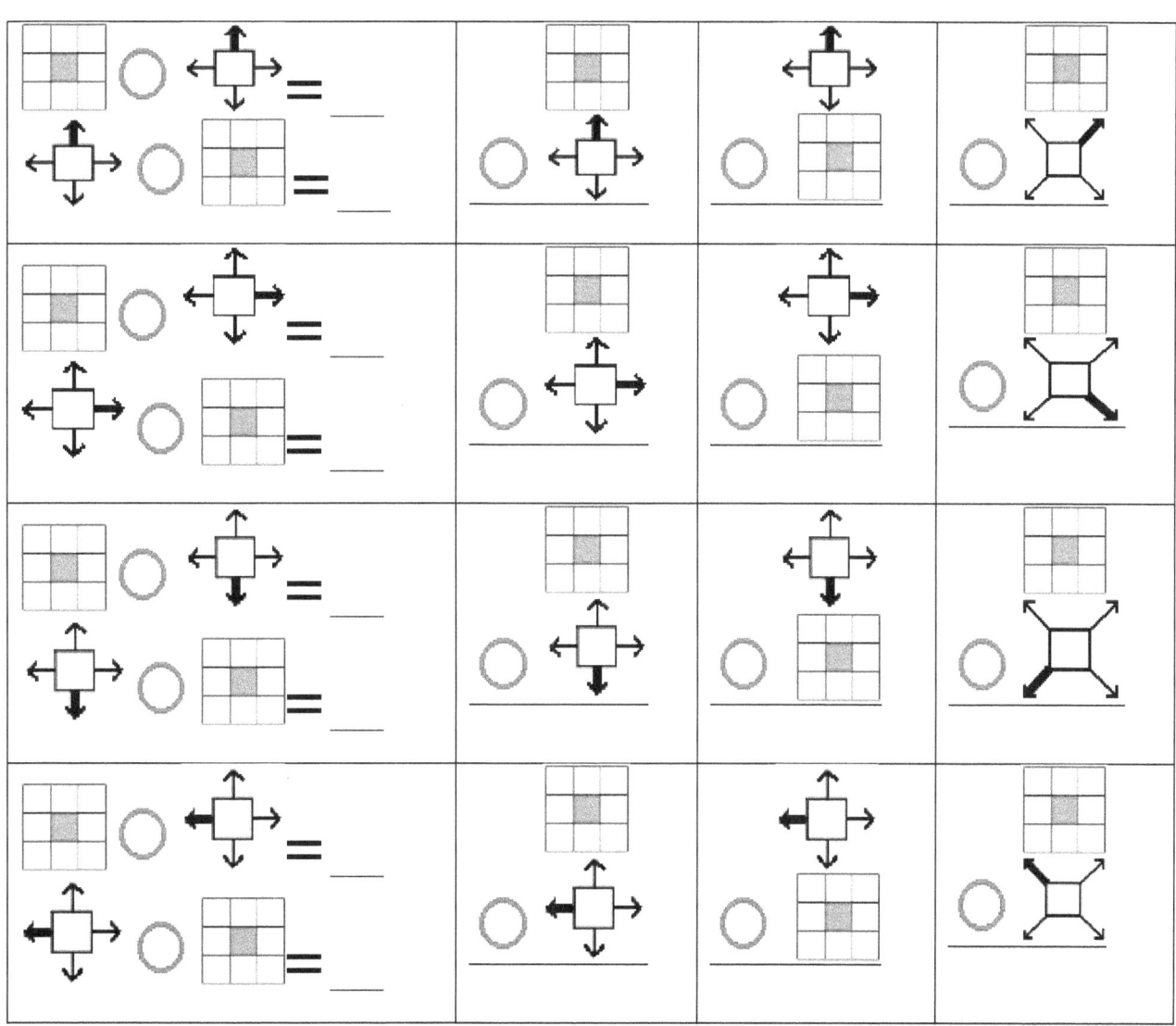

Ho Math Chess — Pre-K and Kindergarten Math

何数棋谜　棋谜式幼儿健脑思维趣味数学

© 2012 – 2021 Frank Ho, Amanda Ho, Canada copyright 1095661, Trademark 771400

Adding up to 9

You are a chess piece located at a square indicated by a shaded square.

4	4	4	5
5	5	5	6
6	6	6	4
5	5	5	5

answer

○ = +

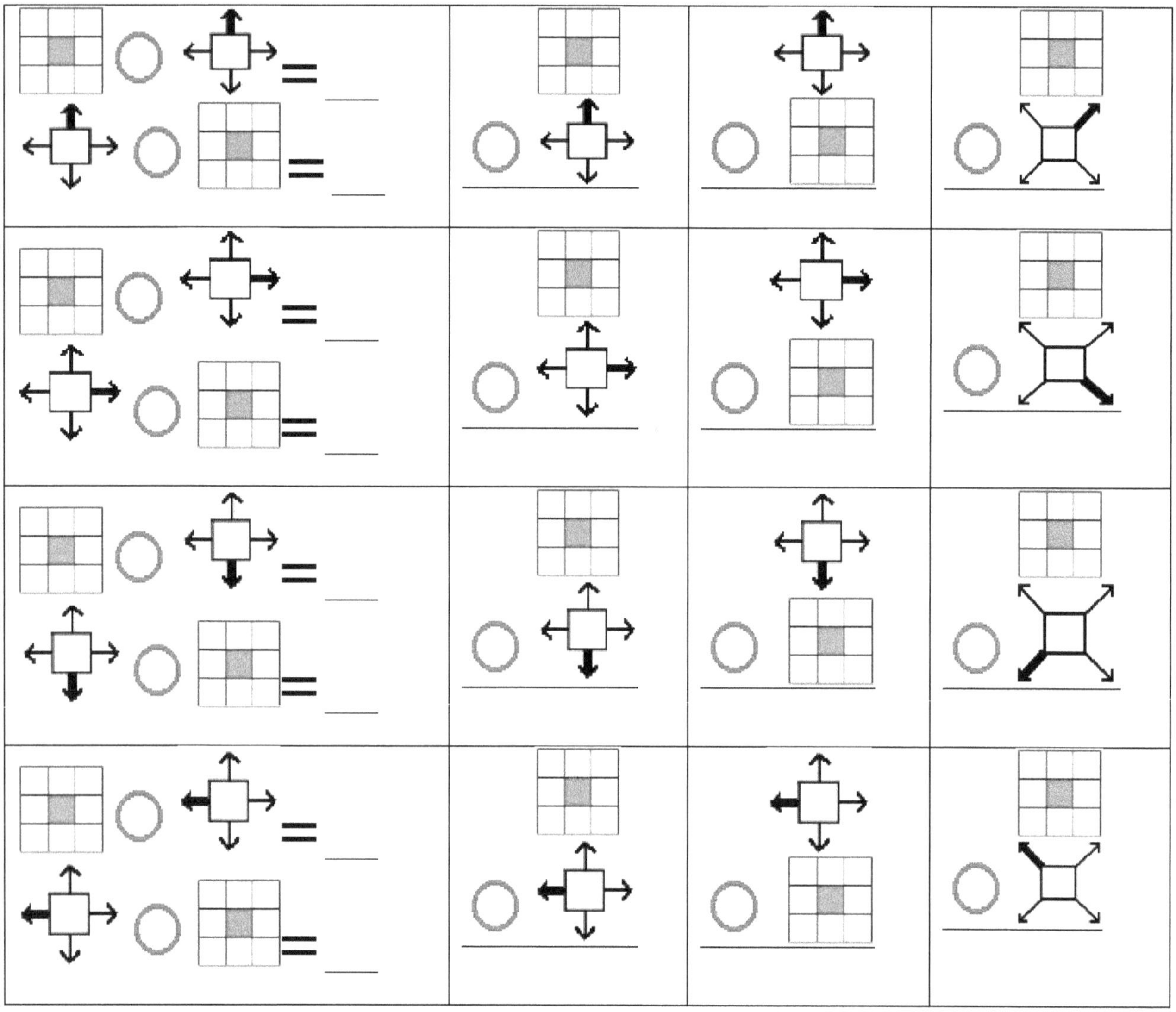

www.homathchess.com   328

Adding up to 9

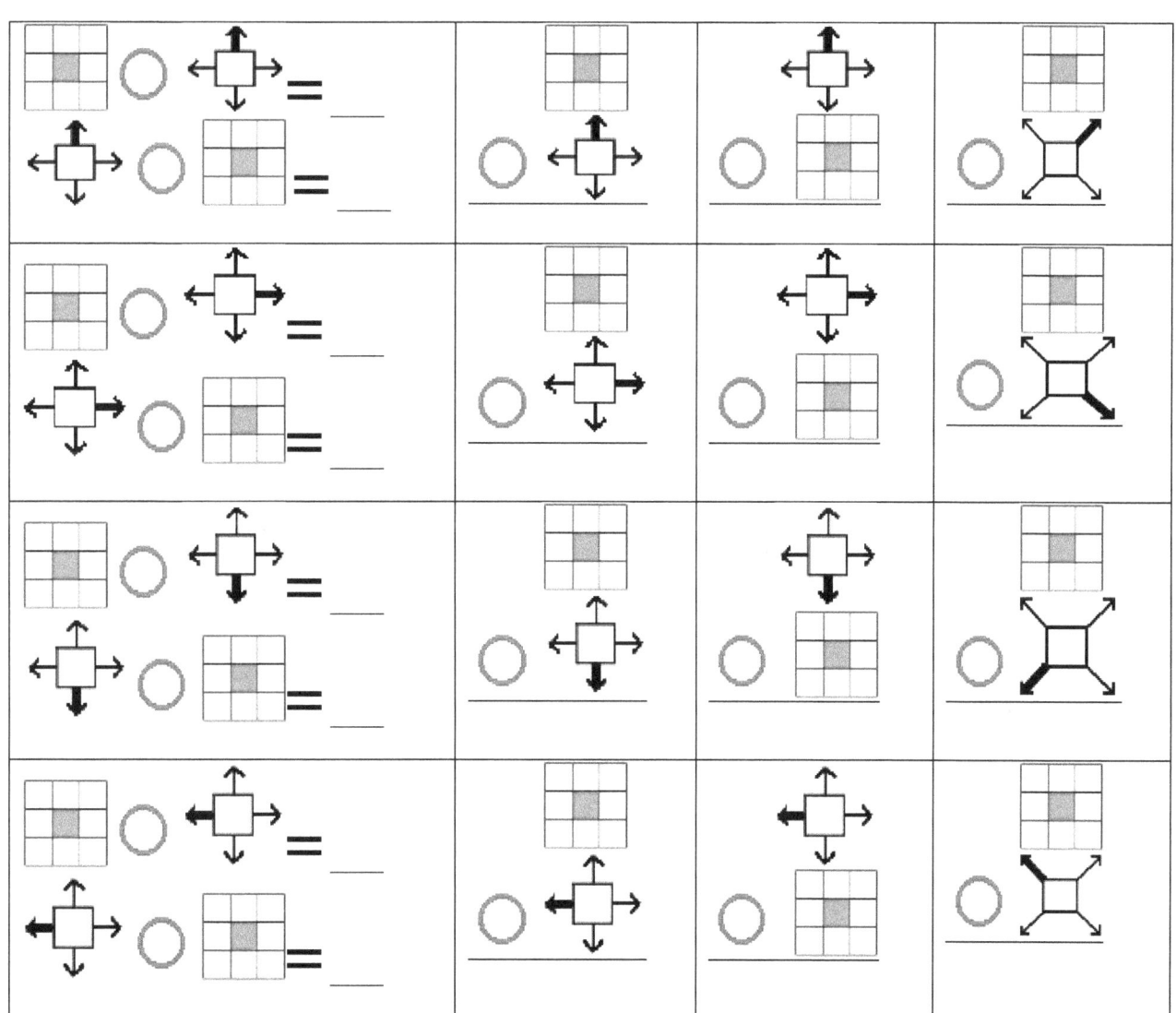

Ho Math Chess — Pre-K and Kindergarten Math

何数棋谜　棋谜式幼儿健脑思维趣味数学

© 2012 – 2021 Frank Ho, Amanda Ho, Canada copyright 1095661, Trademark 771400

Adding up to 9

You are a chess piece located at a square indicated by a shaded square.

6	6	6	7
7	7	7	8
8	8	8	6
7	7	7	7

answer

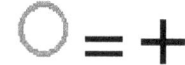

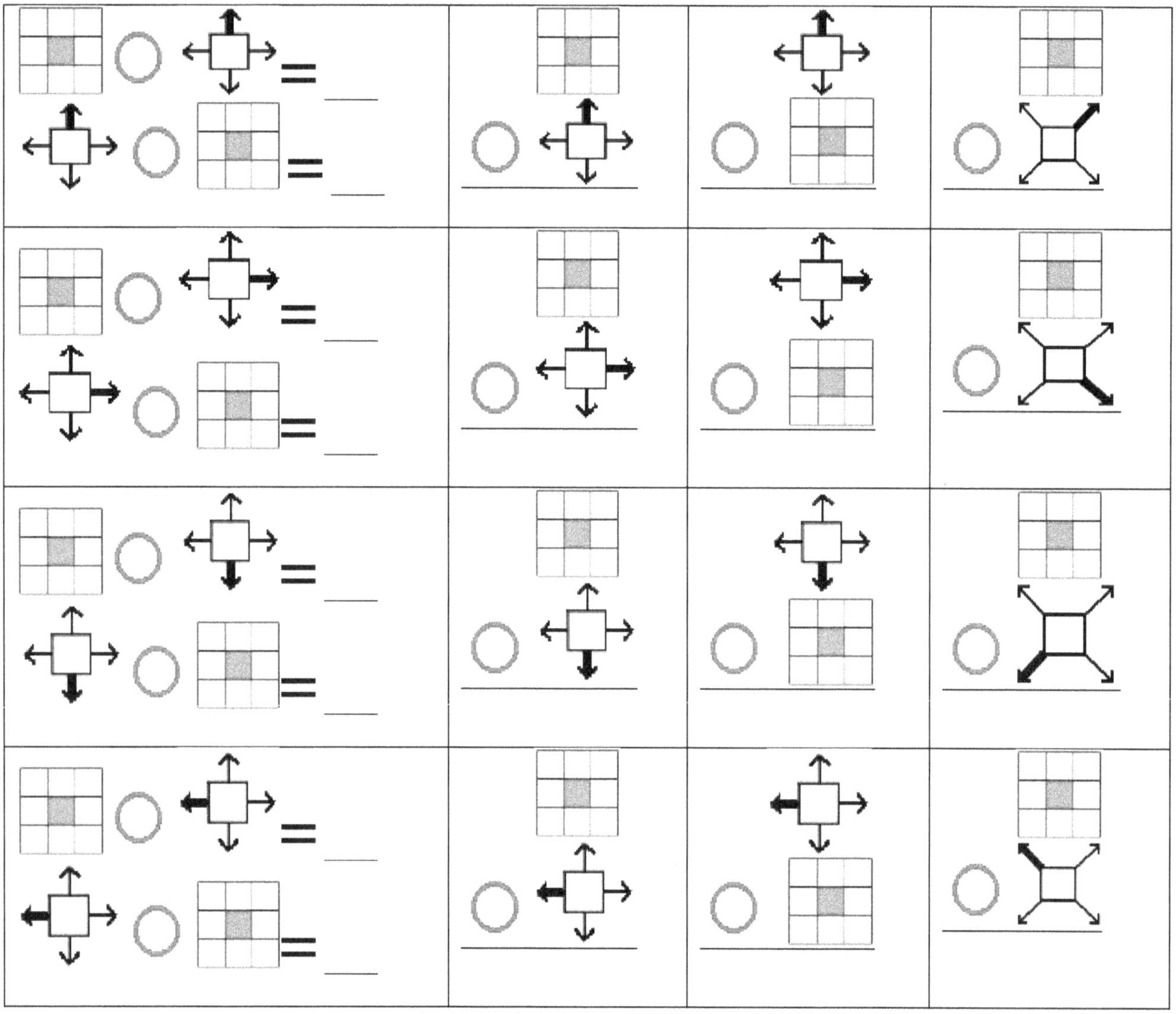

Adding up to 9

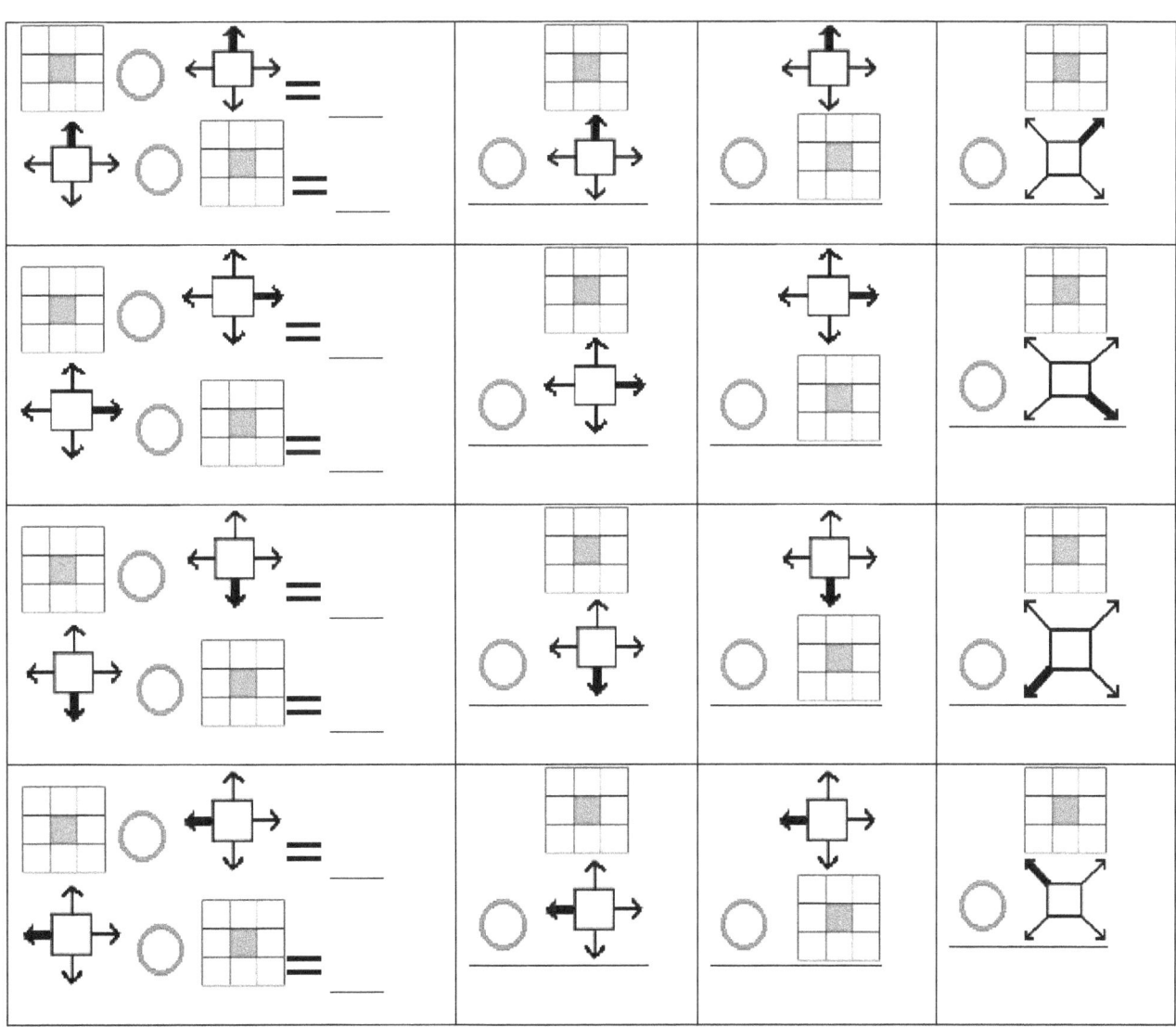

Ho Math Chess — Pre-K and Kindergarten Math

何数棋谜　棋谜式幼儿健脑思维趣味数学

© 2012 – 2021 Frank Ho, Amanda Ho, Canada copyright 1095661, Trademark 771400

Adding up to 9

You are a chess piece located at a square indicated by a shaded square.

4	4	4	5
5	5	5	6
6	6	6	4
5	5	5	5

answer

 = +

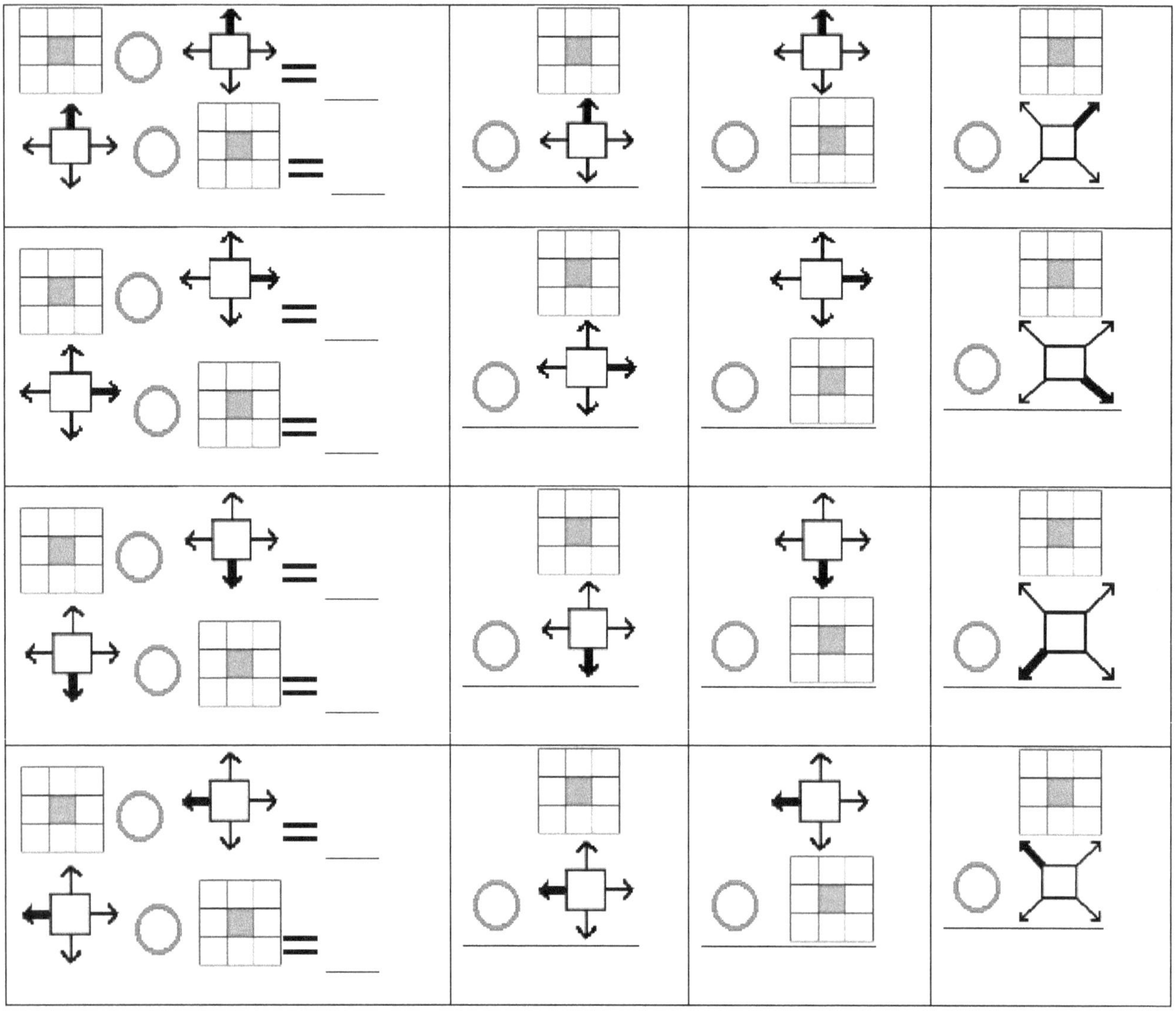

www.homathchess.com　　332

Adding up to 9

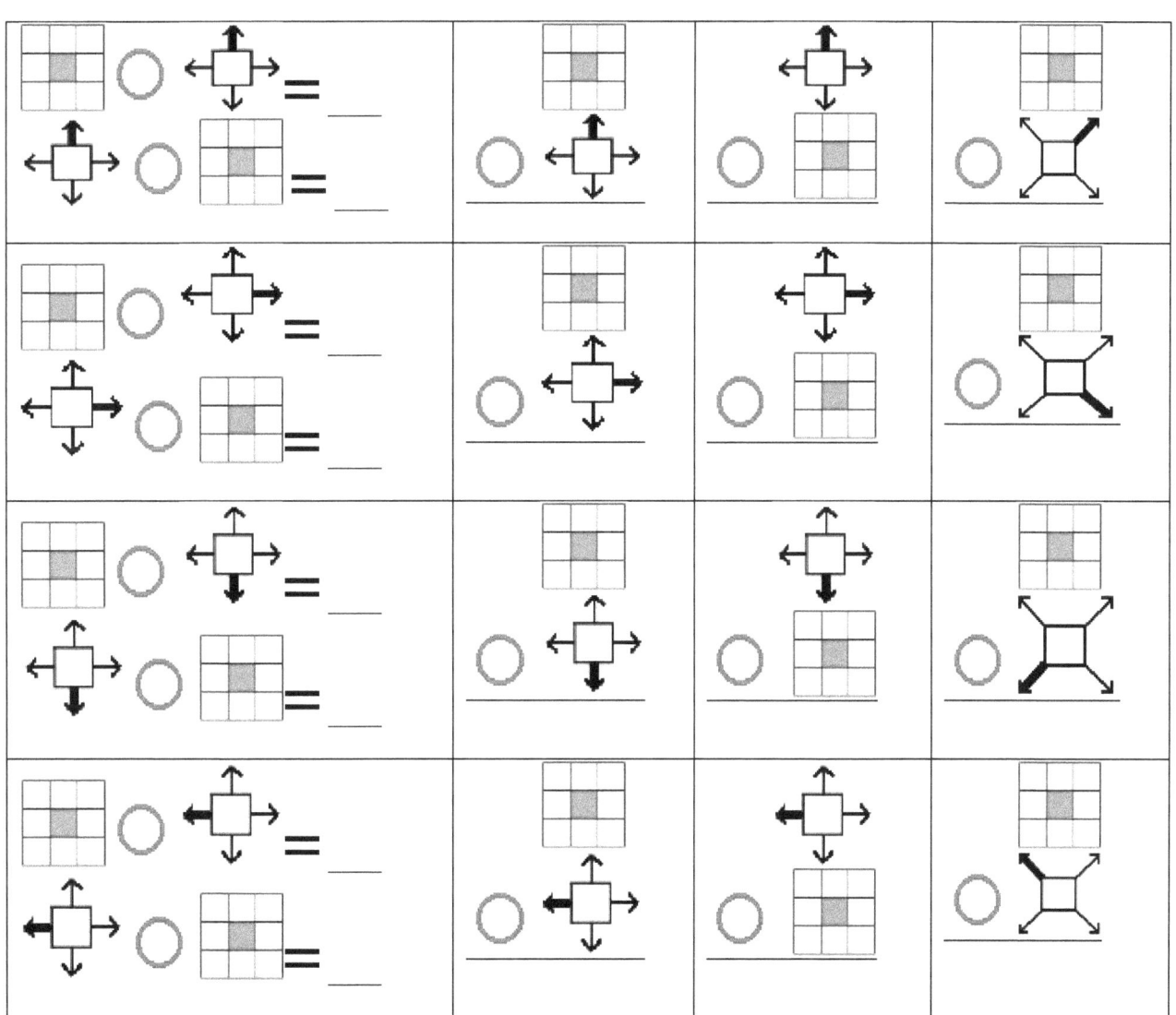

Ho Math Chess — Pre-K and Kindergarten Math

Adding up to 9

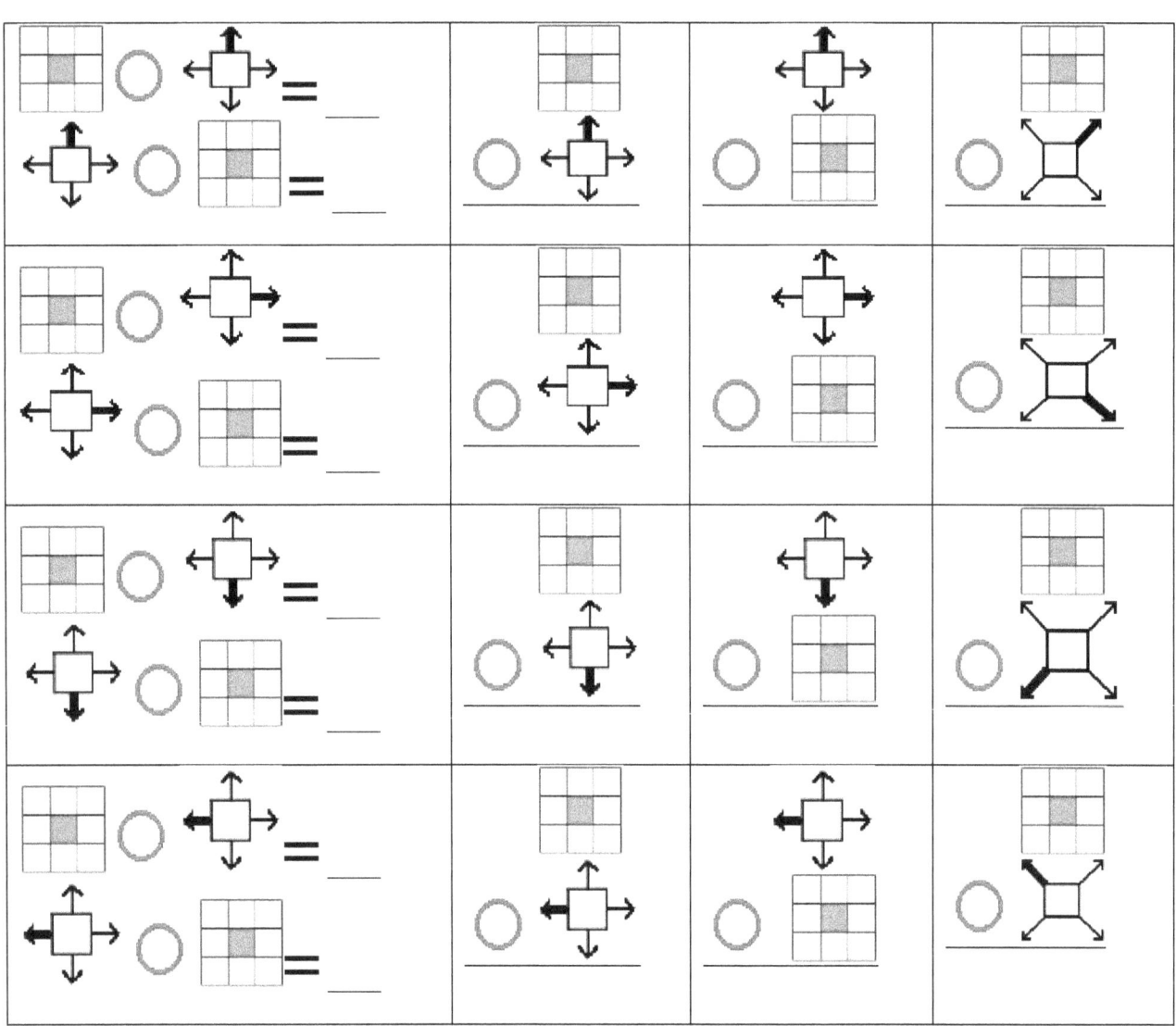

Adding up to 9

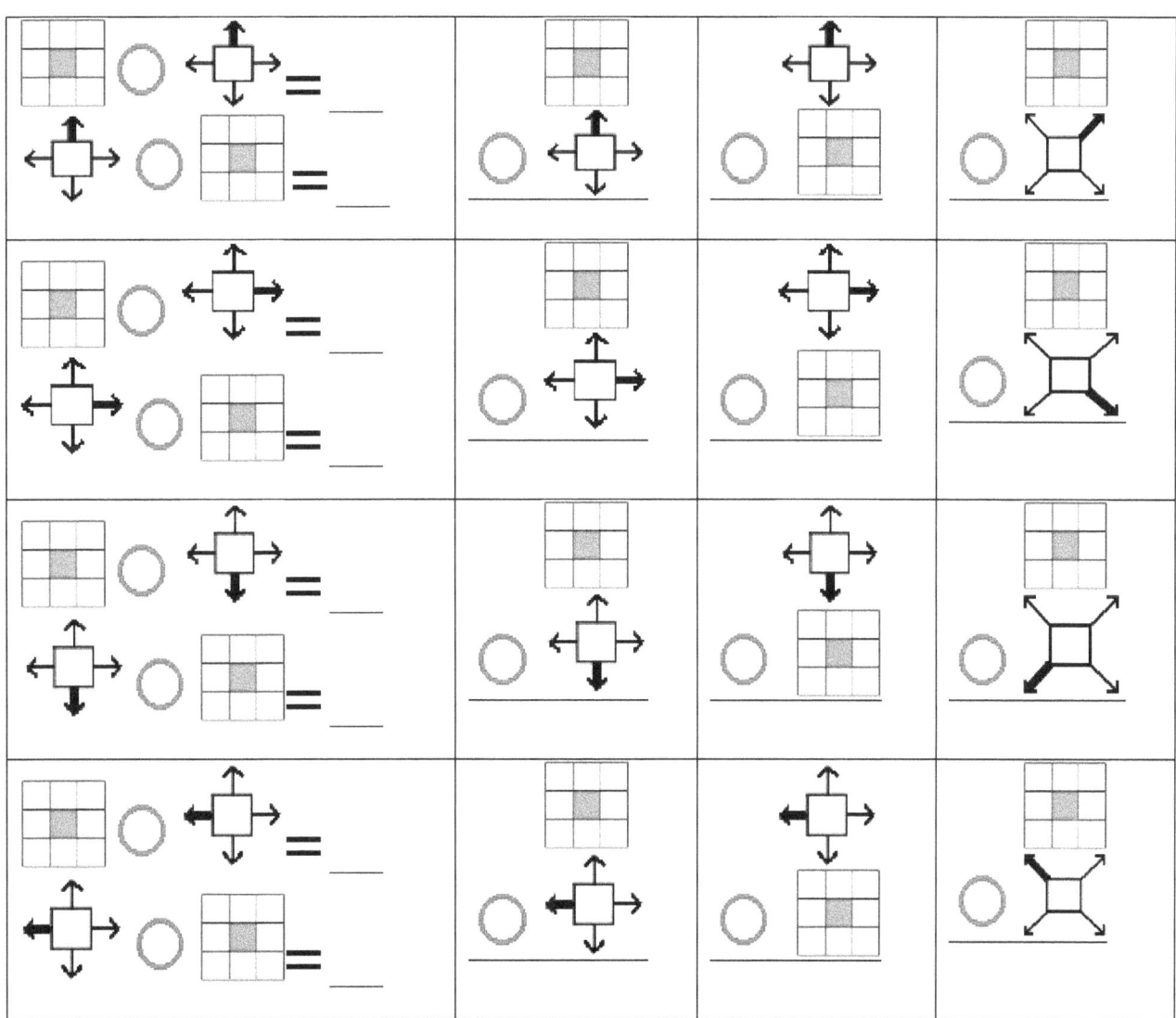

Subtracting up 9

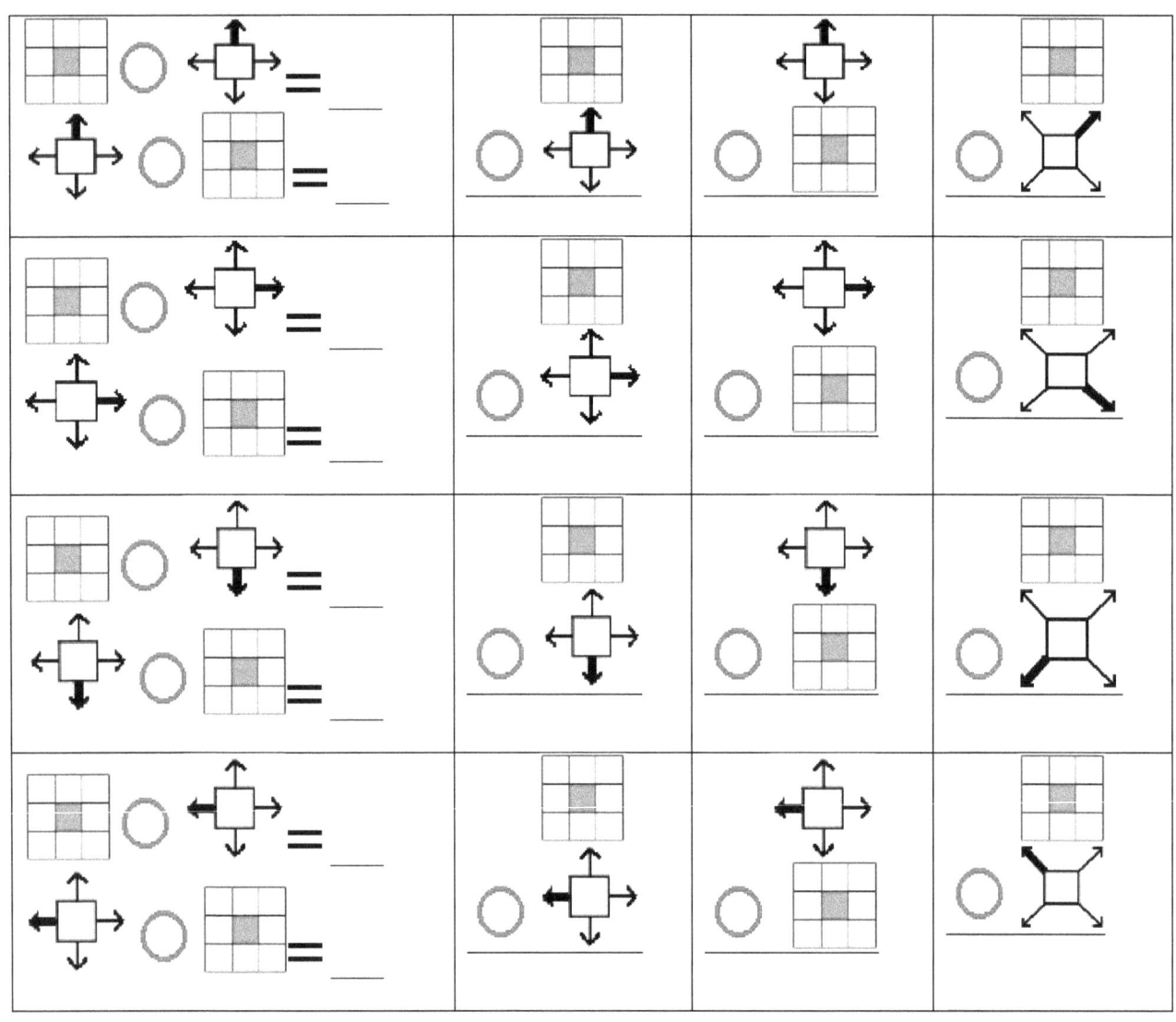

Subtracting and adding up 9

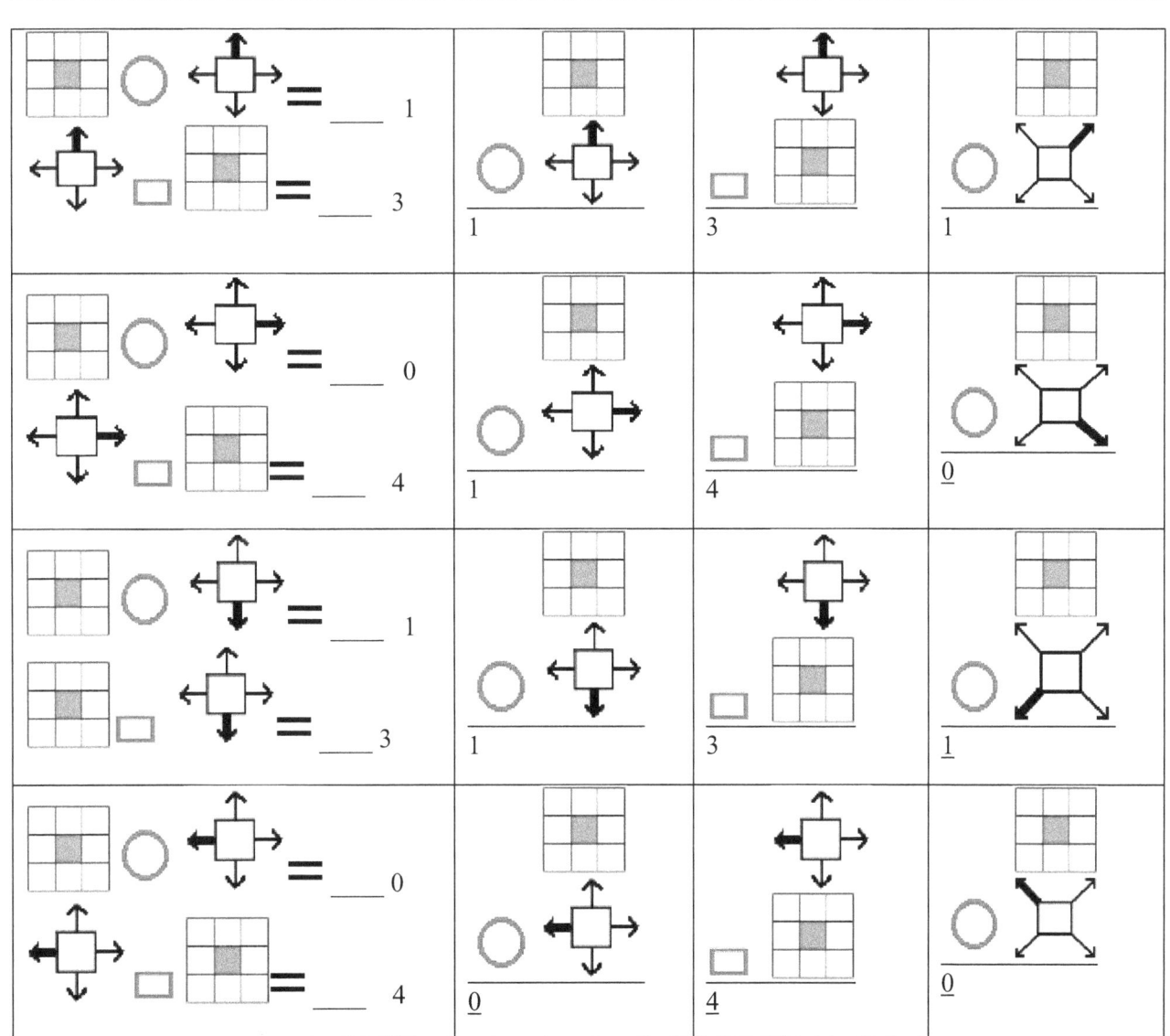

Subtracting and adding up to 9

2	1	1
2	3	2
1	1	2

You are a chess piece located at a square indicated by a shaded square.
24152415
2121
4545
2121
○ = −, □ = +

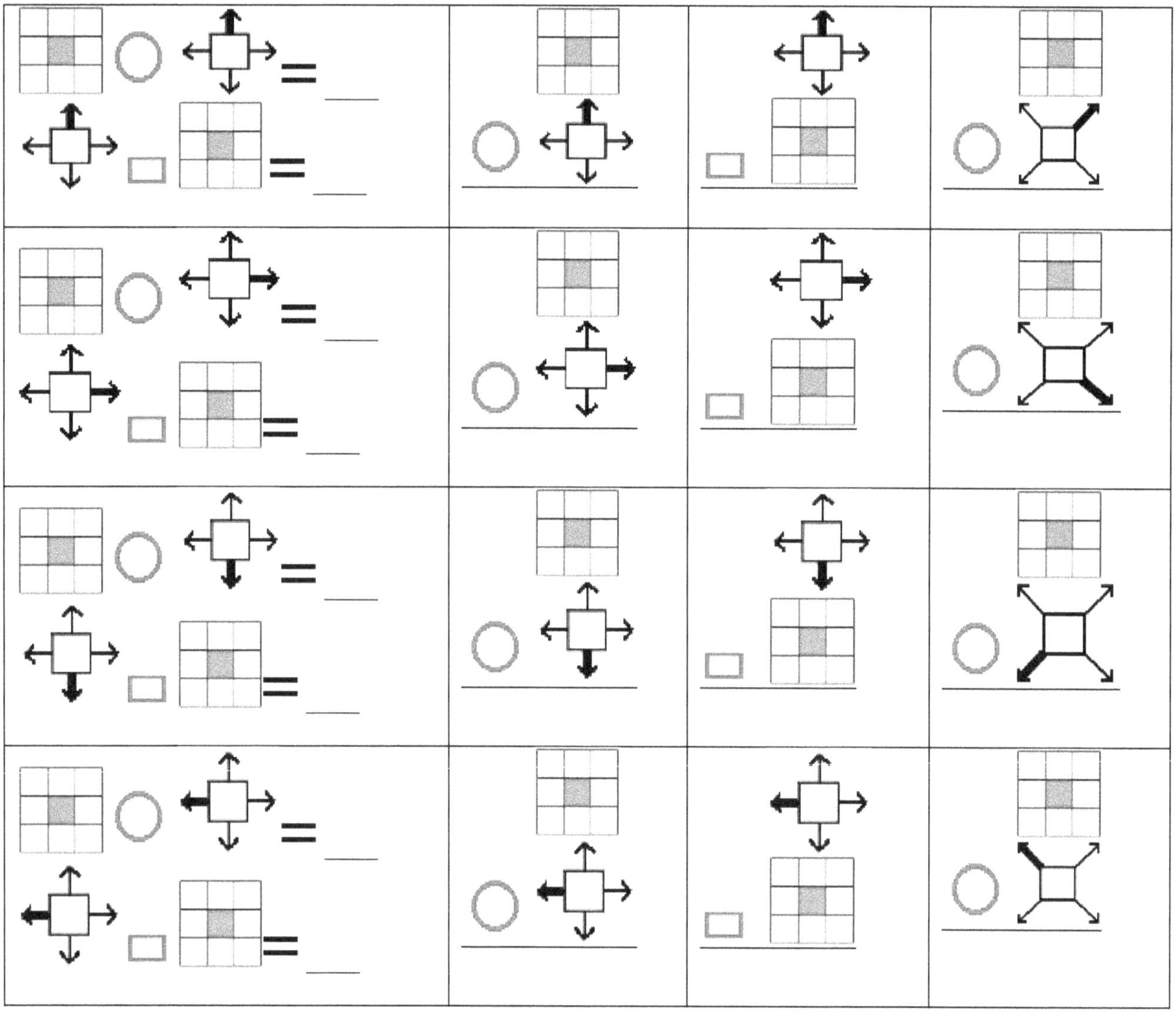

Subtracting and adding up to 9

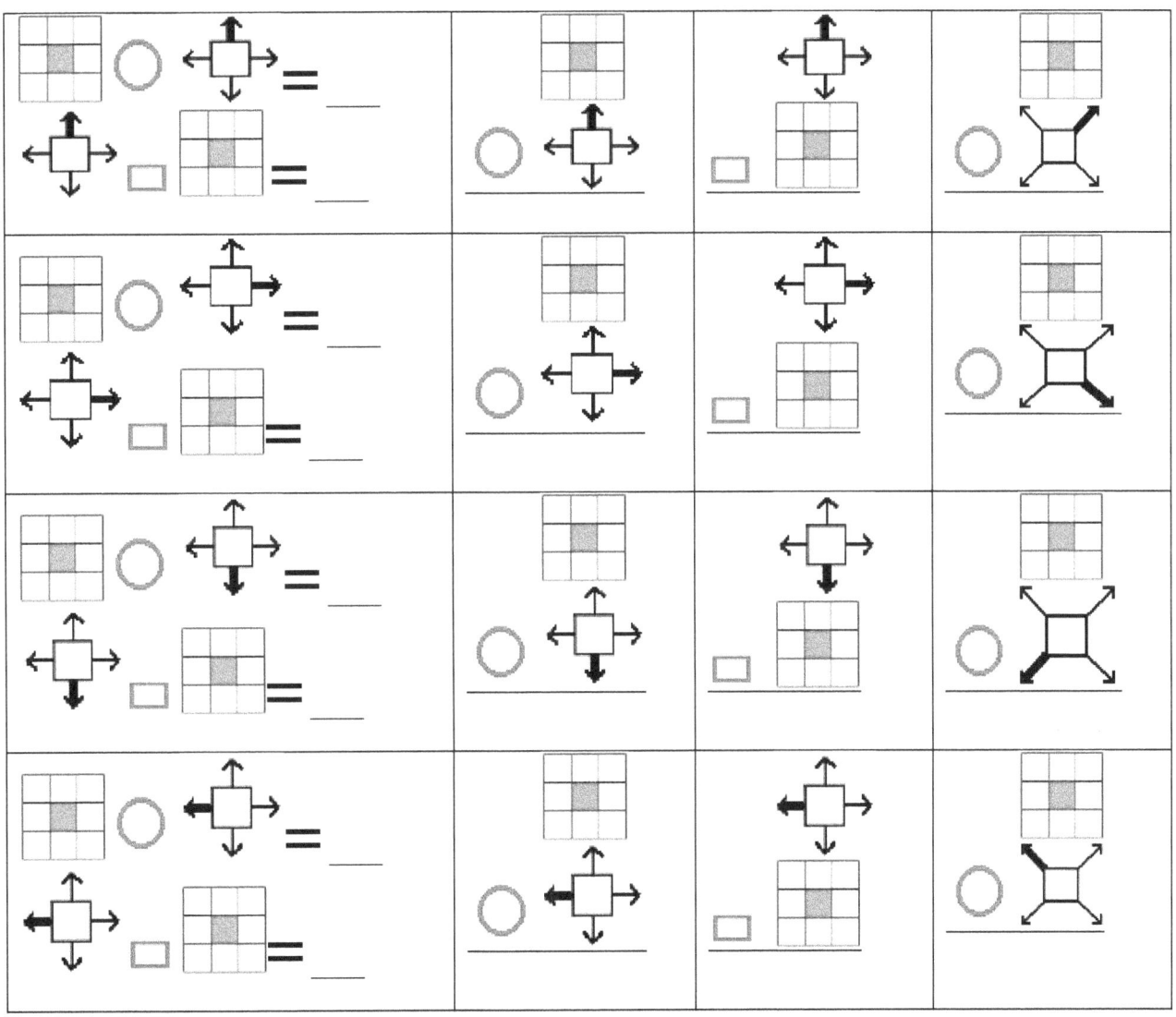

Ho Math Chess   Pre-K and Kindergarten Math
何数棋谜   棋谜式幼儿健脑思维趣味数学
© 2012 - 2021 Frank Ho, Amanda Ho, Canada copyright 1095661, Trademark 771400

Subtracting and adding up to 9

You are a chess piece located at a square indicated by a shaded square.
24556633
2102
4564
2101

○ = −, □ = +

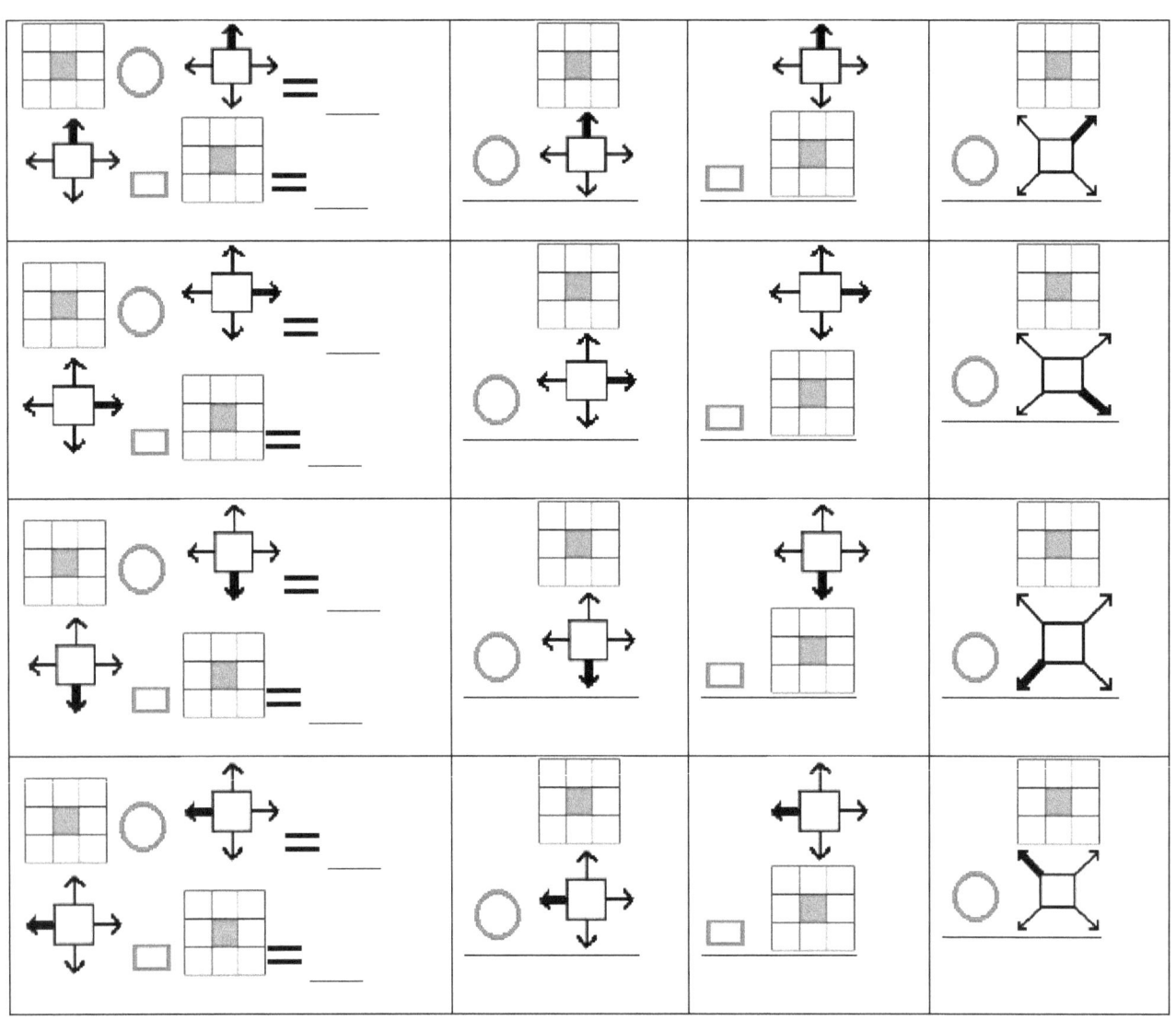

Ho Math Chess — Pre-K and Kindergarten Math

Subtracting and adding up to 9

You are a chess piece located at a square indicated by a shaded square.
35667654
3213
5675
3212

○ = −, □ = +

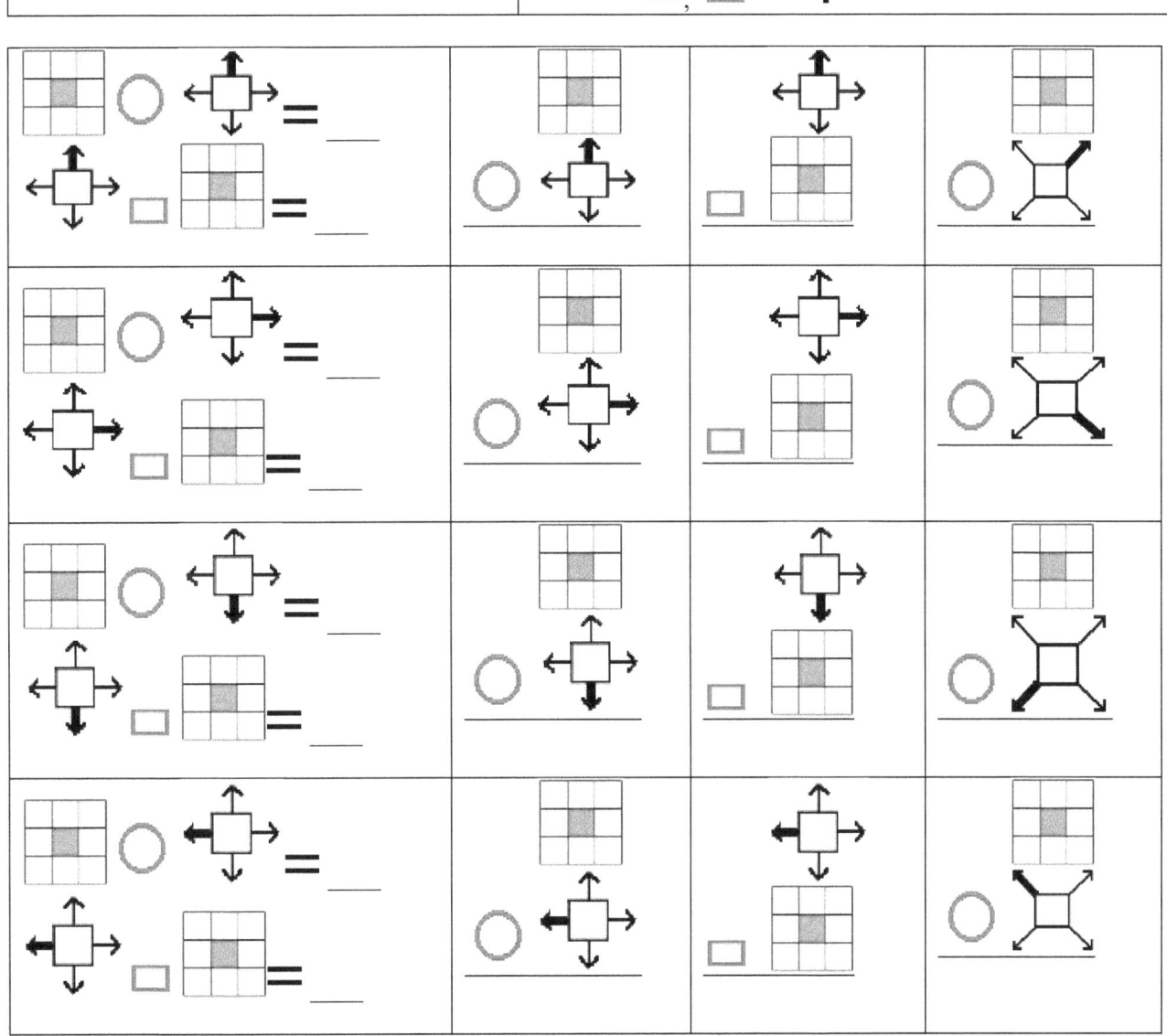

Subtracting and adding up to 9

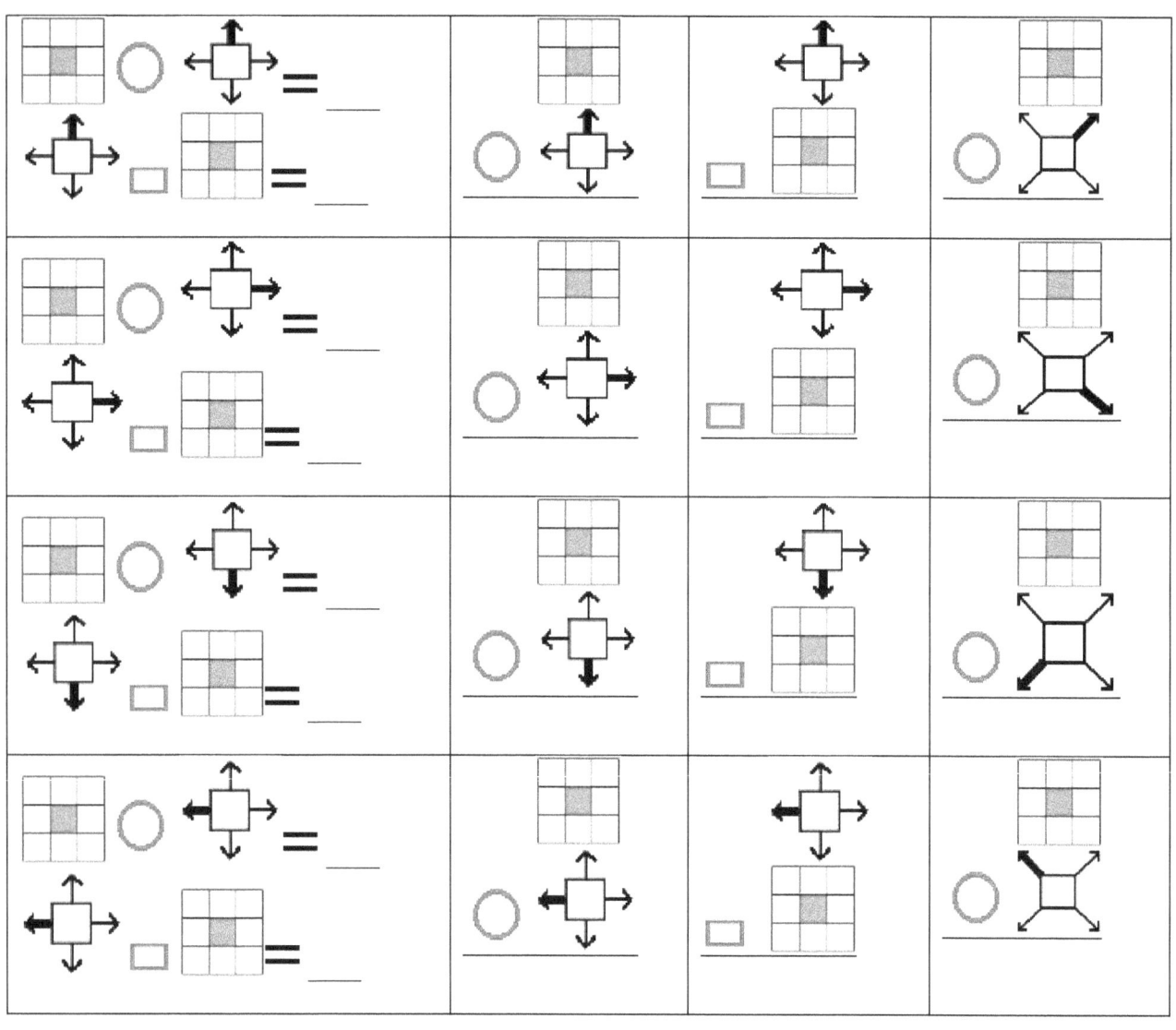

Subtracting and adding up to 9

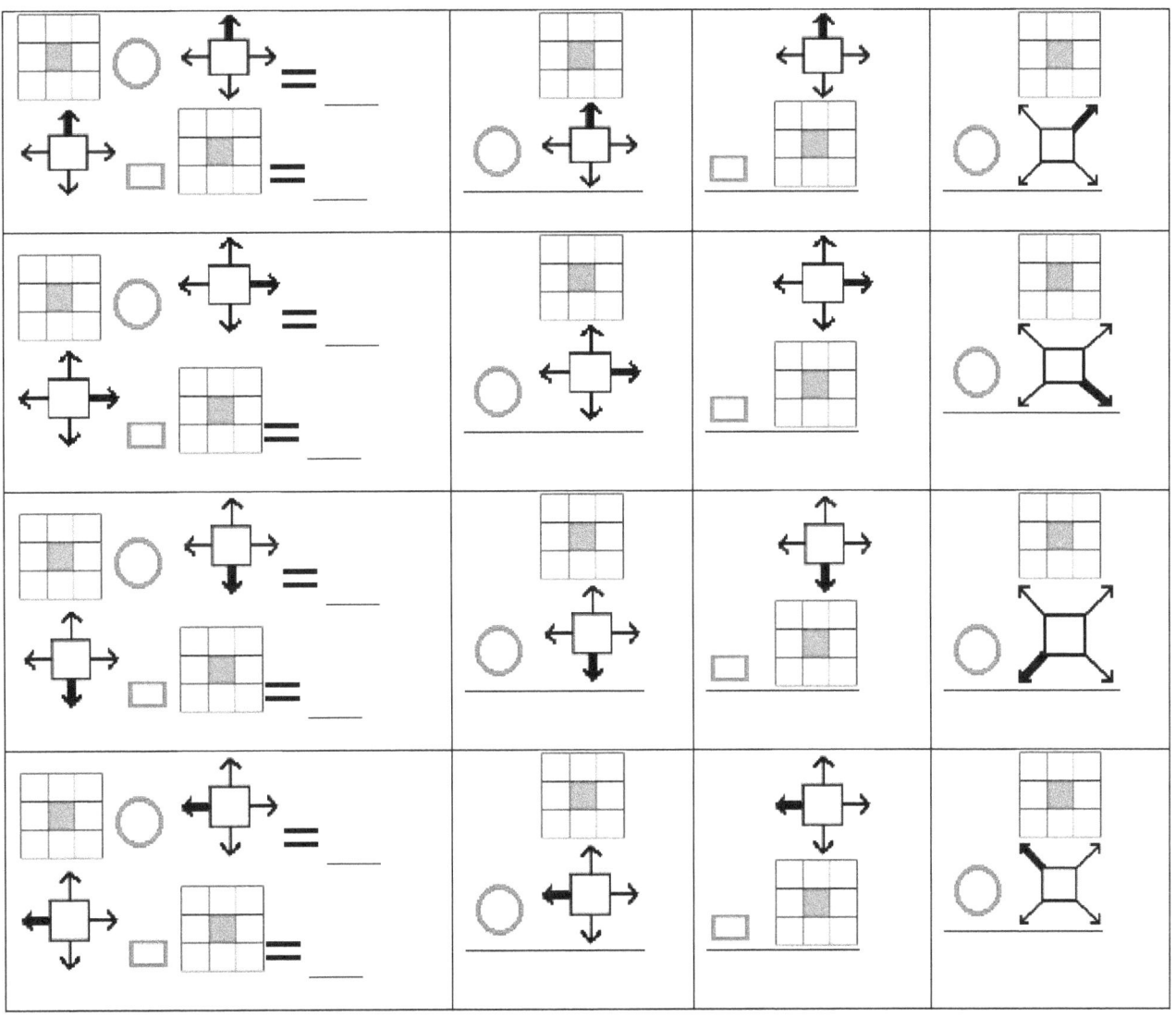

Subtracting and adding up to 9

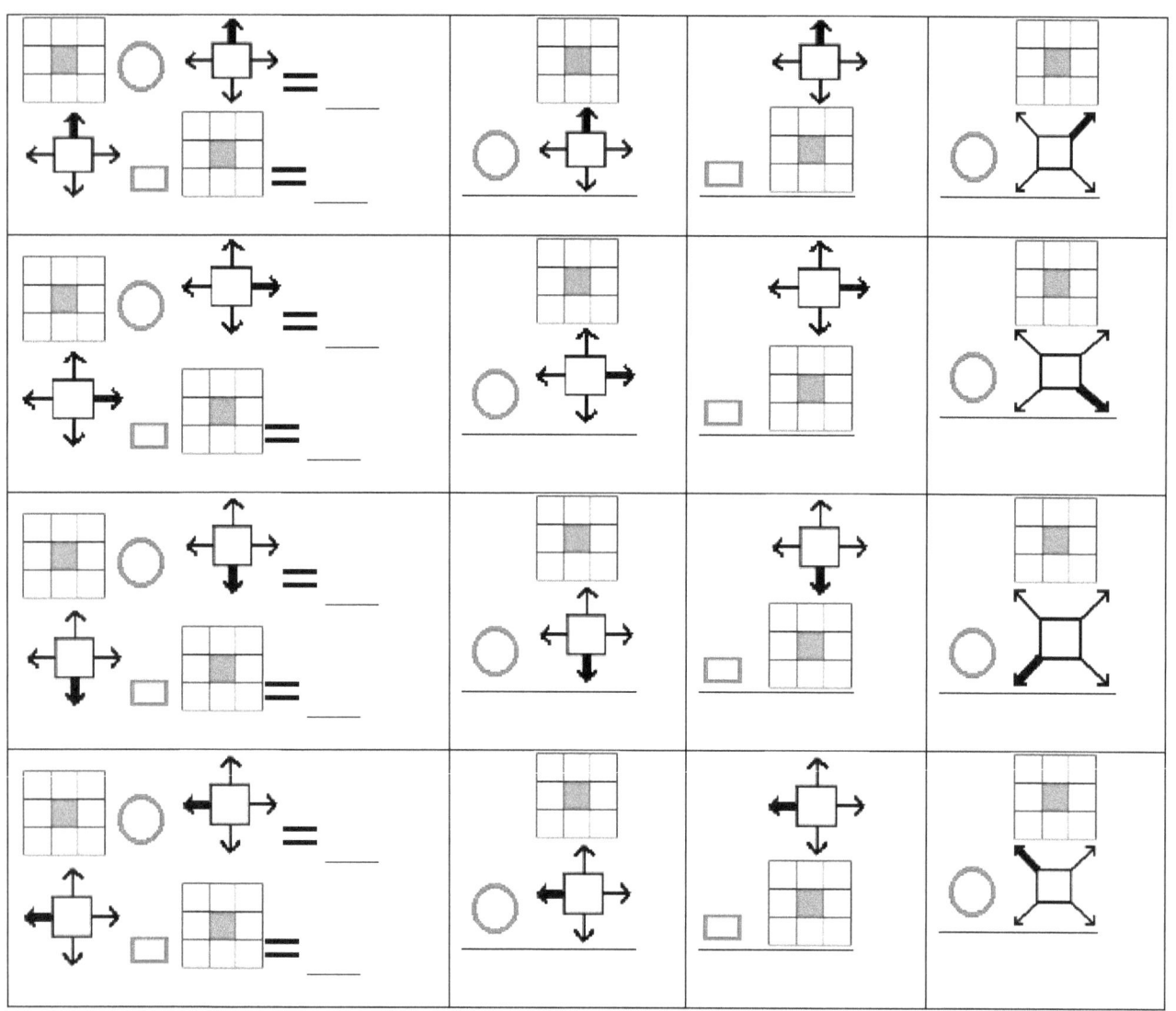

Subtracting and adding up to 9

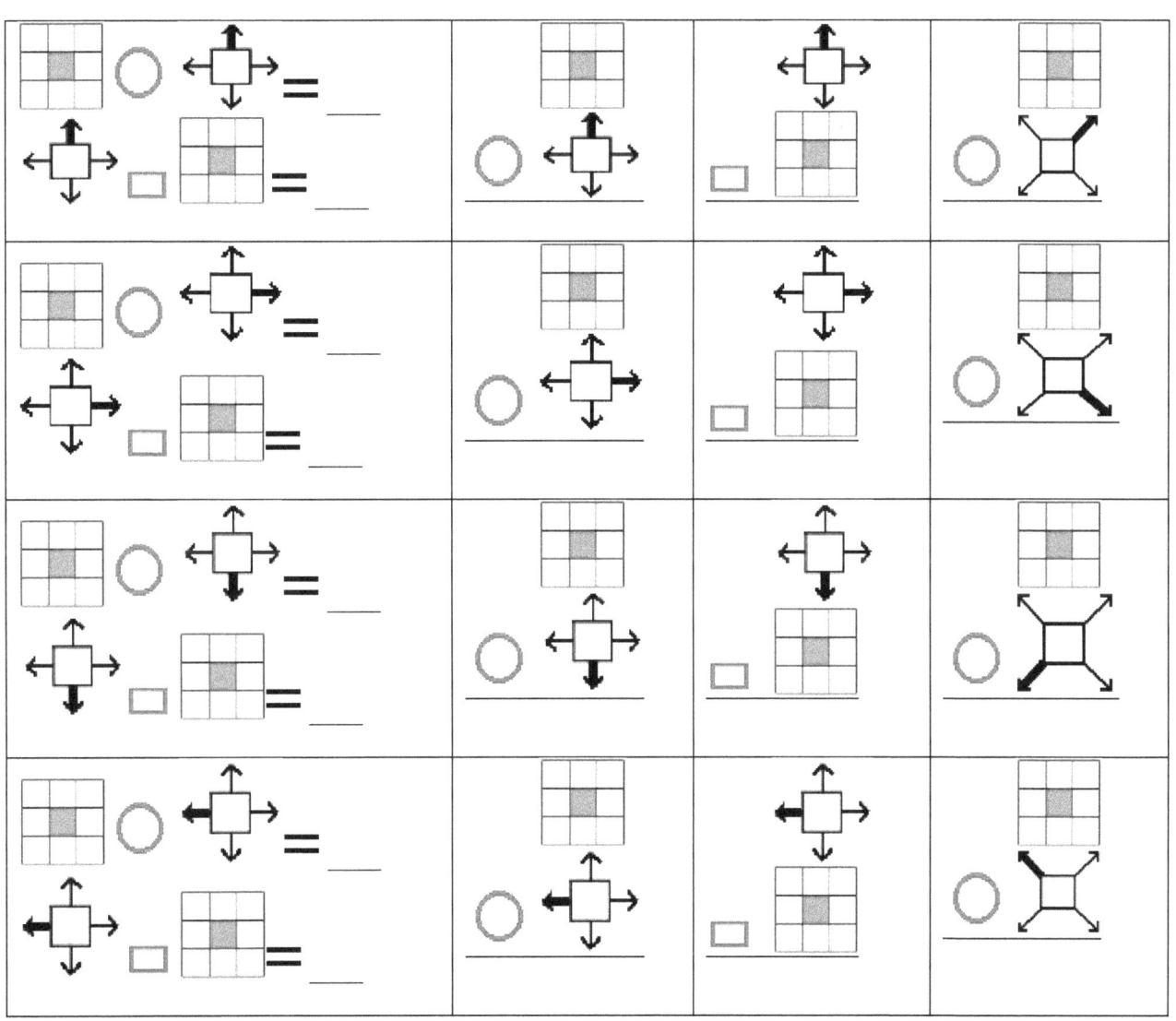

Subtracting and adding up to 9

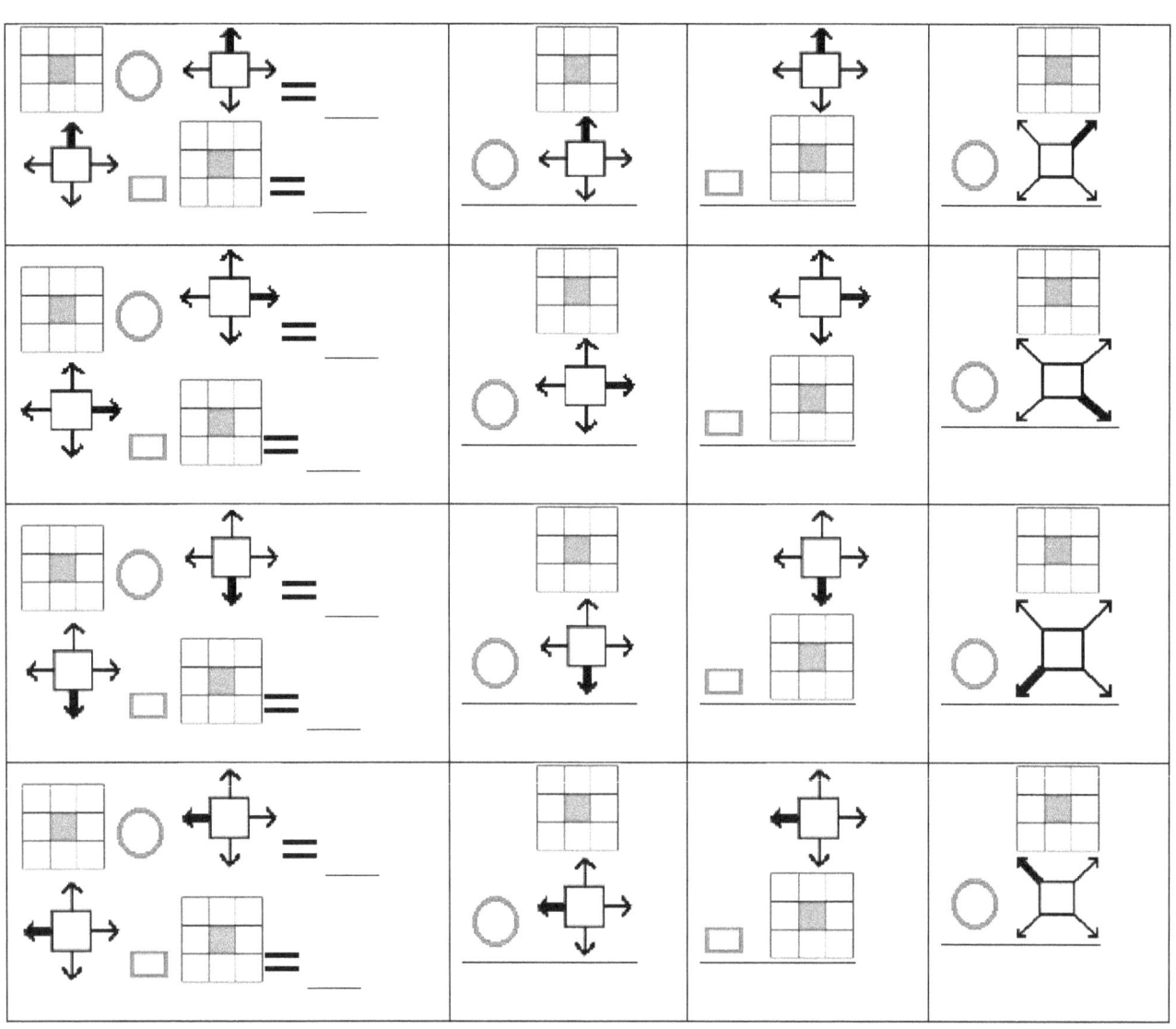

Ho Math Chess   Pre-K and Kindergarten Math

何数棋谜   棋谜式幼儿健脑思维趣味数学

© 2012 – 2021 Frank Ho, Amanda Ho, Canada copyright 1095661, Trademark 771400

Subtracting and adding up to 9

You are a chess piece located at a square indicated by a shaded square.
57483957
5435
7897
5434

○ = −, □ = +

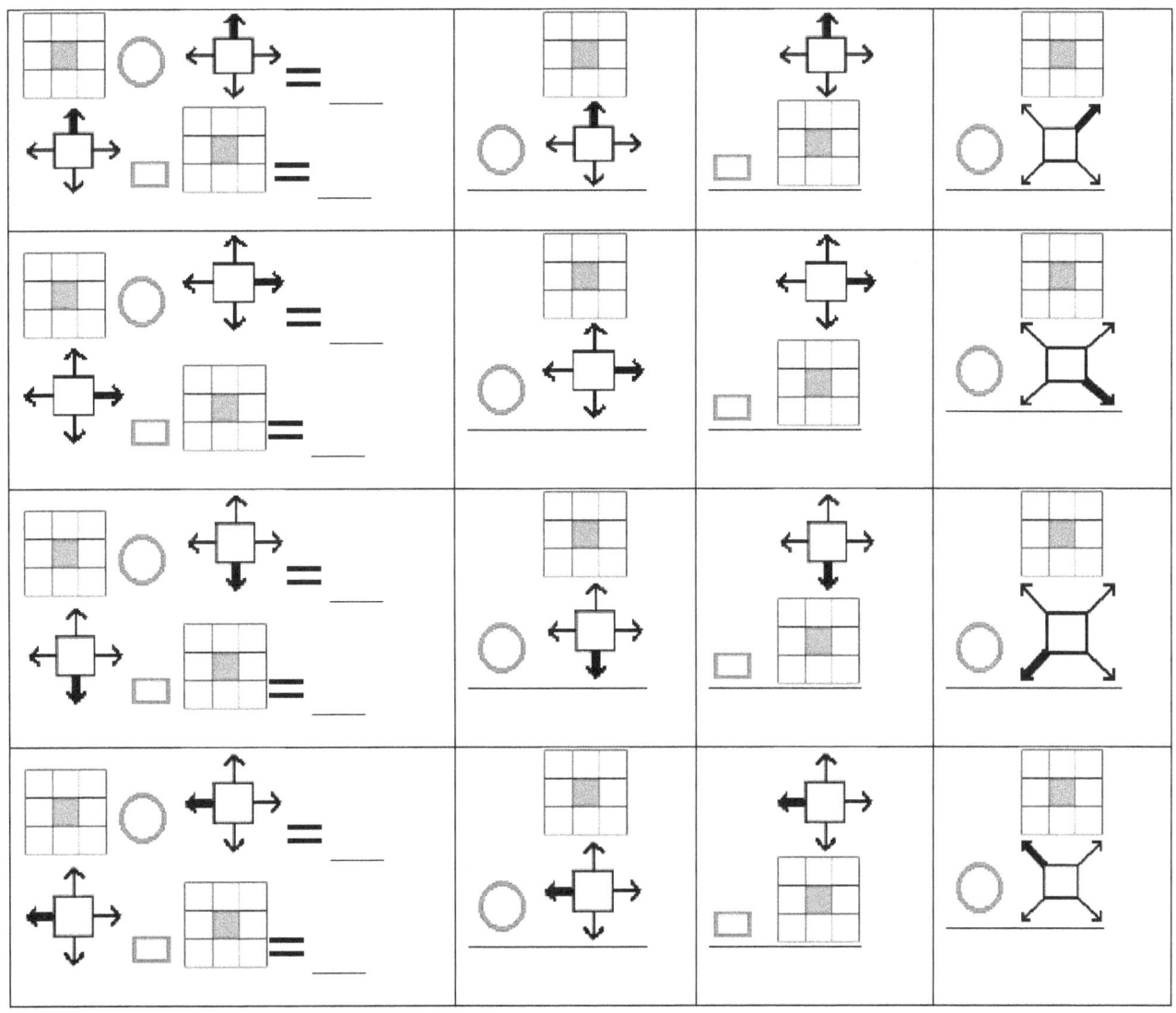

Ho Math Chess — Pre-K and Kindergarten Math

何数棋谜　棋谜式幼儿健脑思维趣味数学

© 2012 – 2021 Frank Ho, Amanda Ho, Canada copyright 1095661, Trademark 771400

Subtracting and adding up to 9

You are a chess piece located at a square indicated by a shaded square.

1	1	1	1
2	2	2	2
3	3	3	3
1	1	1	2

answer ○ = +

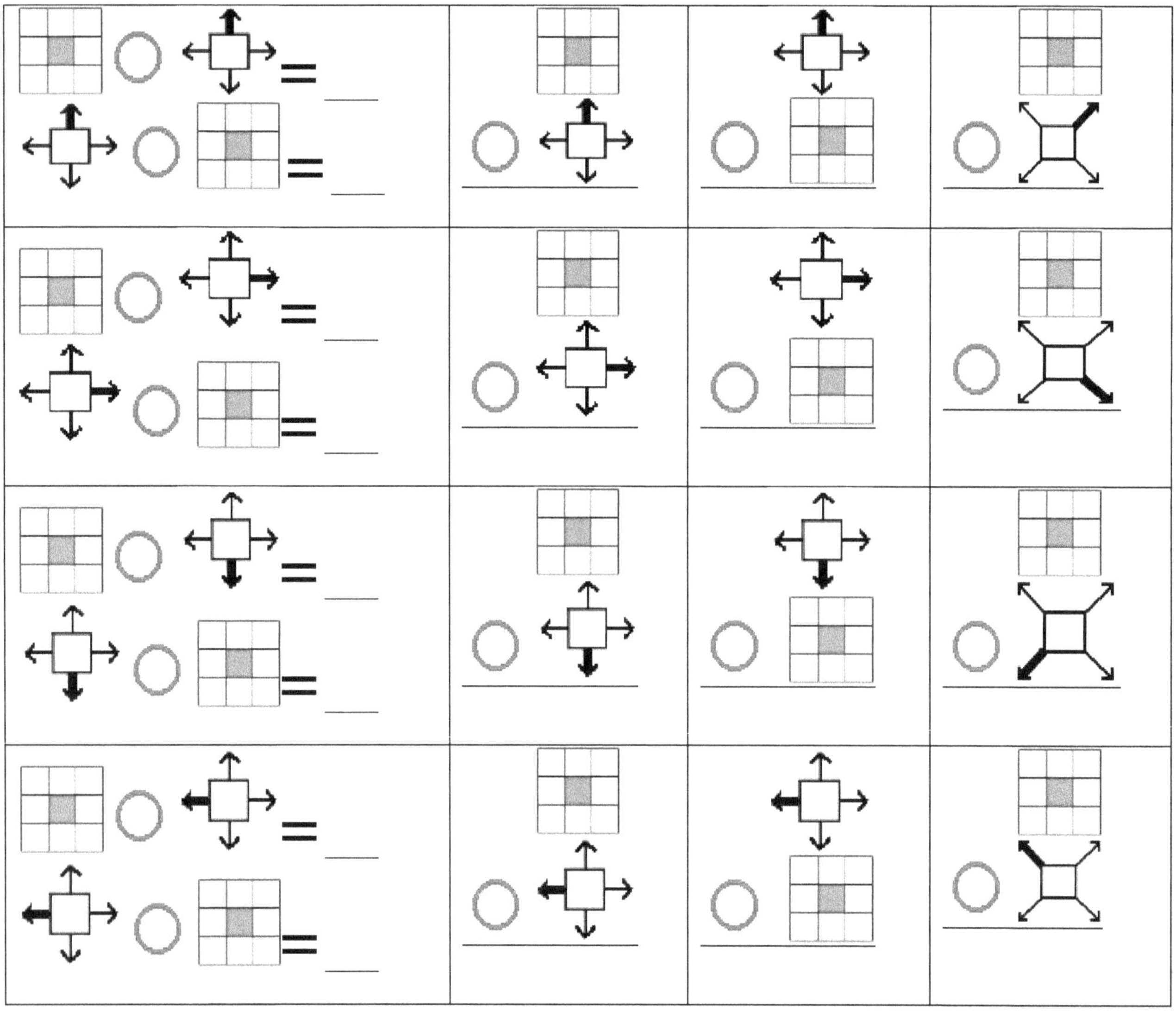

www.homathchess.com   348

Subtracting and adding up to 9

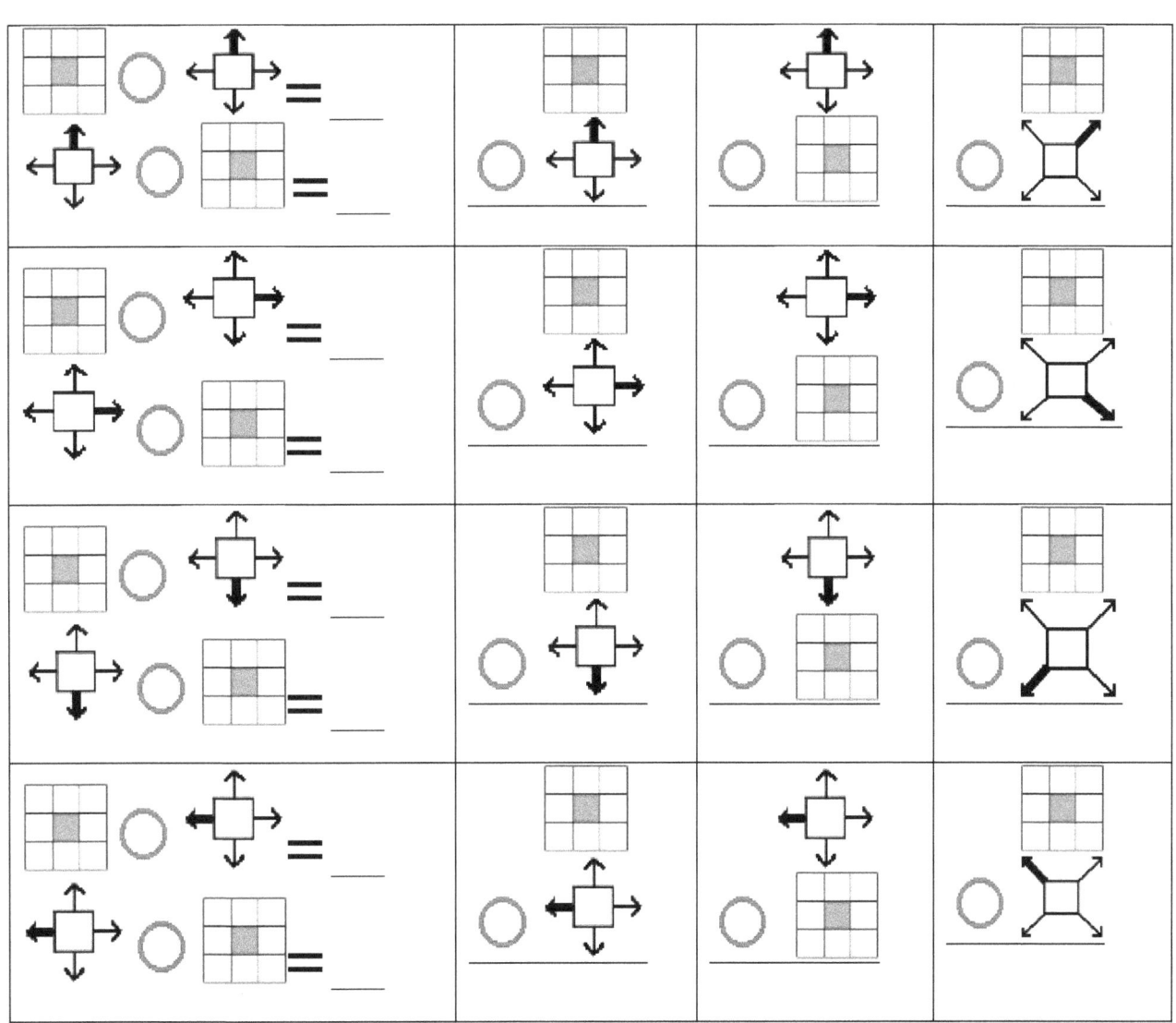

Subtracting and adding up to 9

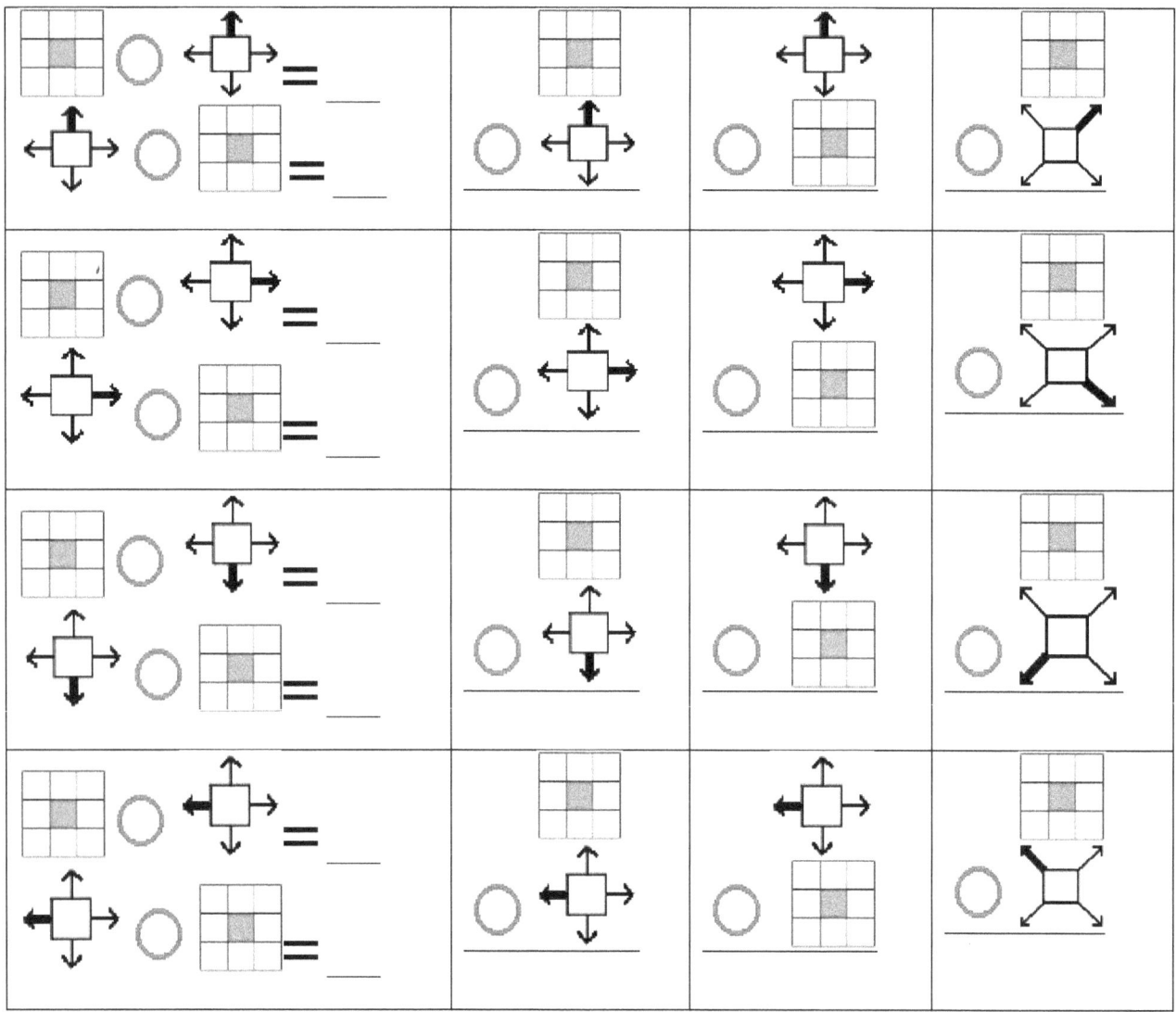

Subtracting and adding up to 9

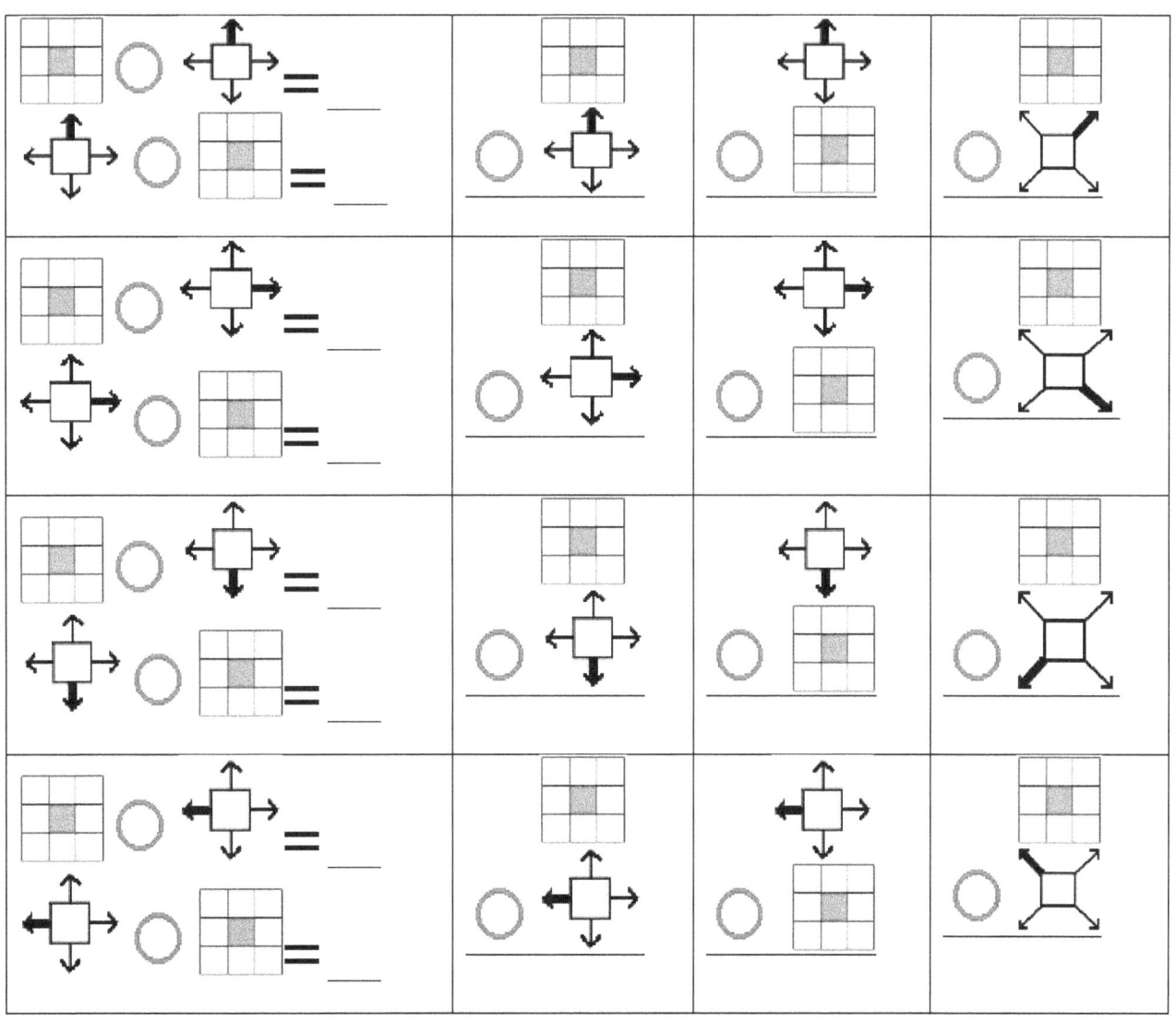

Subtracting and adding up to 9

2	1	1
1	3-2	2
3	3	2

You are a chess piece located at a square indicated by a shaded square.

2			2
3			3
4			4
2			3

answer

○ = +

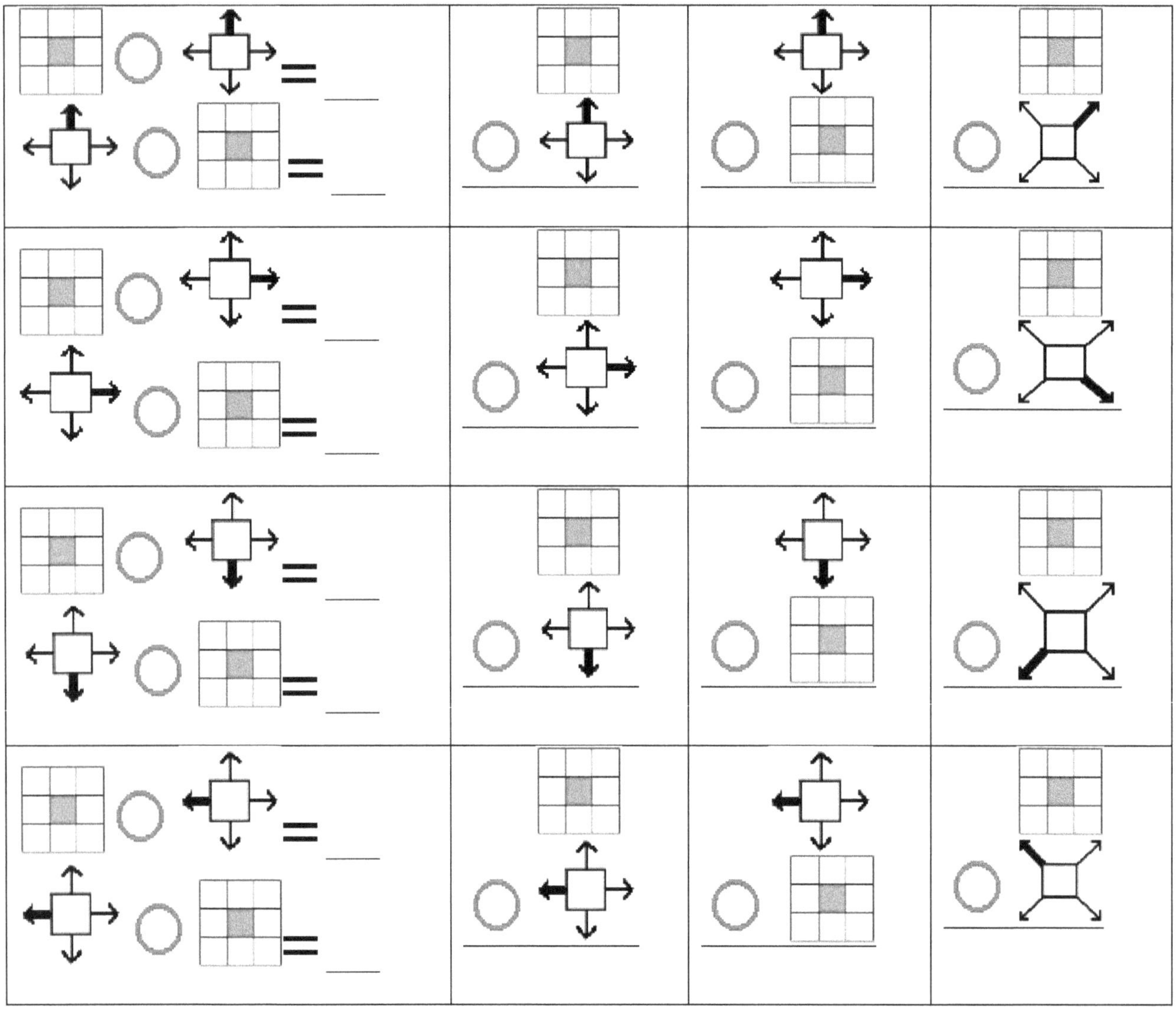

Ho Math Chess — Pre-K and Kindergarten Math

何数棋谜　棋谜式幼儿健脑思维趣味数学

© 2012 – 2021 Frank Ho, Amanda Ho, Canada copyright 1095661, Trademark 771400

Subtracting and adding up to 9

You are a chess piece located at a square indicated by a shaded square.

1		1
2		2
3		3
1		2

answer

◯ = +

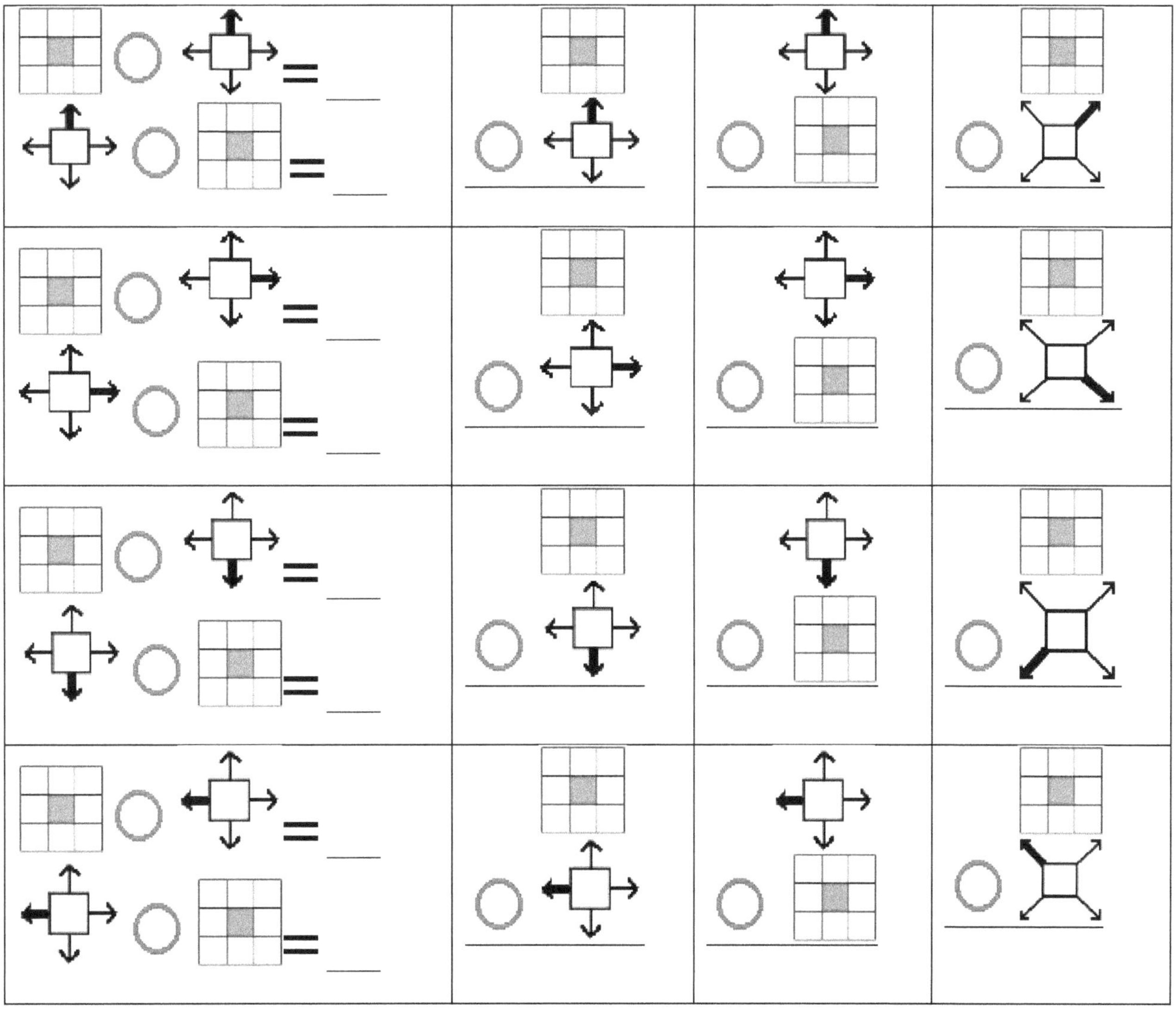

www.homathchess.com

Subtracting and adding up to 9

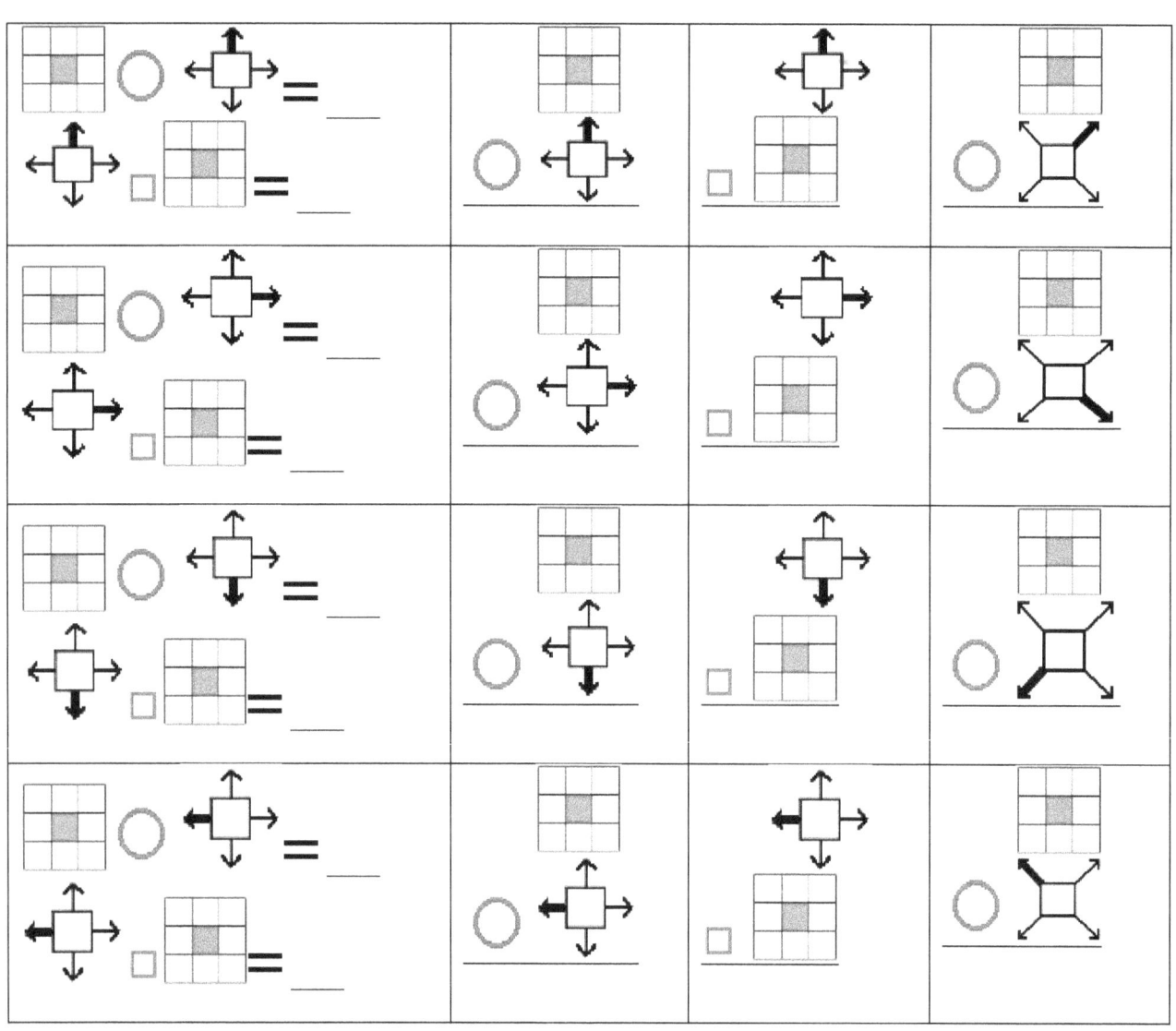

Subtracting and adding up to 9

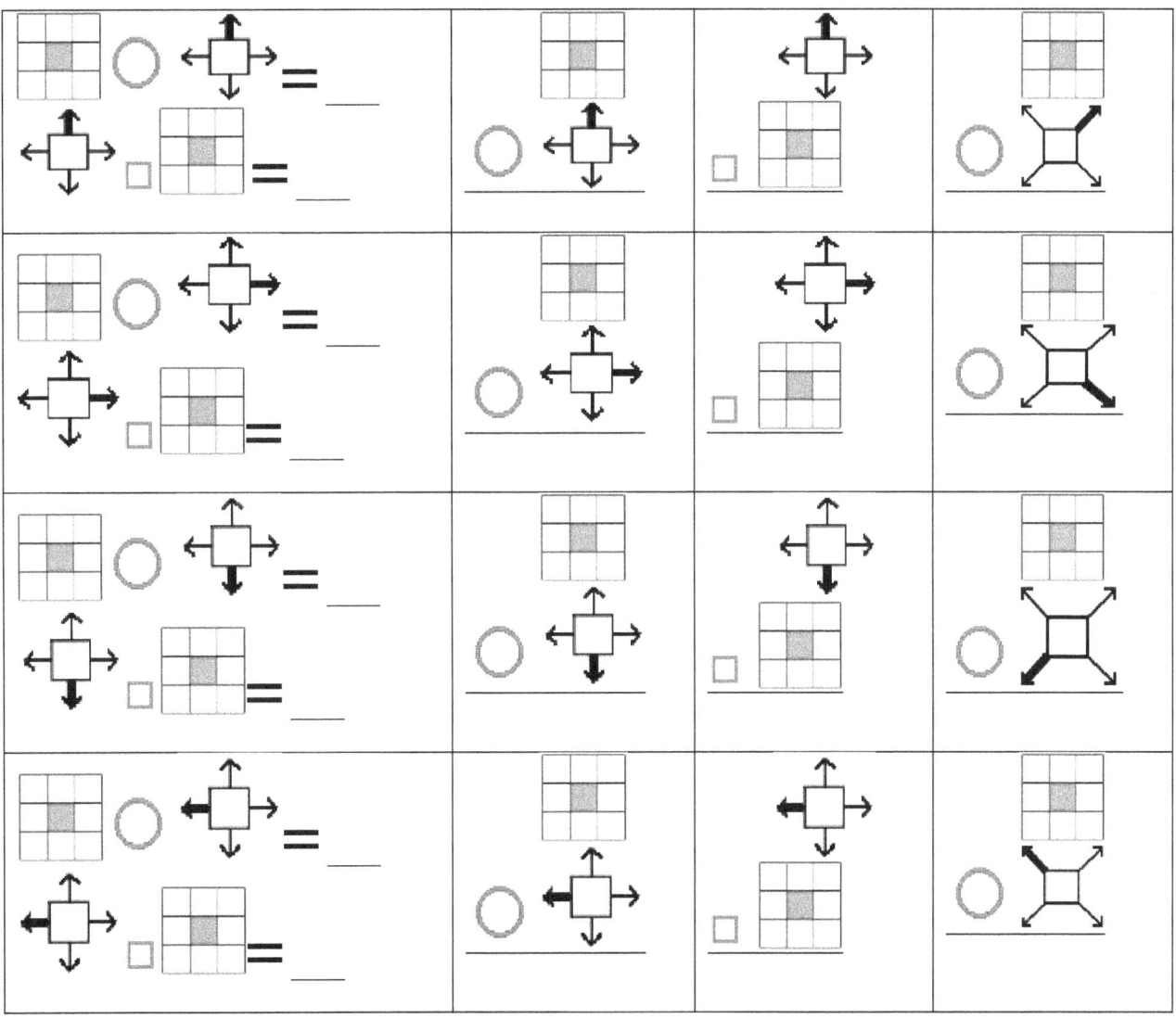

Ho Math Chess — Pre-K and Kindergarten Math

Memory and computation training

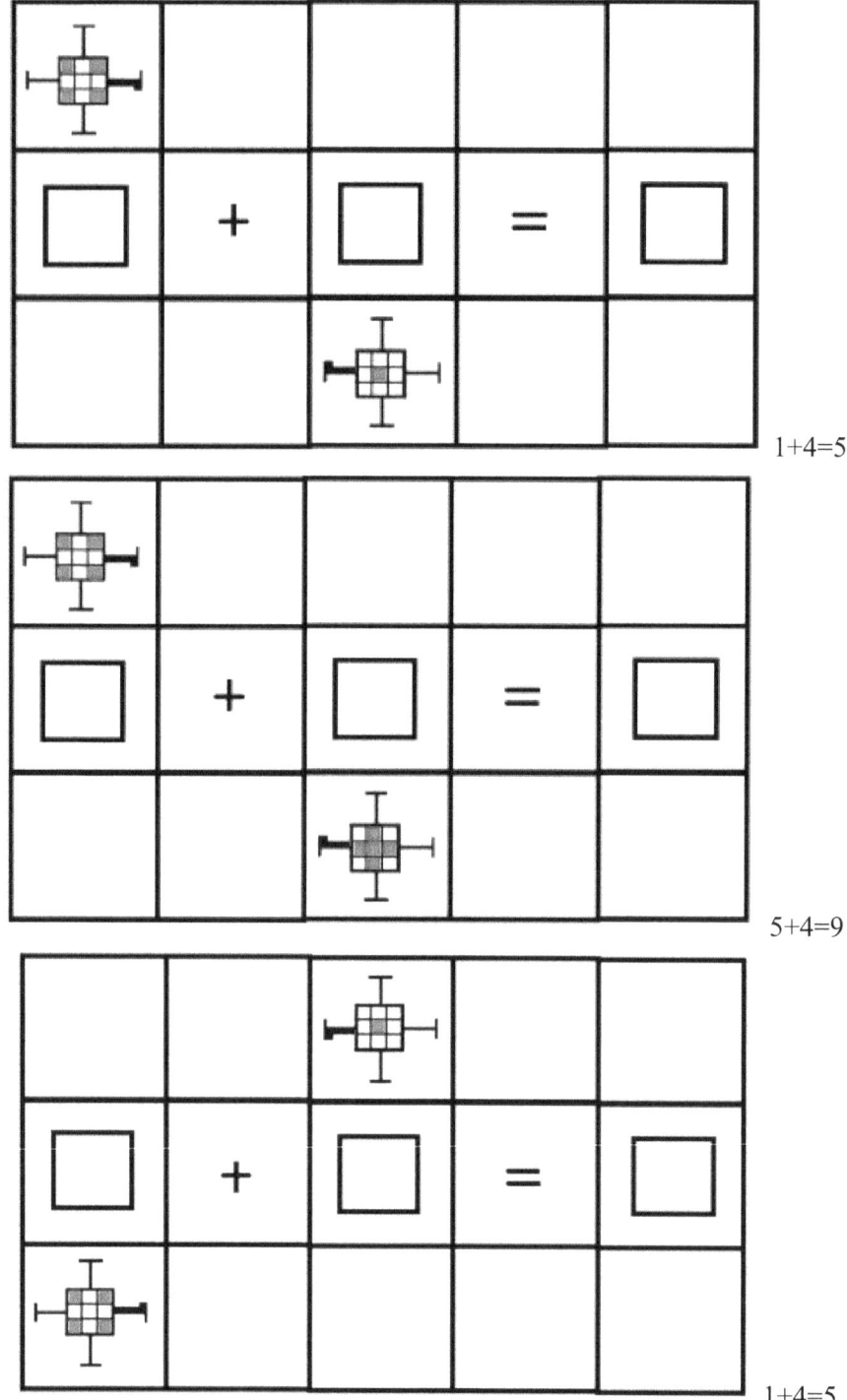

1+4=5

5+4=9

1+4=5

Memory and computation training

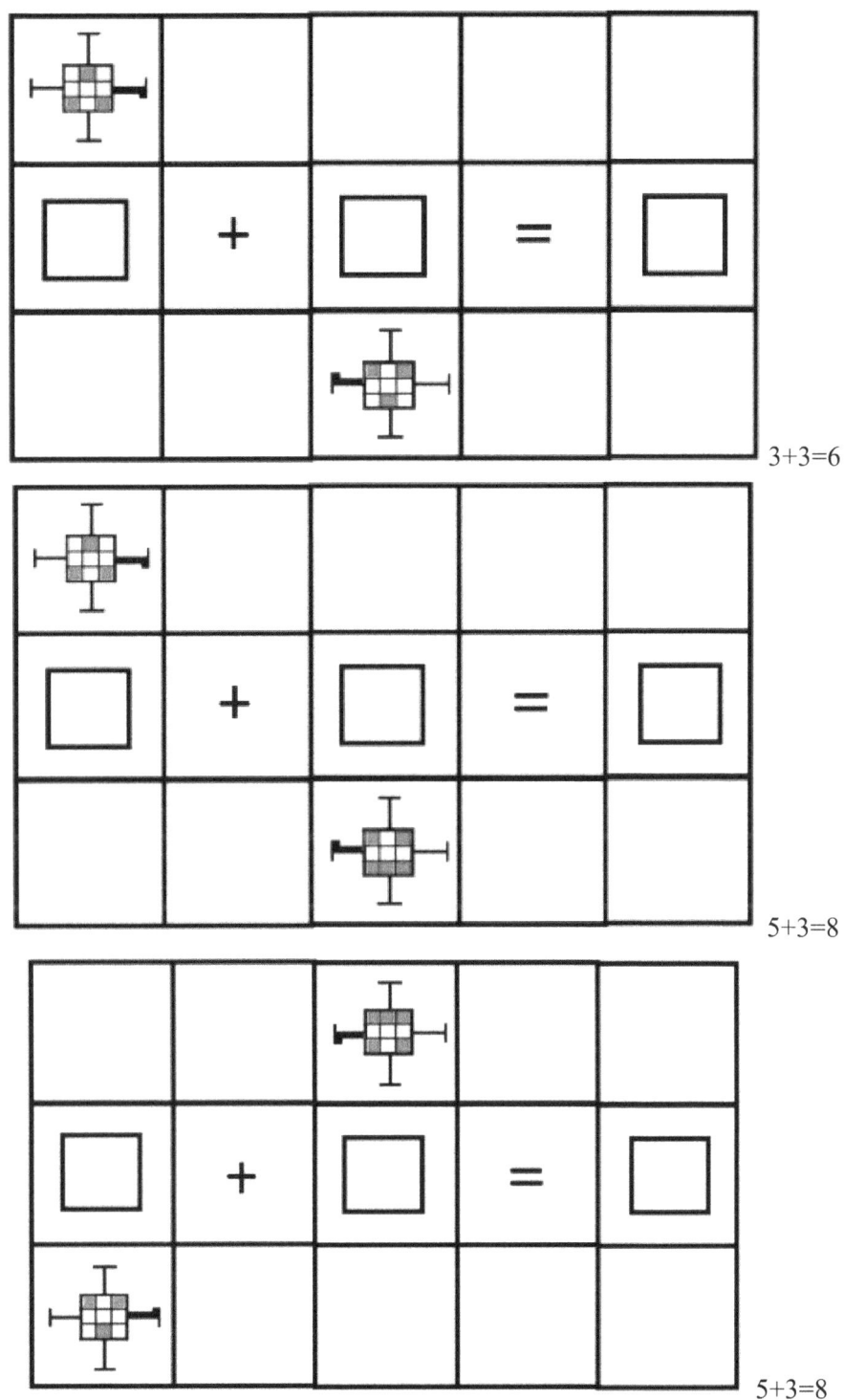

3+3=6

5+3=8

5+3=8

Memory and computation training

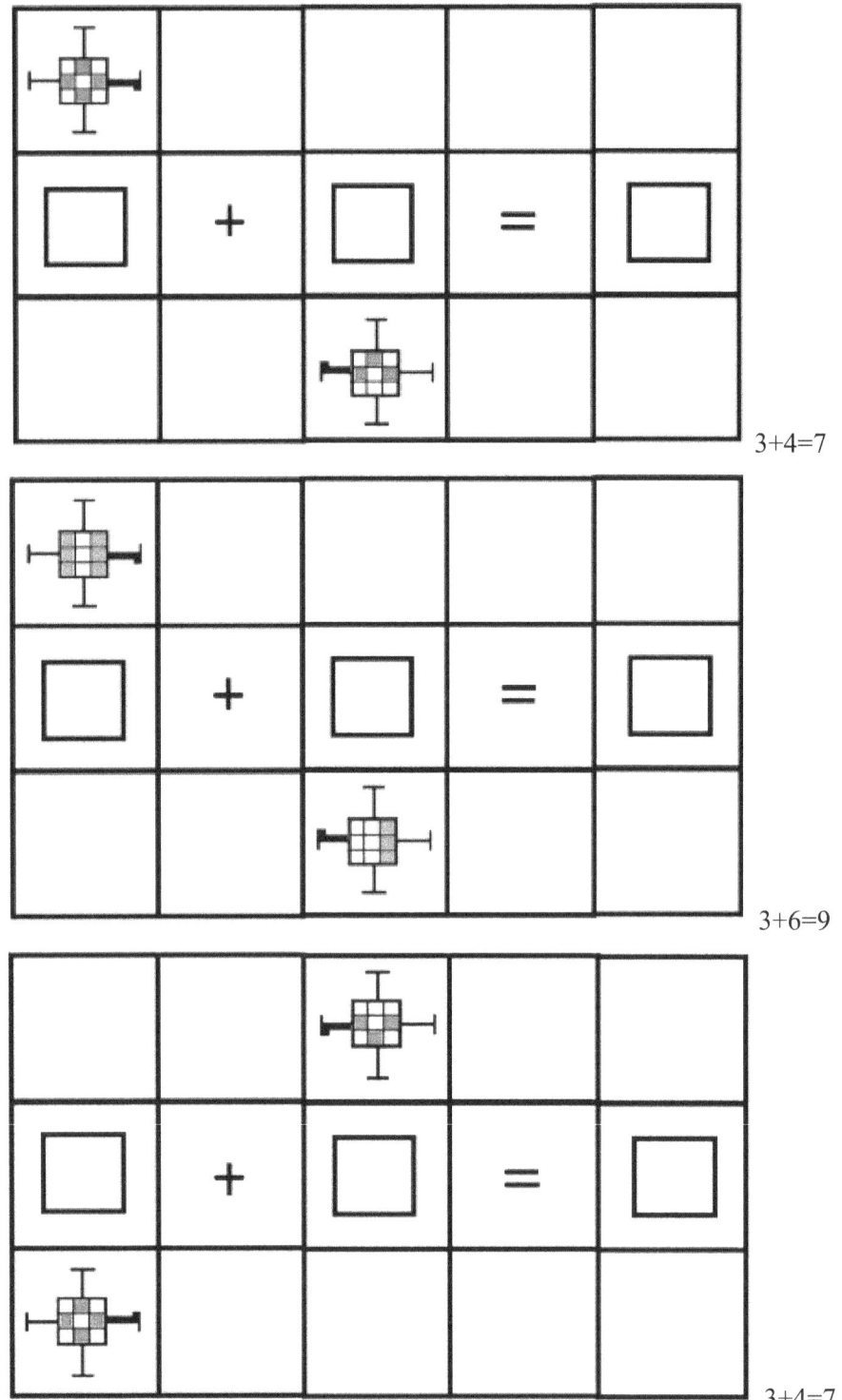

3+4=7

3+6=9

3+4=7

Memory and computation training

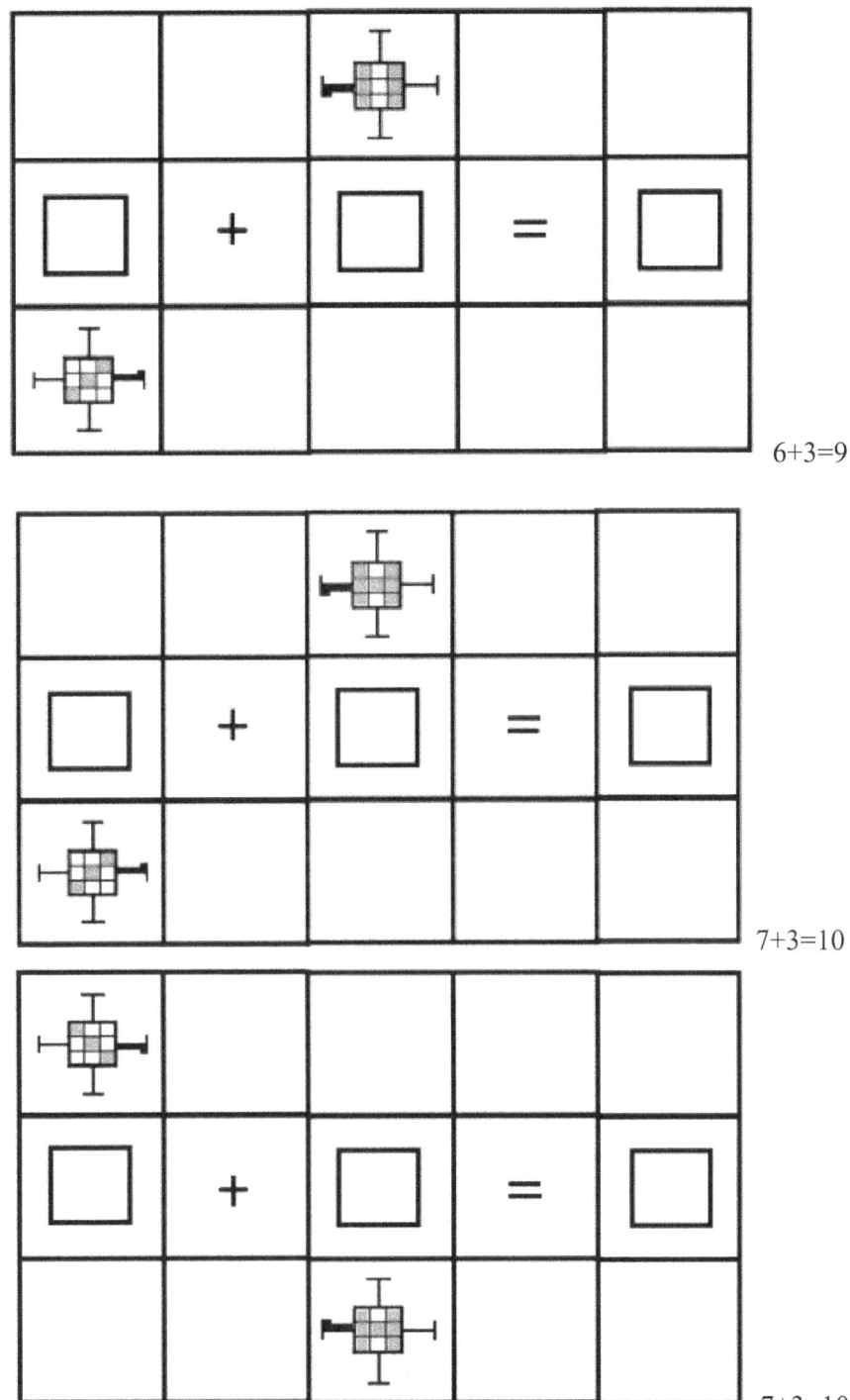

6+3=9

7+3=10

7+3=10

Memory and computation training

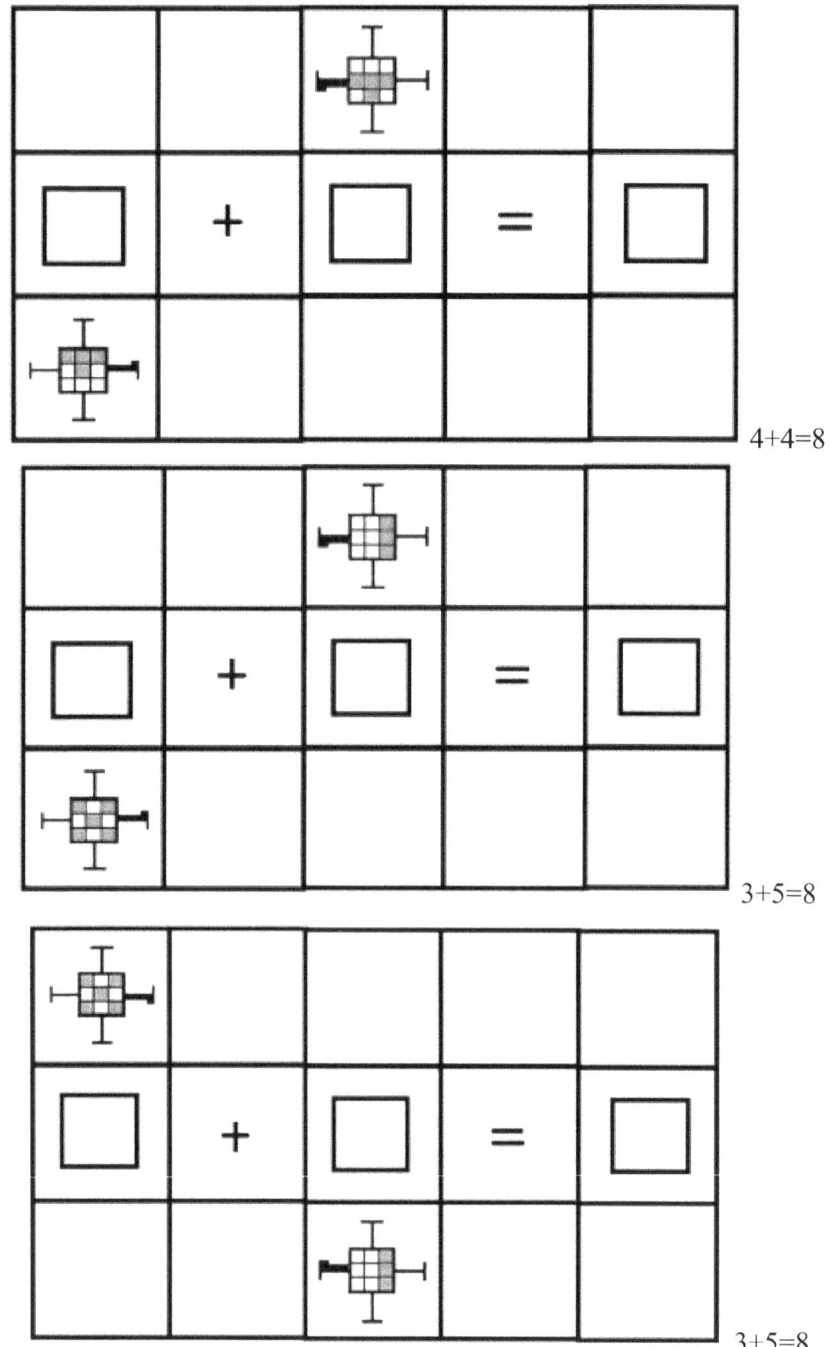

4+4=8

3+5=8

3+5=8

Memory and computation training

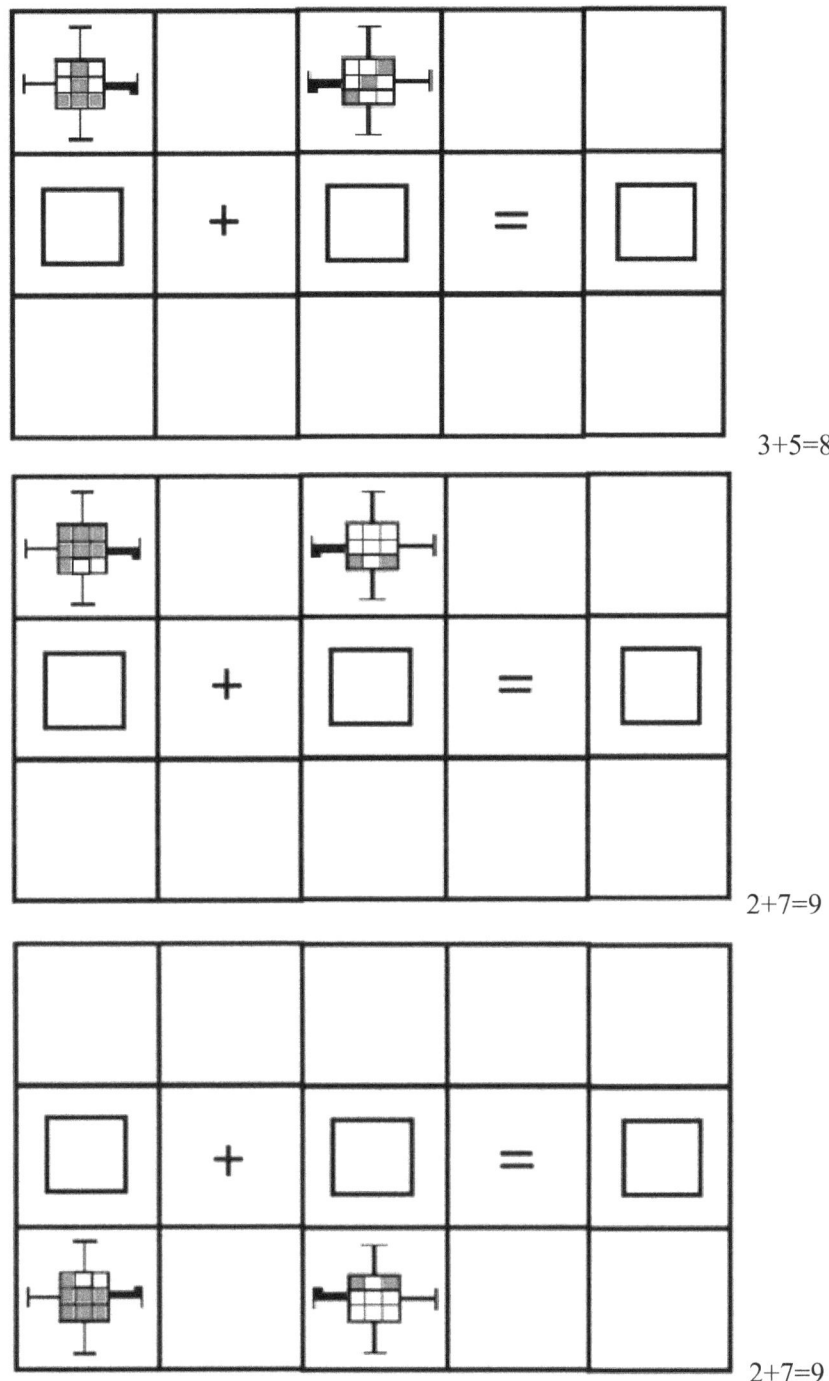

3+5=8

2+7=9

2+7=9

Memory and computation training

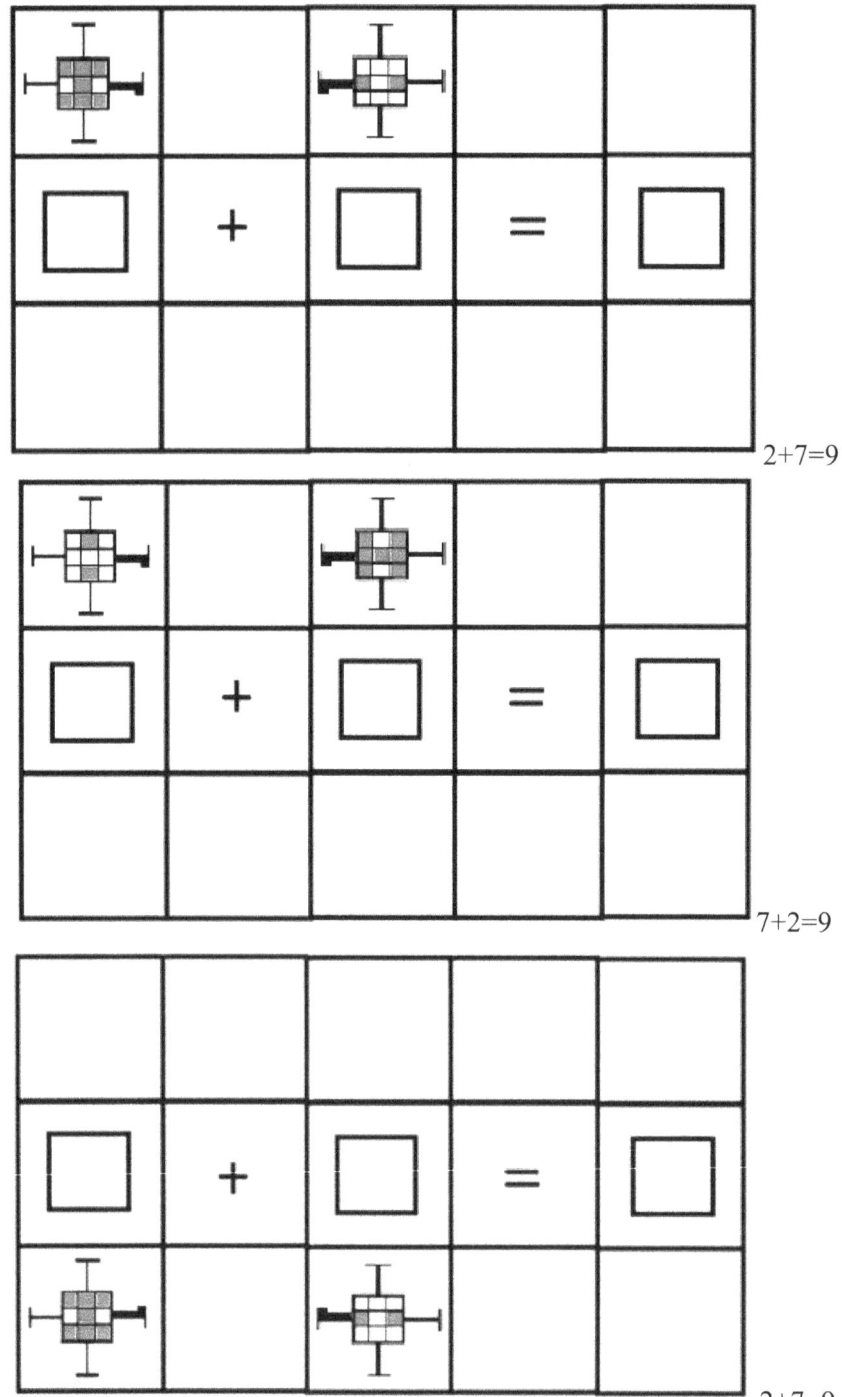

2+7=9

7+2=9

2+7=9

Memory and computation training

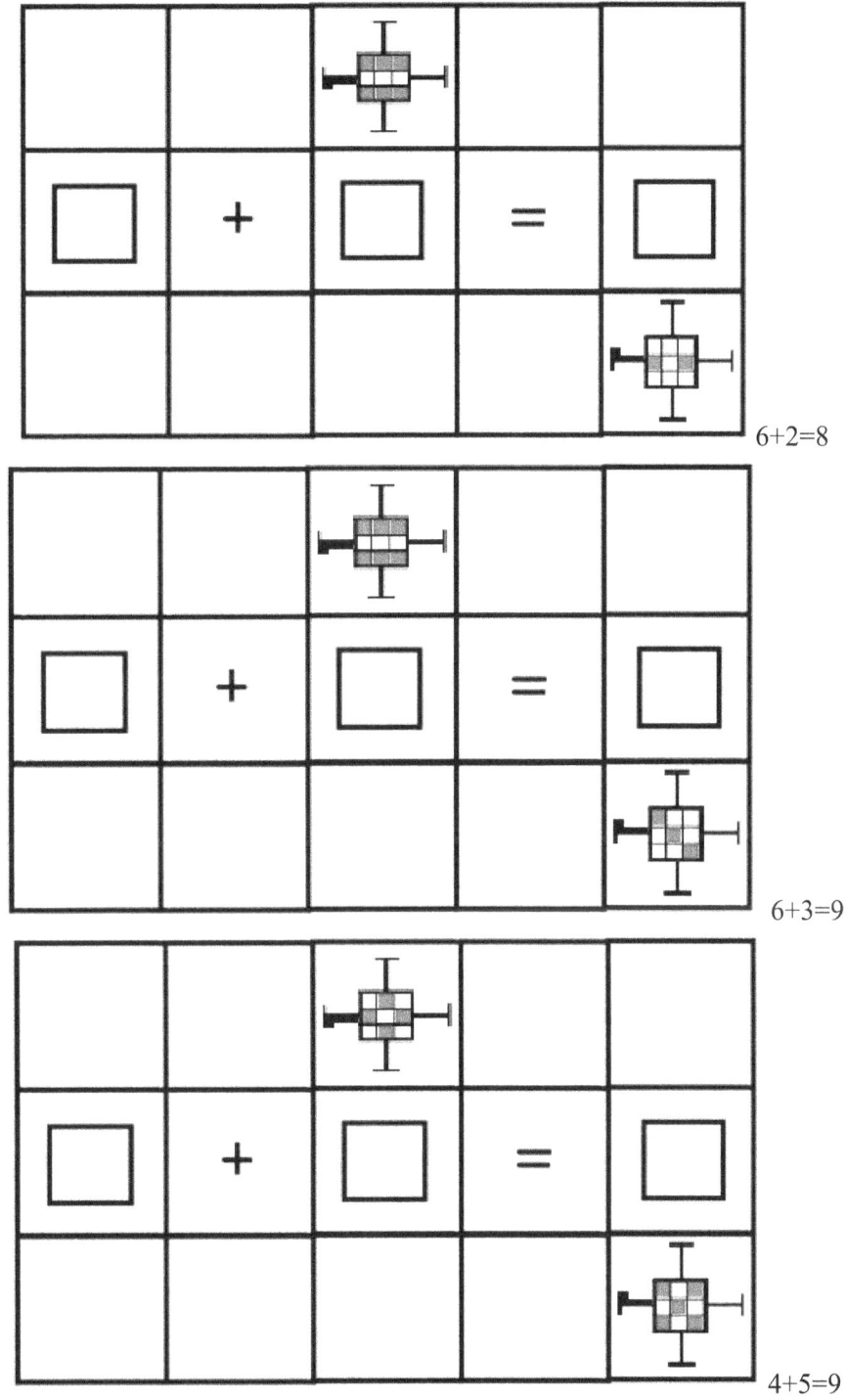

6+2=8

6+3=9

4+5=9

Memory and computation training

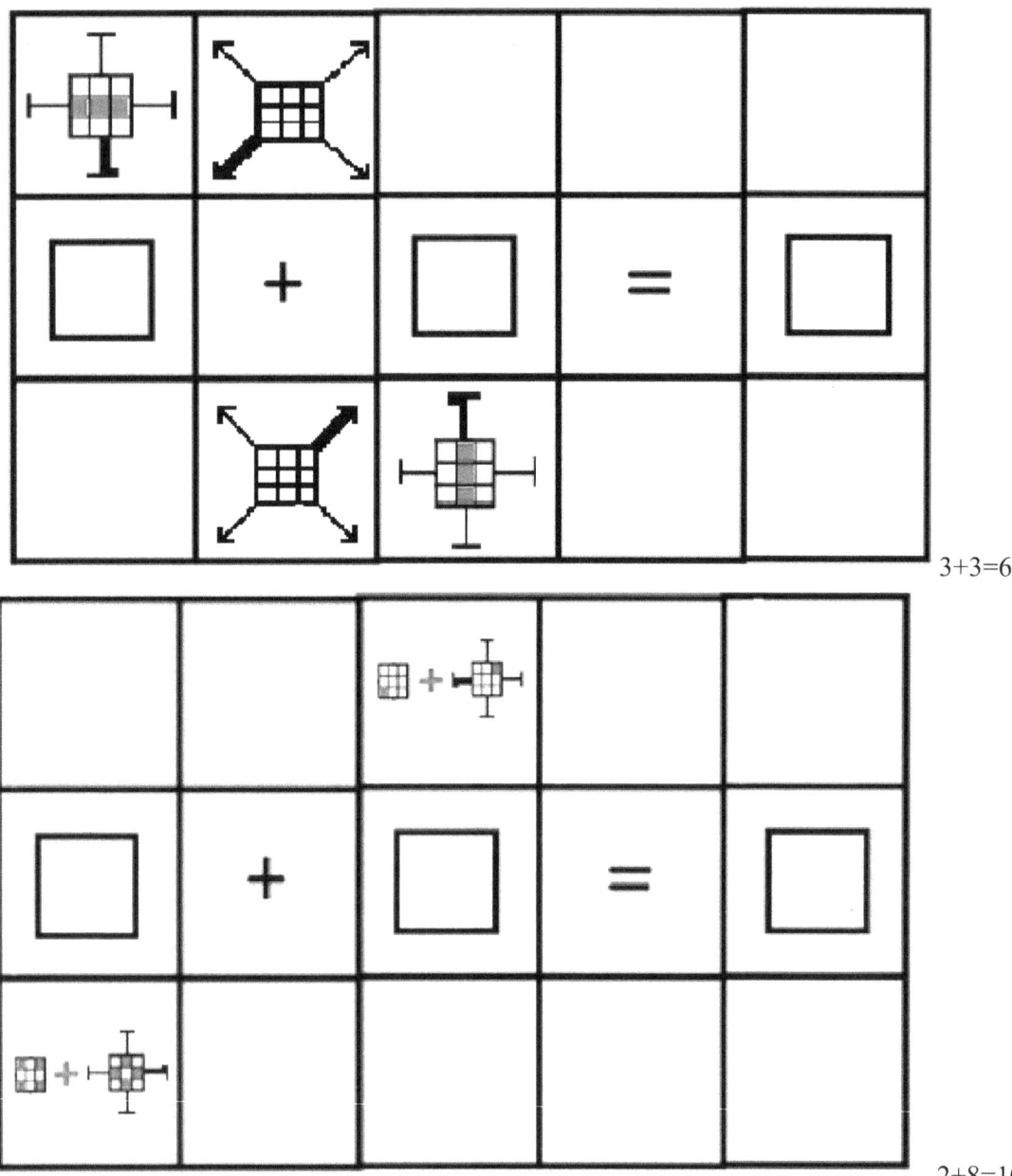

3+3=6

2+8=10

Memory and computation training

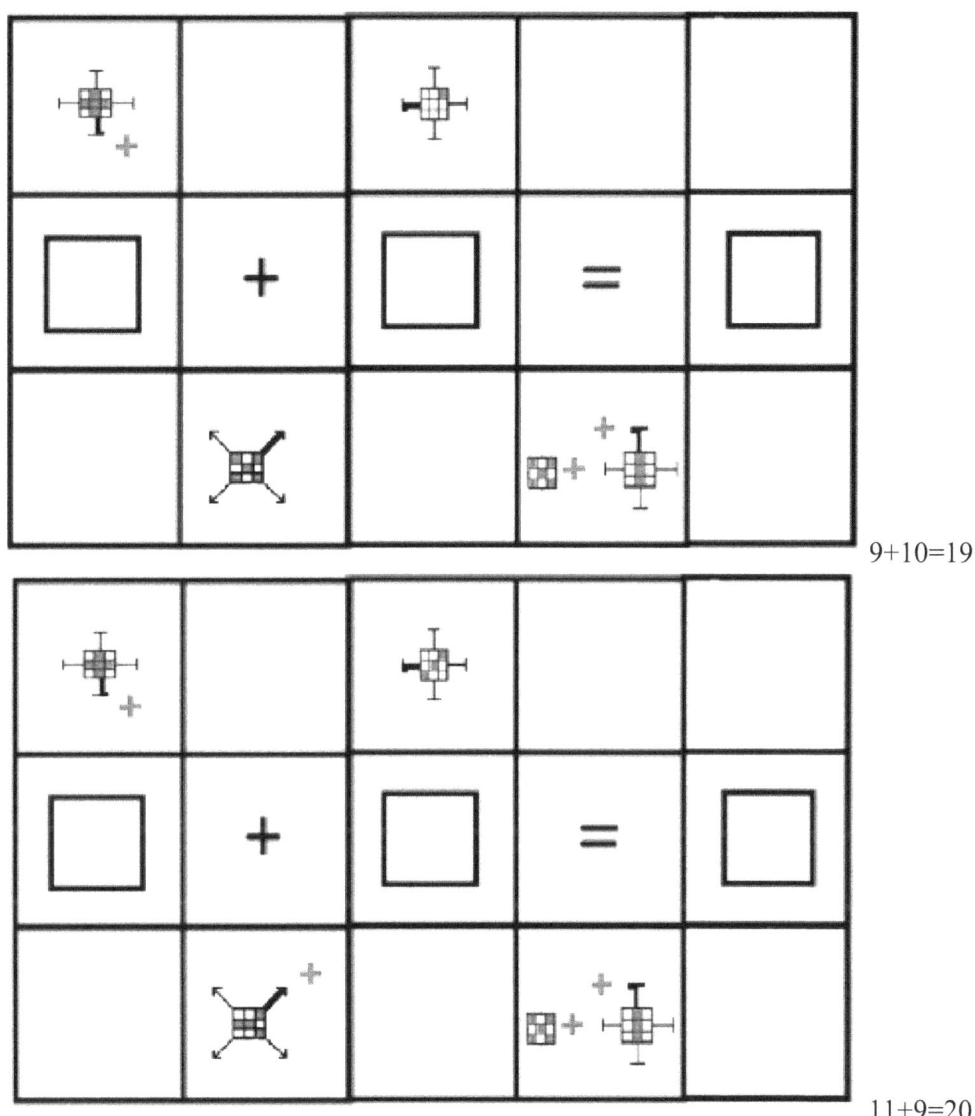

9+10=19

11+9=20

Number ranking puzzles

数的大小比较

Arrange the numbers 1, 2, 3, and 4 so that each number appears only once in each row.

☐ > ☐

☐ > ☐

☐ > ☐

☐ > ☐

☐ > ☐

☐ > ☐

☐ > ☐ > ☐

☐ > ☐ > ☐

☐ > ☐ > ☐

☐ > ☐ > ☐

4>3	4>3>2
4>2	4>3>1
4>1	4>2>1
3>2	3>2>1
3>1	
2>1	

Ho Math Chess — Pre-K and Kindergarten Math

何数棋谜　棋谜式幼儿健脑思维趣味数学

© 2012 – 2021 Frank Ho, Amanda Ho, Canada copyright 1095661, Trademark 771400

Number ranking puzzles

Arrange the numbers 0, 1, 2, 3, and 4 so that each inequality condition is true.

$4 > \square > \square > \square > \square$

$4 > \square > \square$

$4 > \square > \square$

$4 > \square > \square$

$4 > \square > \square$

$4 > \square > \square$

$4 > \square > \square$

$3 > \square > \square$

$3 > \square > \square$

$3 > \square > \square$

$3 > \square$

$3 > \square$

$3 > \square$

4>3>2>1>0	3>2>1	3>2
4>3>2	3>2>0	3>1
4>3>1	3>1>0	3>0
4>3>0		
4>2>1		
4>2>0		
4>1>0		

www.homathchess.com　367

Number ranking puzzles

Arrange the numbers 0, 1, 2, 3, and 4 so that each inequality condition is true.

☐2☐ > ☐, ☐2☐ > ☐ 2>1, 2>0 ☐ > ☐3☐, ☐3☐ < ☐ 4>3, 3<4 ☐ > ☐2☐, ☐2☐ < ☐ 4>2, 2<4 ☐ > ☐1☐, ☐1☐ < ☐ 4>1, 1<4 ☐ > ☐0☐, ☐0☐ < ☐ 4>0, 0<4 ☐ > ☐2☐, ☐2☐ < ☐ 3>2, 2<3 ☐ > ☐1☐, ☐1☐ < ☐ 3>1, 1<3 ☐ > ☐0☐, ☐0☐ < ☐ 3>0, 0<3 ☐ > ☐1☐, ☐1☐ < ☐ 2>1, 1<2 ☐ > ☐0☐, ☐0☐ < ☐ 2>0, 0<2 ☐ > ☐0☐, ☐0☐ < ☐ 1>0, 0<1	☐1☐ > ☐ 1>0 ☐ > ☐1☐, ☐1☐ < ☐ 4 4 ☐ > ☐1☐, ☐1☐ < ☐ 3 3 ☐ > ☐1☐, ☐1☐ < ☐ 2 2 ☐ > ☐0☐, ☐0☐ < ☐ 1 1

Number ranking puzzles

Arrange the numbers 1, 2, 3, and 4 so that each number appears only once in each row and column. Answers may vary.

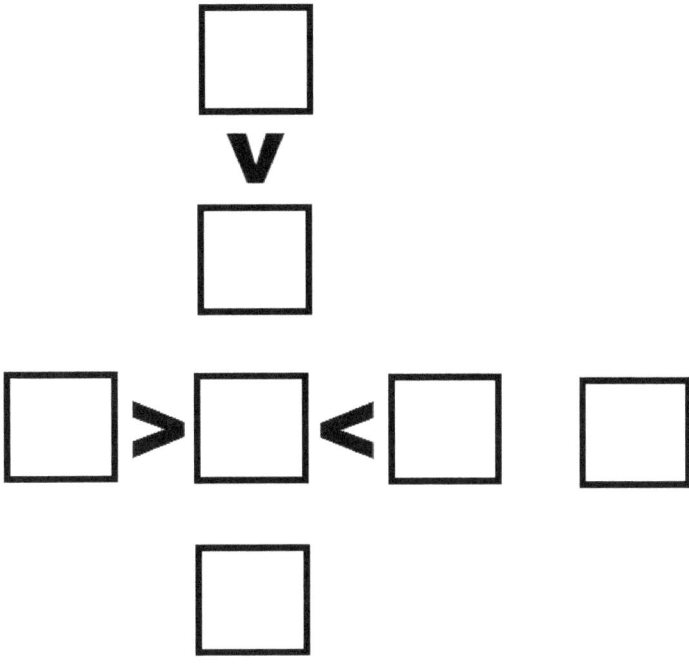

4
3
4123
2

Ho Math Chess — Pre-K and Kindergarten Math

何数棋谜　棋谜式幼儿健脑思维趣味数学

© 2012 – 2021 Frank Ho, Amanda Ho, Canada copyright 1095661, Trademark 771400

Number ranking puzzles

Arrange the numbers 1, 2, 3, and 4 so that each number appears only once in each row and column.

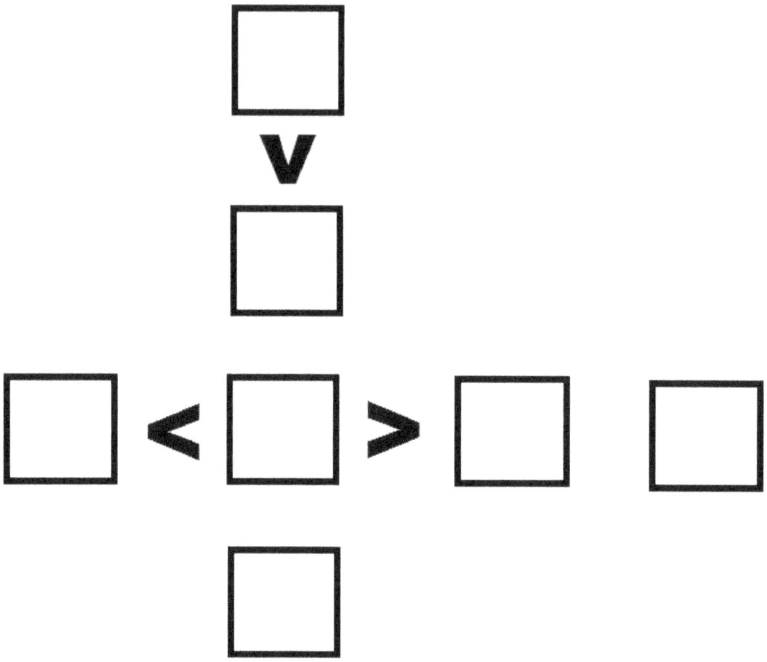

2
1
1432
3

Number ranking puzzles

Arrange the numbers 1, 2, 3, and 4 so that each number appears only once in each row and column.

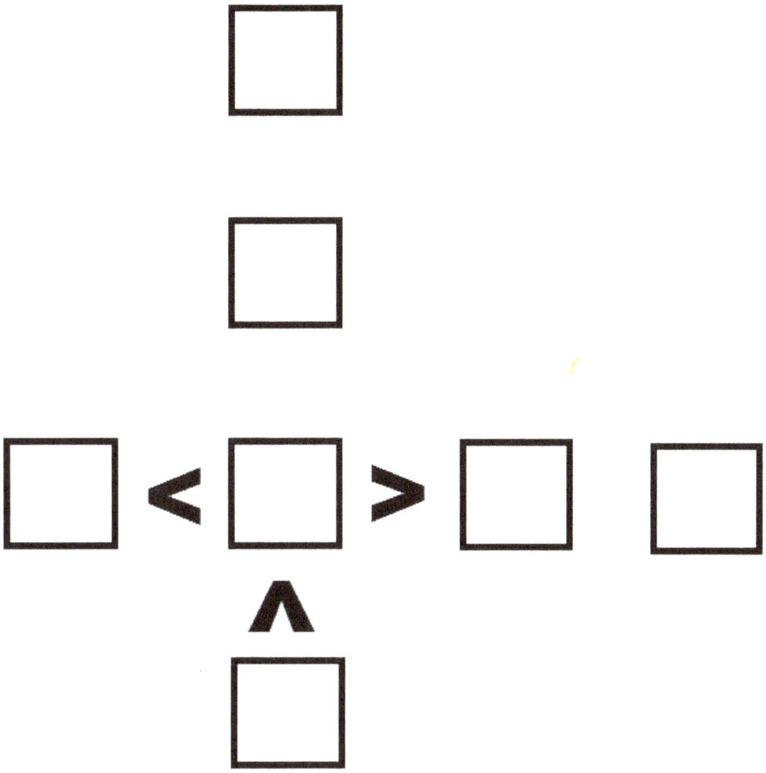

1
4
1324
2

Ho Math Chess — Pre-K and Kindergarten Math

何数棋谜　棋谜式幼儿健脑思维趣味数学

© 2012 – 2021 Frank Ho, Amanda Ho, Canada copyright 1095661, Trademark 771400

Number ranking puzzles

Arrange the numbers 1, 2, 3, and 4 so that each number appears only once in each row and column.

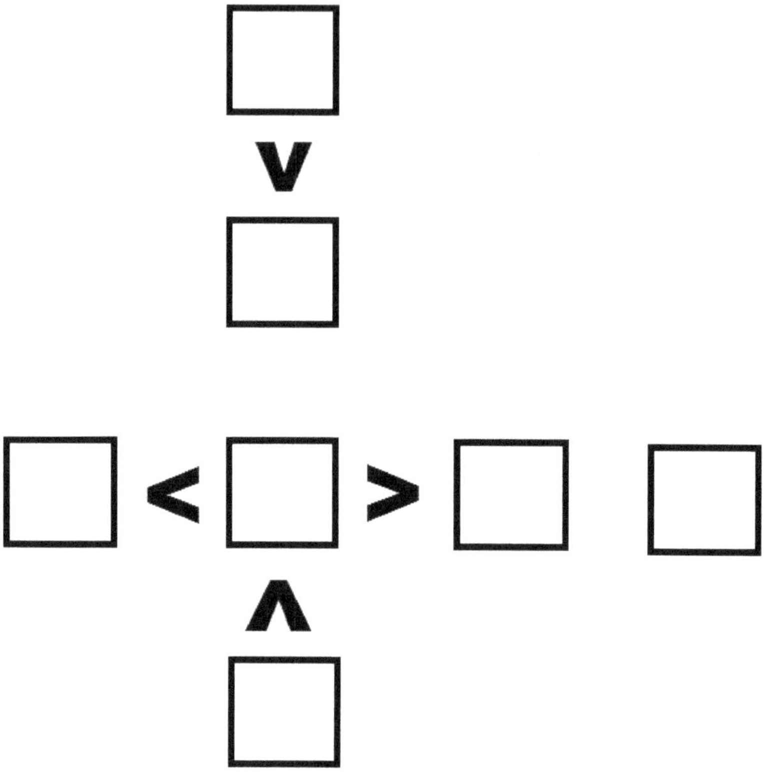

2
1
2314
4

Ho Math Chess — Pre-K and Kindergarten Math

何数棋谜　棋谜式幼儿健脑思维趣味数学

© 2012 – 2021 Frank Ho, Amanda Ho, Canada copyright 1095661, Trademark 771400

Number ranking puzzles

Arrange the numbers 1, 2, 3, and 4 so that each number appears only once in each row and column.

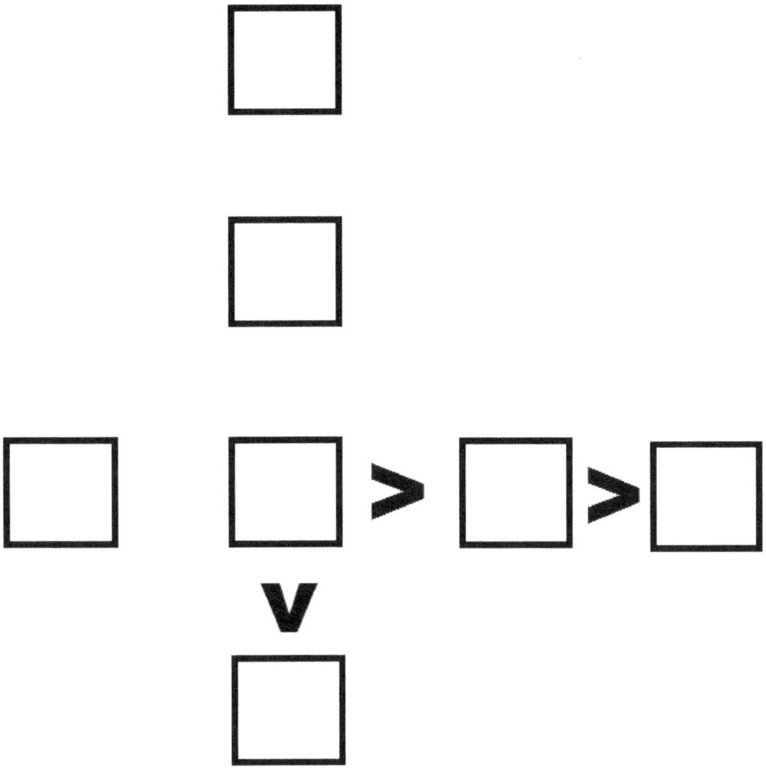

1
2
1432
3

www.ingramcontent.com/pod-product-compliance
Lightning Source LLC
Chambersburg PA
CBHW081125170426
43197CB00017B/2749